THE PSYCHOLOGY OF INSECURITY

Insecurity is an inevitable part of being human. Although life is insecure for every organism, humans alone are burdened by *knowing* that this is so. This ground-breaking volume features contributions by leading international researchers exploring the social psychology of insecurity and how existential, metaphysical, and social uncertainty influence human social behaviour.

Chapters in the book investigate the psychological origins of insecurity, evolutionary theorizing about the functions of insecurity, the motivational strategies people adopt to manage insecurity, self-regulation strategies, the role of insecurity in the formation and maintenance of social relationships, and the influence of insecurity and uncertainty on the organization of larger social systems and public affairs. The chapters also discuss how insecurity influences many areas of contemporary social life, highlighting the applied implications of this line of research. Topics covered include the role of insecurity in social communication, social judgments, decision-making, group identification, morality, interpersonal behaviour, relationships, attitudes, and many applied aspects of social life and politics where understanding the psychology of insecurity is of critical importance.

This accessible and engaging book will be of interest to students, researchers, and practitioners as a textbook or reference book in behavioural and social science fields, as well as to a broad spectrum of intelligent lay audiences seeking to understand one of the most intriguing issues that shapes human social life.

Joseph P. Forgas is a Scientia Professor at the University of New South Wales. His research focuses on affective influences on social cognition and behaviour. For his work, he received the Order of Australia and the

Distinguished Scientific Contribution Award, and he has been elected Fellow of the Australian and Hungarian Academies of Science.

William D. Crano is Oskamp Professor of Psychology at Claremont Graduate University. He was Liaison Scientist for the US Office of Naval Research, NATO Senior Scientist, and Fulbright Senior Scholar. His research focuses on attitude development and attitude change and their applications.

Klaus Fiedler is Professor of Psychology at the University of Heidelberg and Fellow of the German Academies of Science, the Association for Psychological Sciences, and the Society for Personality and Social Psychology. His research focuses on social cognition, language judgments, and decision-making. He received several awards, including the Leibniz Award, and he is on the editorial boards of leading journals.

The Sydney Symposium of Social Psychology Series

This book is Volume 24 in the *Sydney Symposium of Social Psychology* series. The aim of the Sydney Symposia of Social Psychology is to provide new, integrative insights into key areas of contemporary research. Held every year at the University of New South Wales, Sydney, each symposium deals with an important integrative theme in social psychology, and the invited participants are leading researchers in the field from around the world. Each contribution is extensively discussed during the symposium and is subsequently thoroughly revised into book chapters that are published in the volumes in this series. For further details, see the website at www.sydneysymposium.unsw.edu.au.

Previous Sydney Symposium of Social Psychology volumes

SSSP 1. Feeling and Thinking: The role of Affect in Social Cognition★★ ISBN 0-521-64223-X (Edited by Joseph P. Forgas). *Contributors*: Robert Zajonc, Jim Blascovich, Wendy Berry Mendes, Craig Smith, Leslie Kirby, Eric Eich, Dawn Macauley, Len Berkowitz, Sara Jaffee, EunKyung Jo, Bartholomeu Troccoli, Leonard Martin, Daniel Gilbert, Timothy Wilson, Herbert Bless, Klaus Fiedler, Joseph P. Forgas, Carolin Showers, Anthony Greenwald, Mahzarin Banaji, Laurie Rudman, Shelly Farnham, Brian Nosek, Marshall Rosier, Mark Leary, Paula Niedenthal and Jamin Halberstadt.

SSSP 2. The Social Mind: Cognitive and Motivational Aspects of Interpersonal Behavior★★ ISBN 0-521-77092-0 (Edited by Joseph P. Forgas, Kipling D. Williams and Ladd Wheeler). *Contributors*: William & Claire McGuire, Susan Andersen, Roy Baumeister, Joel Cooper, Bill Crano, Garth Fletcher, Joseph P. Forgas, Pascal Huguet, Mike Hogg, Martin Kaplan, Norb Kerr, John Nezlek, Fred Rhodewalt, Astrid Schuetz, Constantine Sedikides, Jeffry Simpson, Richard Sorrentino, Dianne Tice, Kip Williams and Ladd Wheeler.

SSSP 3. Social Influence: Direct and Indirect Processes★ ISBN 1-84169-038-4 (Edited by Joseph P. Forgas and Kipling D. Williams). *Contributors*: Robert Cialdini, Eric Knowles, Shannon Butler, Jay Linn, Bibb Latane,

Martin Bourgeois, Mark Schaller, Ap Dijksterhuis, James Tedeschi, Richard Petty, Joseph P. Forgas, Herbert Bless, Fritz Strack, Eva Walther, Sik Hung Ng, Thomas Mussweiler, Kipling Williams, Lara Dolnik, Charles Stangor, Gretchen Sechrist, John Jost, Deborah Terry, Michael Hogg, Stephen Harkins, Barbara David, John Turner, Robin Martin, Miles Hewstone, Russell Spears, Tom Postmes, Martin Lea and Susan Watt.

SSSP 4. The Social Self: Cognitive, Interpersonal, and Intergroup Perspectives★★ ISBN 1-84169-062-7 (Edited by Joseph P. Forgas and Kipling D. Williams). *Contributors*: Eliot R. Smith, Thomas Gilovich, Monica Biernat, Joseph P. Forgas, Stephanie J. Moylan, Edward R. Hirt, Sean M. McCrea, Frederick Rhodewalt, Michael Tragakis, Mark Leary, Roy F. Baumeister, Jean M. Twenge, Natalie Ciarocco, Dianne M. Tice, Jean M. Twenge, Brandon J. Schmeichel, Bertram F. Malle, William Ickes, Marianne LaFrance, Yoshihisa Kashima, Emiko Kashima, Anna Clark, Marilynn B. Brewer, Cynthia L. Pickett, Sabine Otten, Christian S. Crandall, Diane M. Mackie, Joel Cooper, Michael Hogg, Stephen C. Wright, Art Aron, Linda R. Tropp and Constantine Sedikides.

SSSP 5. Social Judgments: Implicit and Explicit Processes★★ ISBN 0-521-82248-3. (Edited by Joseph P. Forgas, Kipling D. Williams and William Von Hippel). *Contributors*: Herbert Bless, Marilynn Brewer, David Buss, Tanya Chartrand, Klaus Fiedler, Joseph Forgas, David Funder, Adam Galinsky, Martie Haselton, Denis Hilton, Lucy Johnston, Arie Kruglanski, Matthew Lieberman, John McClure, Mario Mikulincer, Norbert Schwarz, Philip Shaver, Diederik Stapel, Jerry Suls, William von Hippel, Michaela Waenke, Ladd Wheeler, Kipling Williams and Michael Zarate.

SSSP 6. Social Motivation: Conscious and Unconscious Processes★★ ISBN 0-521-83254-3 (Edited by J.P. Forgas, K.D. Williams and S.M. Laham). *Contributors*: Henk Aarts, Ran Hassin, Trish Devine, Joseph Forgas, Jens Forster, Nira Liberman, Judy Harackiewicz, Leanne Hing, Mark Zanna, Michael Kernis, Paul Lewicki, Steve Neuberg, Doug Kenrick, Mark Schaller, Tom Pyszczynski, Fred Rhodewalt, Jonathan Schooler, Steve Spencer, Fritz Strack, Roland Deutsch, Howard Weiss, Neal Ashkanasy, Kip Williams, Trevor Case, Wayne Warburton, Wendy Wood, Jeffrey Quinn, Rex Wright and Guido Gendolla.

SSSP 7. The Social Outcast: Ostracism, Social Exclusion, Rejection, and Bullying★ ISBN 1-84169-424-X (Edited by Kipling D. Williams, Joseph P. Forgas and William Von Hippel). *Contributors*: Kipling D. Williams, Joseph P. Forgas, William von Hippel, Lisa Zadro, Mark R. Leary, Roy F. Baumeister, C. Nathan DeWall, Geoff MacDonald, Rachell Kingsbury, Stephanie Shaw,

John T. Cacioppo, Louise C. Hawkley, Naomi I. Eisenberger, Matthew D. Lieberman, Rainer Romero-Canyas, Geraldine Downey, Jaana Juvonen, Elisheva F. Gross, Kristin L. Sommer, Yonata Rubin, Susan T. Fiske, Mariko Yamamoto, Jean M. Twenge, Cynthia L. Pickett, Wendi L. Gardner, Megan Knowles, Michael A. Hogg, Julie Fitness, Jessica L. Lakin, Tanya L. Chartrand, Kathleen R. Catanese, Dianne M. Tice, Lowell Gaertner, Jonathan Iuzzini, Jaap W. Ouwerkerk, Norbert L. Kerr, Marcello Gallucci, Paul A.M. Van Lange and Marilynn B. Brewer.

SSSP 8. Affect in Social Thinking and Behavior* ISBN 1-84169-454-2 (Edited by Joseph P. Forgas). *Contributors*: Joseph P. Forgas, Carrie Wyland, Simon M. Laham, Martie G. Haselton, Timothy Ketelaar, Piotr Winkielman, John T. Cacioppo, Herbert Bless, Klaus Fiedler, Craig A. Smith, Bieke David, Leslie D. Kirby, Eric Eich, Dawn Macaulay, Gerald L. Clore, Justin Storbeck, Roy F. Baumeister, Kathleen D. Vohs, Dianne M. Tice, Dacher Keltner, E.J. Horberg, Christopher Oveis, Elizabeth W. Dunn, Simon M. Laham, Constantine Sedikides, Tim Wildschut, Jamie Arndt, Clay Routledge, Yaacov Trope, Eric R. Igou, Chris Burke, Felicia A. Huppert, Ralph Erber, Susan Markunas, Joseph P. Forgas, Joseph Ciarrochi, John T. Blackledge, Janice R. Kelly, Jennifer R.Spoor, John G. Holmes and Danu B. Anthony.

SSSP 9. Evolution and the Social Mind: Evolutionary Psychology and Social Cognition* ISBN 1-84169-458-0 (Edited by Joseph P. Forgas, Martie G. Haselton and William Von Hippel). *Contributors*: William von Hippel, Martie Haselton, Joseph P. Forgas, R.I.M. Dunbar, Steven W. Gangestad, Randy Thornhill, Douglas T. Kenrick, Andrew W. Delton, Theresa E. Robertson, D. Vaughn Becker, Steven L. Neuberg, Phoebe C. Ellsworth, Ross Buck, Joseph P. Forgas, Paul B.T. Badcock, Nicholas B. Allen, Peter M. Todd, Jeffry A. Simpson, Jonathon LaPaglia, Debra Lieberman, Garth J.O. Fletcher, Nickola C. Overall, Abraham P. Buunk, Karlijn Massar, Pieternel Dijkstra, Mark Van Vugt, Rob Kurzban, Jamin Halberstadt, Oscar Ybarra, Matthew C. Keller, Emily Chan, Andrew S. Baron, Jeffrey Hutsler, Stephen Garcia, Jeffrey Sanchez-Burks, Kimberly Rios Morrison, Jennifer R. Spoor, Kipling D. Williams, Mark Schaller and Lesley A. Duncan.

SSSP 10. Social Relationships: Cognitive, Affective, and Motivational Processes* ISBN 978-1-84169-715-4 (Edited by Joseph P. Forgas and Julie Fitness). *Contributors*: Joseph P. Forgas, Julie Fitness, Elaine Hatfield, Richard L. Rapson, Gian C. Gonzaga, Martie G. Haselton, Phillip R. Shaver, Mario Mikulincer, David P. Schmitt, Garth J.O. Fletcher, Alice D. Boyes, Linda K. Acitelli, Margaret S. Clark, Steven M. Graham, Erin Williams, Edward P. Lemay, Christopher R. Agnew, Ximena B. Arriaga, Juan E. Wilson, Marilynn B. Brewer, Jeffry A. Simpson, W. Andrew Collins, SiSi Tran, Katherine C. Haydon, Shelly L. Gable, Patricia Noller, Susan Conway, Anita Blakeley-Smith,

Julie Peterson, Eli J. Finkel, Sandra L. Murray, Lisa Zadro, Kipling D. Williams and Rowland S. Miller.

SSSP 11. Psychology of Self-Regulation: Cognitive, Affective, and Motivational Processes★ ISBN 978-1-84872-842-4 (Edited by Joseph P. Forgas, Roy F. Baumeister and Dianne M. Tice). *Contributors:* Joseph P. Forgas, Roy F. Baumeister, Dianne M. Tice, Jessica L. Alquist, Carol Sansone, Malte Friese, Michaela Wänke, Wilhelm Hofmann, Constantine Sedikides, Christian Unkelbach, Henning Plessner, Daniel Memmert, Charles S. Carver, Michael F. Scheier, Gabriele Oettingen, Peter M. Gollwitzer, Jens Förster, Nira Liberman, Ayelet Fishbach, Gráinne M. Fitzsimons, Justin Friesen, Edward Orehek, Arie W. Kruglanski, Sander L. Koole, Thomas F. Denson, Klaus Fiedler, Matthias Bluemke, Christian Unkelbach, Hart Blanton, Deborah L. Hall, Kathleen D. Vohs, Jannine D. Lasaleta, Bob Fennis, William von Hippel, Richard Ronay, Eli J. Finkel, Daniel C. Molden, Sarah E. Johnson and Paul W. Eastwick.

SSSP 12. The Psychology of Attitudes and Attitude Change★ ISBN 978-1-84872-908-7 (Edited by Joseph P. Forgas, Joel Cooper and William D. Crano). *Contributors:* William D. Crano, Joel Cooper, Joseph P. Forgas, Blair T. Johnson, Marcella H. Boynton, Alison Ledgerwood, Yaacov Trope, Eva Walther, Tina Langer, Klaus Fiedler, Steven J. Spencer, Jennifer Peach, Emiko Yoshida, Mark P. Zanna, Allyson L. Holbrook, Jon A. Krosnick, Eddie Harmon-Jones, David M. Amodio, Cindy Harmon-Jones, Michaela Wänke, Leonie Reutner, Kipling D. Williams, Zhansheng Chen, Duane Wegener, Radmila Prislin, Brenda Major, Sarah S.M. Townsend, Frederick Rhodewalt, Benjamin Peterson, Jim Blascovich and Cade McCall.

SSSP 13. The Psychology of Social Conflict and Aggression★ ISBN 978-1-84872-932-2 (Edited by Joseph P. Forgas, Arie W. Kruglanski and Kipling D. Williams). *Contributors:* Daniel Ames, Craig A. Anderson, Joanna E. Anderson, Paul Boxer, Tanya L. Chartrand, John Christner, Matt DeLisi, Thomas F. Denson, Ed Donnerstein, Eric F. Dubow, Chris Eckhardt, Emma C. Fabiansson, Eli J. Finkel, Gráinne M. Fitzsimons, Joseph P. Forgas, Adam D. Galinsky, Debra Gilin, Georgina S. Hammock, L. Rowell Huesmann, Arie W. Kruglanski, Robert Kurzban, N. Pontus Leander, Laura B. Luchies, William W. Maddux, Mario Mikulincer, Edward Orehek, Deborah South Richardson, Phillip R. Shaver, Hui Bing Tan, Mark Van Vugt, Eric D. Wesselmann, Kipling D. Williams and Lisa Zadro.

SSSP 14. Social Thinking and Interpersonal Behavior★ ISBN 978-1-84872-990-2 (Edited by Joseph P. Forgas, Klaus Fiedler and Constantine Sedikides). *Contributors:* Andrea E. Abele, Eusebio M. Alvaro, Mauro Bertolotti, Camiel J. Beukeboom, Susanne Bruckmüller, Patrizia Catellani, Cindy K. Chung, Joel Cooper, William D. Crano, István Csertő, John F. Dovidio,

Bea Ehmann, Klaus Fiedler, Joseph P. Forgas, Éva Fülöp, Jessica Gasiorek, Howard Giles, Liz Goldenberg, Barbara Ilg, Yoshihisa Kashima, Mikhail Kissine, Olivier Klein, Alex Koch, János László, Anne Maass, Andre Mata, Elisa M. Merkel, Alessio Nencini, Andrew A. Pearson, James W. Pennebaker, Kim Peters, Tibor Pólya, Ben Slugoski, Caterina Suitner, Zsolt Szabó, Matthew D. Trujillo and Orsolya Vincze.

SSSP 15. Social Cognition and Communication* ISBN 978-1-84872-663-5 (Edited by Joseph P. Forgas, Orsolya Vincze and János László). *Contributors*: Andrea E. Abele, Eusebio M. Alvaro, Maro Bertolotti, Camiel J. Beukeboom, Susanne Bruckmüller, Patrizia Catellani, István Cserto, Cindy K. Chung, Joel Coooper, William D. Crano, John F. Dovidio, Bea Ehmann, Klaus Fiedler, J.P. Forgas, Éva Fülöp, Jessica Gasiorek, Howard Giles, Liz Goldenberg, Barbara Ilg, Yoshihisa Kahima, Mikhail Kissine, Alex S. Koch, János László, Olivier Klein, Anne Maass, André Mata, Elisa M. Merkel, Alessio Nencini, Adam R. Pearson, James W. Pennebaker, Kim Peters, Tibor Pólya, Ben Slugoski, Caterina Suitner, Zsolt Szabó, Matthew D. Trujillo and Orsolya Vincze.

SSSP 16. Motivation and Its Regulation: The Control Within* ISBN 978-1-84872-562-1 (Edited by Joseph P. Forgas and Eddie Harmon-Jones). *Contributors*: Emily Balcetis, John A. Bargh, Jarik Bouw, Charles S. Carver, Brittany M. Christian, Hannah Faye Chua, Shana Cole, Carsten K.W. De Dreu, Thomas F. Denson, Andrew J. Elliot, Joseph P. Forgas, Alexandra Godwin, Karen Gonsalkorale, Jamin Halberstadt, Cindy Harmon-Jones, Eddie Harmon-Jones, E. Tory Higgins, Julie Y. Huang, Michael Inzlicht, Sheri L. Johnson, Jonathan Jong, Jutta Joormann, Nils B. Jostmann, Shinobu Kitayama, Sander L. Koole, Lisa Legault, Jennifer Leo, C. Neil Macrae, Jon K. Maner, Lynden K. Mile, Steven B. Most, Jaime L. Napier, Tom F. Price, Marieke Roskes, Brandon J. Schmeichel, Iris K. Schneider, Abigail A. Scholer, Julia Schüler, Sarah Strübin, David Tang, Steve Tompson, Mattie Tops and Lisa Zadro.

SSSP 17. Social Psychology and Politics* ISBN 978-1-13882-968-8 (Edited by Joseph P. Forgas, Klaus Fiedler and William D. Crano). *Contributors*: Stephanie M. Anglin, Luisa Batalha, Mauro Bertolotti, Patrizia Catellani, William D. Crano, Jarret T. Crawford, John F. Dovidio, Klaus Fiedler, Joseph P. Forgas, Mark G. Frank, Samuel L. Gaertner, Jeremy Ginges, Joscha Hofferbert, Michael A. Hogg, Hyisung C. Hwang, Yoel Inbar, Lee Jussim, Lucas A. Keefer, Laszlo Kelemen, Alex Koch, Tobias Krüger, Mark J. Landau, Janos Laszlo, Elena Lyrintzis, David Matsumoto, G. Scott Morgan, David A. Pizarro, Felicia Pratto, Katherine J. Reynolds, Tamar Saguy, Daan Scheepers, David O. Sears, Linda J. Skitka, Sean T. Stevens, Emina Subasic, Elze G. Ufkes, Robin R. Vallacher, Paul A.M. Van Lange, Daniel C. Wisneski, Michaela Wänke, Franz Woellert and Fouad Bou Zeineddine.

SSSP 18. The Social Psychology of Morality* ISBN 978-1-138-92907-4 (Edited by Joseph P. Forgas, Lee Jussim and Paul A.M. Van Lange). *Contributors*: Stephanie M. Anglin, Joel B. Armstrong, Mark J. Brandt, Brock Bastian, Paul Conway, Joel Cooper, Chelsea Corless, Jarret T. Crawford, Daniel Crimston, Molly J. Crockett, Jose L. Duarte, Allison K. Farrell, Klaus Fiedler, Joseph P. Forgas, Rebecca Friesdorf, Jeremy A. Frimer, Adam D. Galinsky, Bertram Gawronski, William G. Graziano, Nick Haslam, Mandy Hütter, Lee Jussim, Alice Lee, William W. Maddux, Emma Marshall, Dale T. Miller, Benoît Monin, Tom Pyszczynski, Richard Ronay, David A. Schroeder, Simon M. Laham, Jeffry A. Simpson, Sean T. Stevens, William Von Hippel and Geoffrey Wetherell.

SSSP 19. The Social Psychology of Living Well* ISBN 978-0-8153-6924-0 (Edited by Joseph P. Forgas and Roy F. Baumeister). *Contributors*: Yair Amichai-Hamburger, Peter Arslan, Roy F. Baumeister, William D. Crano, Candice D. Donaldson, Elizabeth W. Dunn, Ryan J. Dwyer, Shir Etgar, Allison K. Farrell, Klaus Fiedler, Joseph P. Forgas, Barbara L. Fredrickson, Megan M. Fritz, Shelly L. Gable, Karen Gonsalkorale, Alexa Hubbard, Chloe O. Huelsnitz, Felicia A. Huppert, David Kalkstein, Sonja Lyubomirsky, David G. Myers, Constantine Sedikides, James Shah, Kennon M. Sheldon, Jeffry A. Simpson, Elena Stephan, Yaacov Trope, William Von Hippel and Tom Wildschut.

SSSP 20. The Social Psychology of Gullibility: Conspiracy Theories, Fake News and Irrational Beliefs* ISBN 978-0-3671-8793-4 (Edited by Joseph P. Forgas and Roy F. Baumeister). *Contributors*: Stephanie M. Anglin, Joseph J. Avery, Roy F. Baumeister, Aleksandra Chicoka, Joel Cooper, Karen Douglas, David Dunning, Anthony M. Evans, Johanna K. Falbén, Klaus Fiedler, Joseph P. Forgas, Nicholas Fox, Marius Golubickis, Nathan Honeycutt, Lee Jussim, Alex Koch, Joachim I. Krueger, Spike W.S. Lee, C. Neil Macrae, Jessica A. Maxwell, Ruth Mayo, David Myers, Juliana L. Olivier, Daphna Oyserman, Jan-Willem van Prooijen, Norbert Schwarz, Sean T. Stevens, Fritz Strack, Robbie M. Sutton, Geoffrey P. Thomas, Christian Unkelbach, Kathleen D. Vohs and Claudia Vogrincic-Haselbacher.

SSSP 21. Applications of Social Psychology: How Social Psychology Can Contribute to the Solution of Real-World Problems* ISBN 978-0-367-41833-5 (Edited by Joseph P. Forgas, William D. Crano and Klaus Fidler). *Contributors*: Dana Atzil-Slomin, Hilary B. Bergsieker, H. Blanton, Shannon T. Brady, Pablo Brinol, Christopher N. Burrows, Emily Butler, Akeela Careem, Susannah Chandhook, William D. Crano, Lianne De Vries, Suzanne Dikker, Klaus Fiedler, Joseph P. Forgas, William M. Hall, Nathan Honeycutt, Lee Jussim, Sander L. Koole, Margaret Bull Kovera, Dorottya Lantos, Norman P. Li, Mario Mikulincer, Esther Papies, Richard E. Petty, Timothy Regan, Andrea L. Ruybal, Toni Schmader, Philip R. Shaver, Anna Stefaniak, Sean T. Stevens,

Wolfgang Tschacher, Mark Van Vugt, Gregory M. Walton, Tom Wilderjans and Michael J.A. Wohl.

SSSP 22. The Psychology of Populism: The Tribal Challenge to Liberal Democracy* ISBN 978-0-367-52381-7 (Edited by Joseph P. Forgas, William D. Crano and Klaus Fiedler). *Contributors*: Peter H. Ditto, Cristian G. Rodriguez, Daniel Bar-Tal, Tamir Magal, Michael Bang Petersen, Mathias Osmundsen, Alexander Bor, George E. Marcus, Agnieszka Golec de Zavala, Dorottya Lantos, Oliver Keenan Goldsmiths, Joachim I. Krueger, David J. Grüning, Jan-Willem van Prooijen, Arie W. Kruglanski, Erica Molinario, Gilda Sensales, Klaus Fiedler, Michael A. Hogg, Oluf Gøtzsche-Astrup, Joseph P. Forgas, Péter Krekó, Eotvos Lorand, Leonie Huddy, Alessandro Del Ponte, Michele J. Gelfand, Rebecca Lorente, Amber M. Gaffney, Joel Cooper, Joseph Avery, Robin R. Vallacher, Eli Fennell, Stanley Feldman and William D. Crano.

SSSP 23. The Psychology of Sociability: Understanding Human Attachment* ISBN 978-1-032-19307-6 (Edited by Joseph P. Forgas, William D. Crano and Klaus Fiedler). *Contributors*: Roy F. Baumeister, William D. Crano, Elizabeth W. Dunn, Molly Ellenberg, Klaus Fiedler, Alan Page Fiske, Joseph P. Forgas, Amber M. Gaffney, David J. Grüning, Michael A. Hogg, Mandy Hütter, Guy Itzchakov, Joachim I. Krueger, Arie W. Kruglanski, Karisa Y. Lee, Iris Lok, Sonja Lyubomirsky, Heather M. Maranges, Mario Mikulincer, Radmila Prislin, Annie Regan, Harry T. Reis, Yan Ruan, Phillip R. Shaver, Nicholas M.A. Smith, Tanushri Sudnar and William von Hippel.

SSSP 24. The Psychology of Insecurity: Seeking Certainty Where None Can Be Found* ISBN 978-1-032-32986-4 (Edited by Joseph P. Forgas, William D. Crano and Klaus Fiedler). *Contributors*: Ximena Arriaga, Joel Cooper, William D. Crano. Molly Ellenberg, Julie R. Eyink, Klaus Fiedler, Danica Finkelstein, Alan Page Fiske, Joseph P. Forgas. Amber M. Gaffney, David Gruening, Samantha Heiman, Gilad Hirschberger, Edward R. Hirt, Michael A. Hogg, Zachary Hohman, Lee Jussim, Péter Krekó, Joachim I. Krueger, Arie Kruglanski, Madoka Kumashiro, Veronica M. Lamarche, Linda McCaughey, John Merakovsky, Mario Mikulincer, Sandra L. Murray, Logan Pearce, Tom Pyszczynsky, Sean Sevens, Phillip R. Shaver, Jonathan Sundby, Kees Van den Bos, Jan-Willem Van Prooijen and William von Hippel.

*Published by Routledge
**Published by Cambridge University Press

For more information about this series, please visit: www.routledge.com/Sydney-Symposium-of-Social-Psychology/book-series/TFSE00262

THE PSYCHOLOGY OF INSECURITY

Seeking Certainty Where
None Can Be Found

*Edited by Joseph P. Forgas, William D. Crano,
and Klaus Fiedler*

NEW YORK AND LONDON

Designed cover image: Jacob Peeter Gowy, *The Fall of Icarus* (c.1636–1638)

First published 2023
by Routledge
605 Third Avenue, New York, NY 10158

and by Routledge
4 Park Square, Milton Park, Abingdon, Oxon, OX14 4RN

Routledge is an imprint of the Taylor & Francis Group, an informa business

© 2023 selection and editorial matter, Joseph P. Forgas, William D. Crano, and Klaus Fiedler; individual chapters, the contributors

The right of Joseph P. Forgas, William D. Crano, and Klaus Fiedler to be identified as the authors of the editorial material, and of the authors for their individual chapters, has been asserted in accordance with sections 77 and 78 of the Copyright, Designs and Patents Act 1988.

All rights reserved. No part of this book may be reprinted or reproduced or utilised in any form or by any electronic, mechanical, or other means, now known or hereafter invented, including photocopying and recording, or in any information storage or retrieval system, without permission in writing from the publishers.

Trademark notice: Product or corporate names may be trademarks or registered trademarks, and are used only for identification and explanation without intent to infringe.

Library of Congress Cataloging-in-Publication Data
Names: Forgas, Joseph P, editor. | Crano, William D., 1942- editor. | Fiedler, Klaus, 1951- editor.
Title: The psychology of insecurity: seeking certainty where none can be found / edited by Joseph P. Forgas, William D. Crano and Klaus Fiedler.
Description: 1 Edition. | New York, NY: Routledge, 2023. | Series: Sydney symposium of social psychology | Includes bibliographical references and index.
Identifiers: LCCN 2022056056 (print) | LCCN 2022056057 (ebook) | ISBN 9781032329864 (hardback) | ISBN 9781032323954 (paperback) | ISBN 9781003317623 (ebook)
Subjects: LCSH: Social psychology. | Human security.
Classification: LCC HM1033 .P845 2023 (print) | LCC HM1033 (ebook) | DDC 302/.1—dc23/eng/20221202
LC record available at https://lccn.loc.gov/2022056056
LC ebook record available at https://lccn.loc.gov/2022056057

ISBN: 978-1-032-32986-4 (hbk)
ISBN: 978-1-032-32395-4 (pbk)
ISBN: 978-1-003-31762-3 (ebk)

DOI: 10.4324/9781003317623

Typeset in Bembo
by codeMantra

CONTENTS

List of Contributors   xvii

PART I
The Nature and Sources of Insecurity   1

1 Understanding the Psychology of Insecurity:
Evolutionary, Cognitive, and Cultural Perspectives   3
Joseph P. Forgas

2 The Evolution of Insecurity   21
William von Hippel and John Merakovsky

3 The Interactive Role of Death, Uncertainty, and the Loss
of Shared Reality on Societal and Individual Insecurity   35
Tom Pyszczynski and Jonathan Sundby

4 The Uncertainty Challenge: Escape It, Embrace It   54
Arie W. Kruglanski and Molly Ellenberg

5 Insecurity Can Be Beneficial: Reflections on Adaptive
Strategies for Diverse Trade-Off Settings   74
Klaus Fiedler and Linda McCaughey

PART II
Managing Individual Insecurity 93

6 The Arc of Dissonance: From Drive to Uncertainty 95
   Joel Cooper and Logan Pearce

7 Persuasion as a Sop to Insecurity 108
   William D. Crano and Zachary Hohman

8 Self-Handicapping in the Face of Uncertainty: The
   Paradox That Most Certainly Is 130
   *Edward R. Hirt, Samantha L. Heiman, Julie R. Eyink, and
   Sean McCrea*

9 Strategy, Trust, and Freedom in an Uncertain World 150
   Joachim I. Krueger and David J. Grüning

10 Seeking Moral Meaning in Misfortune: Assigning Blame,
   Without Regard for Causation 170
   Alan Page Fiske

PART III
The Role of Insecurity in Social Relationships 185

11 Attachment Security and Coping with Existential
   Concerns: Studying Security Dynamics in Dyadic,
   Group, Sociopolitical, and Spiritual/Religious Relationships 187
   Mario Mikulincer and Phillip R. Shaver

12 Beyond Dyadic Interdependence: Romantic
   Relationships in an Uncertain Social World 205
   Sandra L. Murray and Veronica M. Lamarche

13 Adult Attachment Insecurity During the COVID
   Pandemic: Heightened Insecurity and Its Undoing 224
   Ximena B. Arriaga and Madoka Kumashiro

14 Social Identity Dynamics in the Face of Overwhelming
   Uncertainty 244
   Michael A. Hogg and Amber M. Gaffney

15 From Individual Insecurity to Collective Security: The Group Survival Motivation 265
Gilad Hirschberger

PART IV
The role of Insecurity and Uncertainty in Politics and Public Life 285

16 Trust in Social Institutions: The Role of Informational and Personal Uncertainty 287
Kees van den Bos

17 The Politics of Insecurity: How Uncertainty Promotes Populism and Tribalism 307
Joseph P. Forgas

18 Uncertainty, Academic Radicalization, and the Erosion of Social Science Credibility 329
Lee Jussim, Danica Finkelstein, and Sean T. Stevens

19 Escape From Uncertainty: To Conspiracy Theories and Pseudoscience 349
Péter Krekó

20 Feelings of Insecurity as a Driver of Anti-Establishment Sentiments 368
Jan-Willem van Prooijen

Index *389*

LIST OF CONTRIBUTORS

Ximena B. Arriaga Purdue University, USA

Joel Cooper Princeton University, New Jersey, USA

William D. Crano Claremont Graduate University, USA

Molly Ellenberg University of Maryland, USA

Julie R. Eyink Indiana University-Bloomington, USA

Klaus Fiedler University of Heidelberg, Germany

Danica Finkelstein Rutgers University, New Jersey, USA

Alan Page Fiske Department of Anthropology, UCLA, USA

Joseph P. Forgas University of New South Wales, Sydney, Australia

Amber M. Gaffney California Polytechnic University Humboldt, USA

David J. Grüning Gesis – Leibniz Institute for the Social Sciences and Heidelberg University, Germany

Samantha L. Heiman Indiana University-Bloomington, USA

Gilad Hirschberger Reichman University, Israel

Edward R. Hirt Indiana University-Bloomington, USA

Michael A. Hogg Claremont Graduate University, University of Kent, USA

Zachary Hohman Texas Tech University, Texas, USA

Lee Jussim Rutgers University, New Jersey, USA

Péter Krekó Eotvos Lorand University, Hungary

Joachim I. Krueger Brown University, Rhode Island, USA

Arie W. Kruglanski University of Maryland, USA

Madoka Kumashiro Goldsmiths, University of London, UK

Veronica M. Lamarche University of Essex, UK

Linda McCaughey University of Heidelberg, Germany

Sean McCrea University of Wyoming, USA

John Merakovsky University of Queensland, Australia

Mario Mikulincer Reichman University (IDC) Herzliya, Israel

Sandra L. Murray University at Buffalo, The State University of New York, USA

Logan Pearce Princeton University, New Jersey, USA

Tom Pyszczynski University of Colorado at Colorado Springs, Colorado, USA

Sean T. Stevens Rutgers University, New Jersey, USA

Phillip R. Shaver University of California, Davis, USA

Jonathan Sundby University of Colorado, Colorado Springs, Colorado, USA

Kees van den Bos Utrecht University, Utrecht, Netherlands

Jan-Willem van Prooijen VU Amsterdam, NSCR, and Maastricht University, Netherlands

William von Hippel University of Queensland, Australia

PART I
The Nature and Sources of Insecurity

1
UNDERSTANDING THE PSYCHOLOGY OF INSECURITY

Evolutionary, Cognitive, and Cultural Perspectives

Joseph P. Forgas

UNIVERSITY OF NEW SOUTH WALES, SYDNEY

Abstract

The experience of psychological insecurity is a defining feature of humans, as we are the only creatures with consciousness and the symbolic ability to imagine, forecast, and manipulate alternative realities. In this introductory chapter, we review the basic nature and sources of insecurity, the way people manage insecurity, and the consequences of insecurity for our personal lives, our relationships, and for society and politics at large. It is argued that while insecurity can play a crucial beneficial role in anticipating and forecasting danger and coordinating responses, it is also an enduring source of anxiety for many. The roles of symbolic consciousness, 'theory of mind', mortality salience, and existential uncertainty in the generation of insecurity are highlighted. Psychological mechanisms for managing insecurity are also considered, and the functions of social support, close personal relationships, and group affiliation in reducing insecurity are reviewed. The consequences of insecurity for political movements and public affairs receive special attention, given the recent growth of various tribal, populist movements. The contributions to the book are summarized and organized into four topic areas dealing with (1) the nature and sources of psychological insecurity, (2) psychological and social strategies in the management of insecurity, (3) the role of insecurity in social relationships, and (4) the role of politics and public affairs.

Insecurity is an inevitable part of the human condition. As philosophers like Hobbes (1651) noted, life is continual fear. We alone, among all species, have consciousness that allows us to be intensely aware of the manifold risks and dangers that are an essential part of being alive. Of course, life is dangerous and unpredictable for every organism, but humans alone *know* that this is so

DOI: 10.4324/9781003317623-2

and feel compelled to prepare and plan for inevitable misfortunes (von Hippel, 2018; see also von Hippel & Merakovsky; Pyszczynsky & Sundby, this volume). Although physical insecurity is universal, psychological insecurity appears uniquely human (Kiernan & Baumeister, 2020; Leary, 2017).

Insecurity is not easily defined. Generally, it refers to a sense of uncertainty, inadequacy, and anxiety about ourselves, our performance, our relationships, our qualities with respect to others, our society, and ultimately, our very existence. This book, and this introductory chapter, in particular, will focus on the sources and nature of insecurity, its management, and its negative and positive consequences for social adjustment, personal relationships, and public life and politics.

Throughout history, most humans lived in abysmal conditions. Physical insecurity was ever present, and psychological insecurity necessarily took the second place. To dwell on psychological insecurity, one must have a modicum of leisure and reflexive ability. Accordingly, concern with psychological insecurity was mostly the domain of philosophers, artists, and writers. The fundamental existential absurdity of human life was noted by many thinkers since antiquity (stoics, epicureans, and hedonists), but paradoxically, it has only become a topic of mainstream concern *after* the rapid material progress of humanity since the industrial revolution (Kant, 1784; Pinker, 2018).

Insecurity often undermines reason and plays an important role in politics and public life, as noted by Plato (1943) more than 2000 years ago. The recent growth of populist ideologies makes understanding the nature and management of psychological insecurity particularly important (see Section 4 of this volume). This introductory chapter seeks to explore (1) the nature and sources of insecurity, (2) the management of insecurity, and (3) its implications and consequences – negative and positive – of insecurity for individuals and societies.

Nature and Sources of Insecurity

The Role of Consciousness

Humans alone have a subjective awareness of ourselves as causal agents, including the abstract ability to symbolically represent and manipulate experiences that are not part of the present situation. Unlike other primates, only humans "understand the world in intentional and causal terms" (Tomasello, 1999, p. 19). We can only speculate why this strange ability evolved in our species. Evolutionary psychologists suggest that the emergence of consciousness was a relatively recent development, probably driven by the intense cognitive demands of coordinating ever-larger social groups (Buss, 2019; Dunbar, 1998; Harari, 2018). Consistent with this idea, there is a close relationship between group size and brain size in higher primates, with humans possessing by far the largest brains, and living in the largest primary social groups (Dunbar, 1998).

The evolution of a 'theory of mind' – the ability to mentally represent others as conscious and predictable agents – represented a huge advantage to the survival of early hominid groups (Buss, 2019; von Hippel & Merakovsky, this volume). Forecasting, planning, and coordinating collaborative behaviors is probably the most powerful adaptive skill, allowing us to become the dominant species on the planet (see also Hirschberger; Hogg & Gaffney, this volume). The mental capacity to represent others as causal agents necessarily implied the ability also to think about *ourselves* as similarly conscious agents. Many of our cognitive habits, as well as cultural practices, serve the purpose of managing insecurity and maintaining a shared sense of social order (see also Cooper & Pearce; Krueger & Gruening; Fiske, this volume). The evolution of consciousness and a theory of mind were great adaptive achievements of our species, but at the same time, representing ourselves as causal agents is also the source of insecurity that is uniquely human (von Hippel, 2018).

The nature of human insecurity may also have something to do with the unusual structure of the human brain, consisting of three different modular units that developed at very different times and may also cast some light on the nature of the human experience of insecurity. Although the idea is still debated, some neuroanatomists suggest that the human brain consists of distinct units (the 'triune' brain) – (1) the basal ganglia (the reptilian brain) that evolved over 400 million years ago, (2) the limbic system (the paleo-mammalian brain) that evolved about 250 million years ago, and the surprisingly recent neocortex, the seat of our symbolic processes, only around 250 thousand years old (MacLean, 1990).

All three brain structures show a natural evolutionary tendency toward focusing on threatening, negative information (Kiernan & Baumeister, 2021). Our often hard-to-control and enduring experiences of insecurity, anxiety, and aggression may be linked to the limited capacity of the integrative, symbolic processing parts of the neocortex to control the negative impulses originating in the reptilian cortex and the amygdala (Hoffman, 2003). Some theorists even argue that the poor integration of these different brain structures, and the lack of conscious control over our archaic impulses, may be responsible for the uniquely troubled and often violent and bloodthirsty history of our species (Koestler, 1967).

It is the inability of symbolically representing but not effectively controlling negative inputs by the pre-frontal cortex that often gives rise to psychological insecurity. Insecurity need not be aversive, and sometimes, coping with uncertainty or risky endeavors has definite adaptive advantages (Fiedler & McCaughey, this volume). So, having consciousness and a 'theory of mind' are important tools for adaptive learning and for survival, but also the main source of psychological insecurity. We shall first look at perhaps the most basic source of insecurity, existential insecurity.

Existential Insecurity

One of the deeply troubling consequences of having self-awareness is that humans, again alone among all living organisms, are intensely conscious of the certainty of their own death. As Gilgamesh's powerful 4,000-year-old lament suggests, the dread of death has appeared since humans first started recording their history. William James famously described our awareness of our own inevitable death as 'the worm at the core' of human existence (1902/1985, p. 119).

Cultural practices throughout history have attempted to deny and conquer death, from the ancient Egyptian practice of mummification to the Day of the Dead festival in Mexico and other rituals (Fiske, this volume). Many religions, mythologies, and legends attempted to deal with the fear of our own mortality. Stoic philosophers like Marcus Aurelius, Seneca, and Epictetus offered cold-eyed advice about the need to accept the inevitability of death and misfortune.

More recently, countless experiments showed that death anxiety is a major source of existential and metaphysical insecurity (see Pyszczynsky & Sundby, this volume). Numerous studies found that mortality salience triggers an avalanche of often subconscious defensive reactions, including a powerful tendency to bolster our identity groups (Hirschberger; Hogg & Gaffney, this volume). Of course, most of the time, people live their lives by suppressing mortality awareness. As my mentor and friend, the late Gordon Bower at Stanford University, stated on his web page, 'Immortal until proven otherwise…'.

Closely related to mortality salience, the all-too-human search for metaphysical meaning is another important source of insecurity. As Shakespeare wrote, *"Life's but a walking shadow, A poor player that struts and frets his hour upon the stage, and then is heard no more: It is a tale told by an idiot, full of sound and fury, signifying nothing" (Macbeth)*. The human ability to be self-reflective and seek meaning may often interfere with our ability to function effectively in the here and now (Leary, 2007). Paradoxically, rational thinking offers little help in resolving these existential insecurities, offering no solution either to mortality, or questions about the meaning of life, as we shall see.

The Challenge of Rationality

The advancement of science brought untold benefits to humanity, and material insecurity − including hunger, fear, illness, and violence − has markedly decreased in Western societies (Acemoglu & Robinson, 2016; Pinker, 2018). At the same time, age-old metaphysical belief systems and religions have become less and less credible as a solution to our existential dilemmas. Ironically, the material progress produced by science and rationality, much of which was prompted by attempts to resolve various insecurities, may have reduced physical insecurity but in some instances increased psychological insecurity.

The existential challenge posed by rationality and science is eloquently illustrated by the epistemological principles of Karl Popper (1945), who argued that certainty and security are logically impossible to achieve in the world of science. All knowledge is temporary, only valid insofar as it can be falsified, and falsification should be accepted and even welcomed as the way forward to new knowledge. This epistemological framework is deeply challenging to the way humans think, seeking security, certainty, and safety rather than uncertainty and falsification. Scientific progress is predicated on the necessity of accepting eternal uncertainty, representing a metaphysical challenge and an unwelcome source of insecurity. Science has undermined the credibility of metaphysical belief systems that have given existence hope and meaning, but science now also insists that scientific knowledge itself must forever remain uncertain and temporary. In many ways, good science is not characterized by the mere acceptance of insecurity, but by its celebration as a positive force for discovery.

Personal Responsibility

Another source of psychological insecurity is our belief that humans possess a unique 'freedom of will' – that we can make choices, and consequently, we bear individual responsibility for our actions. As Dostoyevsky wrote in 'The grand inquisitor', never was there anything more unbearable to humanity than personal freedom. There is now extensive psychological literature documenting how freedom also imposes heavy burdens on people. Every 'free' choice creates insecurity, as (1) its consequences are uncertain, and (2) often desirable alternatives are necessarily eliminated. Research on cognitive dissonance demonstrated the length people will go to minimize post-decisional insecurity (see Cooper & Pearce, this volume).

Attributing freedom of will also complicate our interactions with others, making it far more difficult to predict others' behavioral choices that may often deviate from what might be considered rational. As Krueger and Gruening (this volume) show, interacting with humans can be far more difficult and insecurity-inducing than interacting with systems that are expected to follow rational rules.

Achievement Insecurity: The Trials of Sisyphus

Having the freedom of will also imply a responsibility for our successes and failures. Despite our best endeavors, our efforts will often fail, leading to experiences of insecurity. Many centuries ago, St. Augustine noted the paradox that the human struggle to achieve virtue and redemption is often in vain – but nevertheless there is value in the struggle. Subsequently, Kant (1784) came to a similar conclusion, noting the inevitability of conflict between our aspirations and our actual achievements. Existentialist writers such as Camus in *The Myth of Sisyphus* (1955) and *The Stranger* (1946) wrote eloquently about the paradoxical nature of effort and achievement, suggesting that the process of trying rather than succeeding should be our objective.

The fear of failure and resulting insecurity can often motivate self-defeating strategies to minimize threats to our self-esteem. Self-handicapping – creating reasons to produce and explain expected failure – is one of those paradoxical strategies discussed by Hirt, Eyink, and Heiman in their chapter. Self-handicapping represents a striking illustration of how insecurity can produce maladaptive outcomes. Self-handicapping exemplifies how *perceived* social reality often trumps the *real* world – when succeeding is less important than not being *seen* as a failure. Such distortions could be explained in terms of the paramount evolutionary importance of maintaining social status even at the expense of success and reality (Buss, 2019).

Time Perspective: Past, Present, and Future

One of the most remarkable tricks of consciousness is that it allows us to represent past and future experiences as if they were present. Goldfish have an apparent memory span of about three seconds, so they live in a permanent present. Humans, on the other hand, can retrieve, re-experience, and symbolically manipulate past experiences going back many decades, and imagine events and experiences far into the future. However, the ability to conjure past and future experiences can also be the source of considerable insecurity (Leary, 2007).

Remembering the past is sometimes pleasurable but also can produce insecurity and negative emotions such as nostalgia, sadness, regret, guilt, and shame. These emotions can have important evolutionary functions (Al Shawaf et al., 2015; Forgas, 2022), but also contribute to experiences of insecurity and anxiety.

Mental time travel into the future is similarly problematic. Humans are not particularly good at predicting future events and often suffer from a strong negativity bias (Fazio, Eiser & Shook, 2004; Kiernan & Baumeister, 2019). Our ability to forecast future affective reactions is deeply flawed (Gilbert, 2006). Many clinical psychologists see the inability to live in the present and spend most of our time pre-occupied with the past or future as one of the major sources of insecurity, anxiety, and depression (Forgas & Baumeister, 2019). Learning to live in the present is a common theme in many psychotherapies.

Social Insecurity

Humans have evolved to live in intense, small-scale primary groups, an ancestral social environment where feelings of belonging and identity were a natural consequence of daily social interaction, and a significant defense against feelings of insecurity and uncertainty (Hirschberger; Hogg & Gaffney; van den Bos, this volume). All this changed with the advent of anonymous, impersonal mass societies based on the revolutionary ideas of freedom and individualism. Fundamentally, we are tribal creatures, now living in a world of strangers. The nature of this historical change from communal to individualistic societies was analyzed by Durkheim (1997) as the emergence of a new form of social integration, organic solidarity, replacing the archaic forms of mechanical solidarity that characterized face-to-face human groups since time immemorial.

The rise of insecurity, alienation, anxiety, and even suicide were associated with this change. Toennies (2001) described this process as a move from community (Gemeinschaft) characterized by face-to-face sociability to associations (Gesellschaft) featuring impersonal, rule-based relationships. Max Weber (1947) in his classic analysis of bureaucracy explored the dehumanizing, alienating consequences of applying impersonal bureaucratic rules to individuals. In philosophy, Heidegger coined the term '*Geworfenheit*' (thrown into the world) to describe this insecure state of affairs.

Modern individualist societies are inhabited by WEIRD people – Western, Educated, Industrialized, Rich, and Democratic – who are frequently also lonely and insecure (Henrich, 2020; Zimbardo, 1989). Primary groups provided our ancestors with a natural sense of status, identity, and belonging – needs that now often remain unsatisfied and provide rich pickings for advertisers, politicians, and marketers (Zimbardo, 1989). The scientific study of social behavior – social psychology – probably emerged in individualistic Western societies at the very point when, perhaps for the first time in human history, social interaction became problematic.

The experience of social insecurity and loneliness, although certainly not a new phenomenon, has increased exponentially in the modern age (Berkowitz, 2022). Throughout human history, loneliness was probably a borderline phenomenon suffered only in certain marginal social conditions. It has now become a common everyday experience. Many people feel adrift without a primary group, experiencing life as purposeless and lonely (Zimbardo, 1989). It is this sense of abandonment (*Verlassenheit*) that makes people vulnerable to various tribal narratives offering national pride, ethnic superiority, or religious salvation (see also Section 4 of this book). It is the modern search for meaning and certainty that leads people to accept more comforting, if illusory, world views (Berkowitz, 2022).

Relationship Insecurity

In addition to deriving status, identity, and significance from our primary group, humans also need a pattern of close and nurturing intimate relationships to feel secure. Maintaining rewarding relationships in turn requires a sophisticated mental ability to understand, represent, and predict the behavior of partners. After birth, humans have an exceedingly long period of dependence on others, and during this time, a great deal of knowledge about the world is learned, including basic patterns of how to relate to significant others (Tomasello, 1999).

At this time, life-long predispositions of forming secure, insecure, or avoidant relationships are shaped by the behavior of primary caregivers (Mikulincer & Shaver, this volume). The challenge of forming secure and sustaining relationships is a major source of insecurity throughout life (Murray & Lafranche, this volume). Security challenges from the external world – such as the recent COVID pandemic – can be effectively cushioned by the existence of secure attachments to significant others (Arriaga & Kumashiro, this volume).

However, one can never perfectly know the mind of another person, and so our very security is always at risk from the mysterious workings of another person's mind. An important field of psychological research, attribution theory, is devoted to studying the mental inferences humans engage to understand and predict the behaviors of others as causal agents. Unfortunately, such knowledge is necessarily imperfect. Inferences about others based on observed behaviors are only approximations, leaving a great deal of uncertainty and insecurity in their wake.

Nowhere is this dilemma more acute than in our intimate relationships that are often the bedrock of our emotional security (Reis, 2022). Although the theory of attachment security was originally conceived to focus on close relationships, the basic principles apply across the entire spectrum of our important relationships, including romantic partners, family members, teachers, work associates, and even public figures and political leaders (see Section 3, this volume). The ability to trust and depend on others can have far-reaching consequences not only for psychological insecurity but also for the effective functioning of larger social systems (van den Bos; van Prooijen, this volume). In the face of these challenges, how do people manage to maintain a sense of security?

Managing Insecurity

As the world poses numerous threats to our survival, in the face of endemic insecurity humans have developed a range of sophisticated defensive strategies (Dawkins, 2009). Psychologists have done much to explore the psychology of insecurity and uncertainty and its implications for human reasoning (Gigerenzer, 2015; Kahneman, 2013). As we shall see in this section, some of these strategies involve internal, cognitive processes, some focus on the search for meaningful and supportive social relationships, and some rely on finding comfort in reassuring group affiliations and memberships (Tajfel & Forgas, 2000).

Attempts to manage insecurity also extend to the cultural domain. Misfortunes that produce insecurity often give rise to practices that give moral meaning to adverse events and promote group integration and progress (Fiske; Fiedler & McCaughey, this volume). Insecurity and uncertainty are inevitable features of being human, and shared group narratives are a good way to manage these conditions and at the same time promote group cohesion.

Psychological Strategies

As *Homo sapiens* is a profoundly social creature, maintaining an orderly view of the world and ourselves is a primary requirement for survival. Within social psychology, research on *cognitive dissonance* explored how people seek to maintain consistency among their ideas about the world as a means of managing insecurity. As Cooper and Pearce (this volume) show, dissonance reduction is often motivated by the need to maintain our confidence in consensual social reality.

When inconsistent cognitions challenge our representation of reality, psychological processes are activated to reduce the ensuing uncertainty. In extreme cases, people may even jeopardize their own performance in order to safeguard the appearance of competence by engaging in self-handicapping strategies to manage performance insecurity (Hirt et al., this volume). Seeking upward or downward social comparison is another motivated psychological strategy often designed to manage feelings of insecurity (Festinger, 1954).

In other words, the troubling experiences of insecurity and dissonance are "specifically caused by disruptions to our certainty about social reality" (Cooper & Pearce, this volume). Dissonance reduction processes such as effort justification, selective information seeking, and the like clearly serve a social integrative purpose and not just the maintenance of a positive self-concept. It is not surprising then that dissonance reduction often involves seeking social support, affiliation, and confirmation from meaningful others and groups (see also Crano & Hohman; Hirschberger; Hogg & Gaffney; Mikulincer & Shaver, this volume).

Social Support

The experience of insecurity is often managed by recruiting social support through targeted communication strategies and persuasion (Crano & Hohman, this volume). Social psychology offers convincing evidence for the powerful and universal human need to seek agreement and support from others in times of uncertainty. Sherif's (1936) classic studies on norm formation and norm maintenance, Asch's (1951) conformity experiments, Milgram's (1974) classic work on obedience, and Tajfel's work on ingroup favoritism (Tajfel & Forgas, 2000) all confirm the primary evolutionary motivation for humans to fit in and be accepted in meaningful social groups.

Contemporary social media plays an important role in both the creation and the management of insecurity. The cyberspace is eminently suited to creating insecurity but also promotes the formation of virtual groups that offer identity and belonging. The more radical and extreme the group, the more effective it can be in reducing insecurity (Hirschberger; Hogg & Gaffney; Kreko, this volume). Many of the algorithms that were initially designed to capture attention for advertising purposes also promote the creation of virtual identity groups, magnifying perceived threats to group identity (Fisher, 2002).

Accordingly, insecurity is often managed by becoming part of a collective, group, or political movement that becomes more important than the truth (Part 4 of this volume). Much of the research on motivated reasoning shows that membership in a group will often lead to accepting a collective ideology even in the face of clear contrary factual evidence (Tajfel & Forgas, 2001). Human thinking is less concerned with truth than with maintaining consensual *meaning*, which is the main function of collective narratives (Kahneman, 2014).

Collective Narratives

Humans are fundamentally story-telling animals. Managing insecurity is one of the driving forces of story creation and transmission (Harari, 2018). Collective narratives may often be fallacious and mistaken, but they nevertheless do satisfy a fundamental longing for security, epistemic meaning, and significance. For intensely social creatures, sharing stories with others is a powerful method of establishing a consensual reality and a shared system of norms and expectations within an identity group (see also Fiske; Hirschberger, this volume).

Stories are primarily not about reality – they are tools that promote group cohesion and survival. Humans typically do not think analytically in terms of facts, reality, or statistics (Kahneman, 2013; see also Krueger & Gruening, this volume). Truth comes a distinct second when evaluating the effectiveness of stories and providing meaning. Promoting group cohesion is often far more important than truth and reality (Hogg & Gaffney, this volume). Religions are a good example of time-honored story-making, offering an illusion of security, certainty, understanding, and a mechanism of social control and integration. It is remarkable that the human epistemic motivation to seek understanding eventually still led to the emergence of rationality and science.

Collective fiction succeeds because people prefer simple, consistent, and coherent explanations to the messy complexity of reality. As Hannah Arendt noted, when faced with insecurity, people will often choose the consistent certainty of an ideology, not because they are stupid or wicked, but because this escape grants them a degree of coherence and self-respect (see also Hirschberger; Hogg & Gaffney; van Prooijen; van den Bos, this volume).

Of course, not all stories are created equal. Some stories offer better ideas and strategies that ultimately help to produce better outcomes (Forgas et al., 2016). The current age is based on a set of enlightenment narratives emphasizing universal individuality, freedom, equality, and tolerance (Forgas, Crano & Fiedler, 2020). Empowering the individual at the expense of the collective produced unprecedented improvements in the human condition, yet these enlightenment values are now under challenge from the rise of populist, tribal ideologies both from the political left (woke ideology, cultural Marxism, critical race theory) and from the right (Trump, Erdogan, Orban et al.; see also chapters in Part 4, this volume).

It is remarkable how even well-educated people can be attracted to collectivist ideologies that are clearly fallacious. The lasting influence of Marxism on many left-wing intellectuals is a good example. Marxism offers the benefits of simplicity, certainty, moral superiority, and utopistic idealism to its followers, just as many religions do. Yet the economic fundamentals of Marxism have been known to be false for over a hundred years, and regimes based on marxist ideology have failed catastrophically whenever attempted. It is the simplicity and moral trappings rather than the truth of such ideologies that attract unshakeable support.

Liberty, freedom, and individualism offered a more successful narrative (Mill, 1859) but are now challenged by the growth of populism, tribalism, and political polarization (Albright, 2018; Forgas, 2021; Forgas, Crano & Fiedler, 2021). The resurgence of various collectivist conflict narratives suggests that we may well be on the cusp of a dramatic change in the stories we live by and a likely increase in experiences of psychological insecurity and uncertainty.

Consequences and Implications: Insecurity and Public Life

In this section, we consider some of the consequences of psychological insecurity for public life and politics. Of course, insecurity can be useful, allowing us to anticipate, prepare, and plan for possible challenges. Sometimes, seeking risks and dangers can even be beneficial and help reinforce group cohesion (Fiedler & McCaughey, this volume). In numerous experiments, we found that mild negative moods produce significant cognitive benefits, improving attention, promoting eyewitness memory, reducing common judgmental errors, and improving the effectiveness of verbal and interpersonal strategies (for a recent review, Forgas, 2022). In a similar way, mild dysphoria also improves the detection of deception and reduces gullibility (Forgas, 2019).

It is, however, the problematic social and political consequences of psychological insecurity that call for greater attention here. Throughout history, many political and social movements were driven by insecurity and fear, resulting in an erosion of social trust, mass hysteria, and belief in irrational ideologies and conspiracy theories (see chapters in Section 4 of this book).

Catastrophizing

Excessive concern about terrorism, migrants, new technologies, climatic events, and pandemics can be harnessed for political and commercial purposes. Walker (2003) has written extensively about the political manipulation of psychological insecurity in his book *The United States of Paranoia*. It is not surprising that the powerful propensity to pay selective attention to threatening or negative information augmented by human gullibility is exploited by various experts, forecasters, and the media who can acquire personal benefit from recruiting attention (Forgas, 2019; Forgas & Baumeister, 2019).

It is not surprising that the powerful propensity to pay selective attention to threatening or negative information augmented by human gullibility is exploited by various experts, forecasters, and the media who can acquire personal benefits from recruiting attention (Forgas, 2019; Forgas & Baumeister, 2019). Thomas Malthus, Paul Ehrlich, Al Gore, the Club of Rome, and many others acquired temporary fame by making catastrophic but wildly inaccurate predictions. In the 'Population bomb' (1968) and 'The end of affluence', Ehrlich (1974) predicted imminent social collapse, starvation, and the need for self-sufficiency in apocalyptic terms: "The battle to feed all of humanity is over. In the 1970s and 1980s, hundreds of millions of people will starve to death". He was utterly wrong, and now we may be heading into the opposite problem: not enough people.

Despite its alarmist tone and wildly inaccurate predictions, such messages find ready acceptance in many quarters as a rallying cry against the free-market economic system. Media celebrities, such as Al Gore and Greta Thunberg, attract attention by whipping up public hysteria. Such is the power of insecurity that H. L. Mencken, the essayist, once described public life as 'a combat of crazes': "The aim of practical politics is to keep the populace alarmed (and clamorous to be led to safety) by menacing it with an endless series of hobgoblins, most of them imaginary".

The COVID pandemic revealed how insecurity and catastrophizing can lead to public hysteria and erroneous decisions (Atlas, 2022; Frijters, Foster & Baker, 2021; see also Kreko, this volume). In otherwise liberal and tolerant societies, extremely autocratic and sometimes unnecessarily restrictive policies were adopted and often triggered tribal animosity between supporters and opponents on issues such as vaccine mandates. Some countries, such as Australia, even prohibited citizens from *leaving* the country – a most serious transgression against fundamental civil rights.

In recent years, a new source of insecurity, technophobia, joined the time-worn fears of overpopulation, starvation, and social upheaval. Fear of power lines, vaccination, 5G, and GM foods are just some of the more common such fears. It turns out that similar concerns generating insecurity were already common in the 19th century, with claims that electric lights produced brain damage, or that railway travel produced railway madness due to shaking. The power of scare stories to continue to demand attention is testimony to the human bias to concentrate on negativity (Kiernan & Baumeister, 2020).

Political Implications

The tendency to focus on negativity has important political implications and insecurity has shaped politics since time immemorial. Plato was among the first to argue that humans are too easily influenced by fears, hysteria, and emotions, making rational decision-making as required by democracy deeply problematic. These concerns are now shared by several contemporary writers (Brennan, 2016; Caplan, 2008). Psychological research also confirms that there are clear limits to rational reasoning and analytic thinking (Kahneman, 2013). Many human decisions appear to be biased, self-serving, and occasionally, delusional (Mercier & Sperber, 2017). Although thinking fast and heuristically often allows intuitive processes to produce acceptable results (Gigerenzer, 2015), political decision-making frequently demands slow, effortful, and analytic thinking (Forgas, Fiedler & Crano, 2015).

This matters a great deal considering the growing recent appeal of populism and tribalism (Albright, 2018; Forgas, 2021; Forgas, Crano & Fiedler, 2021; Fukuyama, 2018) and the growth of extremist political groups both on the political right (Trump, Orban, AfD, LePen) and on the left (woke, BLM, Antifa, etc.). Irrespective of their ideologies, these movements represent a serious collectivist challenge to individualism, liberalism, and tolerance (Mill, 1859). Uncertainty-identity theory (see Hogg & Gaffney, this volume) argues that insecurity promotes extreme group identification (see also Hirschberger, van den Bos, this volume).

Insecurity and politics also interact in other ways. During the last few hundred years, the state has emerged as the primary provider of security, displacing earlier systems based on family, kinship, and collective. The introduction of pensions, unemployment benefits, and various support schemes reduced material insecurity, yet psychological insecurity seems undiminished. Indeed, some research suggests a paradoxical pattern: the richer and safer we become, the more insecure we feel (Pinker, 2018). This 'progress paradox' suggests that fear of losing what we already have may have a greater influence on feelings of insecurity than actual, real insecurity.

There are growing demands for the state to go even further in alleviating material insecurity. Despite the obvious benefits, such moves may not alleviate psychological insecurity. The imposition of ever-more complex bureaucracies to administer such schemes may actually make people feel more dependent and insecure than before (Weber, 1947). Citizens happily accept these programs, but by becoming more dependent they lose part of their autonomy and independence. Eventually, state support to reduce insecurity becomes an entitlement, and eventually, an inalienable right.

Commercial Implications

The exploitation of insecurity is also all too common in the commercial sphere. Many advertising messages target people's insecurities, misleadingly promising status, identity, and belonging as associated with purchasing decisions. Such manipulative messages ultimately cannot promote human well-being (Forgas et al., 2019). Certain product categories, such as cosmetics, luxury-branded goods, upmarket cars, and many fashion items, falsely suggest that through consumption we can alleviate insecurity and acquire meaningful symbols of status and social identity.

Other marketing strategies directly exploit insecurity and fear. Many products seek buyers by manipulating irrational paranoia about harmful chemicals, fertilizers, 5G, etc. The market for health products with unproven efficacy, food supplements, vitamins, and fake health devices directly benefits from fueling feelings of insecurity. Many products are now advertised by highlighting the absence of allegedly 'harmful' substances rather than emphasizing their benefits.

Overview of the Book

In considering these issues, the book is divided into four parts of five chapters each, dealing with (1) the nature and functions of human insecurity, (2) managing insecurity, (3) the role of insecurity in social relationships, and (4) the role of insecurity in politics and public life.

In the first part, this introductory essay (Forgas, Chapter 1) offers a general overview of the nature, origins, and varieties of insecurity in human affairs, and its consequences for individuals and public life. In Chapter 2, von Hippel and Merakovsky review the evidence for the evolutionary origins of insecurity as a product of the unique human capacity to imagine and forecast alternative futures. Pyszczynsky and Sundby (Chapter 3) focus on existential and metaphysical insecurity and death-related anxiety, suggesting that death awareness fundamentally challenges our sense of agency and security. In Chapter 4, Kruglanski and Ellenberg discuss the costs and benefits of insecurity, suggesting that reactions to insecurity depend on a wide variety of psychological variables. Fiedler and McCaughey, in Chapter 5, offer a psychological analysis of why the challenges of insecurity are often advantageous and can be the driver of better outcomes in a variety of trade-off settings.

Understanding the Psychology of Insecurity   **17**

The second part of the volume looks at how humans manage psychological insecurity. Cooper and Pearce, in Chapter 6, offer an impressive extension of cognitive dissonance theory and suggest that dissonance reduction is driven by the need to maintain a consensual sense of reality. In Chapter 7, Crano and Hohman analyze how insecurity promotes the need to obtain support through social influence mechanisms and persuasion. Hirt, Eyink, and Heiman, in Chapter 8, review the fascinating literature and their research on self-handicapping, a paradoxical strategy of reducing uncertainty by compromising actual performance. Krueger and Gruening, in Chapter 9, discuss how freedom of the will can be a source of existential insecurity and show how people might manage the intrinsic uncertainty when interacting with others using a game theoretical approach. Chapter 10 by Fiske offers a fascinating cross-cultural perspective on how insecurity caused by misfortune can promote social practices designed to find moral meaning in adversity.

The third section of the book looks at the role insecurity plays in social relationships. Mikulincer and Shaver (Chapter 11) review attachment theory and discuss how the search for attachment security is a major factor in many kinds of human relationships. In Chapter 12, Murray and Lamarche show how close relationships represent a source of security and how relationship security is linked to sources of insecurity in the wider world around us. Arriaga and Kumashiro, in Chapter 13, discuss how attachment security interacts with insecurity created by pandemic conditions, suggesting that public insecurity can be a challenge, but also an opportunity for developing attachment security in close relationships. Hogg and Gaffney, in Chapter 14, describe their influential *uncertainty-identity theory* and argue that self-uncertainty can motivate extreme forms of group identification. In Chapter 15, Hirschberger suggests that group survival motivation is a major consequence of insecurity, often best served by ideological diversity, accounting for the often-intractable nature of intergroup animosities.

In the final, fourth section of the book, we turn to the social, cultural, and political consequences of psychological insecurity. In Chapter 16, van den Bos argues that psychological insecurity and uncertainty play a key role in the erosion of trust in social and political institutions. Forgas, in Chapter 17, analyzes the role of insecurity in the growth of populist autocracies, illustrated through the loss of democracy in countries such as Hungary. Jussim, Finkelstein, and Stevens (Chapter 18) discuss the alarming growth of doctrinaire, authoritarian left-wing political movements in universities and argue that totalitarian ideologies, such as critical race theory, cancel culture, and woke movements, can be directly linked to psychological insecurity. Kreko, in Chapter 19, explores the role of insecurity in the recent spread of conspiracy theories and pseudoscience, within the context of the recent COVID epidemic. Finally, in Chapter 20, van Prooijen offers a theory and empirical evidence demonstrating the role of insecurity in producing anti-establishment sentiments and support for radical political movements.

Conclusion and Acknowledgments

The experience of psychological insecurity is one of the most ubiquitous features of human existence. Our aim with this book is to contribute to a better understanding of the nature and characteristics of psychological insecurity, its sources, management, and social and individual consequences. Insecurity is closely linked to the unique human ability to represent fictional realities and our ability to see ourselves and others as causal agents. Understanding the nature of insecurity is particularly important at this time, when populist political movements can easily exploit the current longing for security and certainty by proffering misleading tribal ideologies (Forgas, Crano & Fiedler, 2021). At its extreme, unquestioning attachment to cohesive tribal collectives can produce conflict and violence, as human history amply demonstrates (Harari, 2014; Koestler, 1967).

We hope these chapters will help to highlight the complex patterns of the origins and sources of psychological insecurity, its management, its role in our close relationships, and its influence on politics and public life. Recent years produced genuine breakthroughs in our understanding of human insecurity and its consequences. This introductory chapter tried to give a general background to the volume and anticipate some of the main themes that would be covered.

As editors, we are deeply grateful to all our contributors for accepting our invitation to contribute to this, the 24th volume of the Sydney Symposium of Social Psychology Series, and sharing their valuable ideas with our readers. We are also grateful to the Australian Research Council and the University of New South Wales for financially supporting this project. We sincerely hope that the insights contained in these chapters will contribute to a better understanding of the crucial role that insecurity plays in shaping us both as individuals and as communities.

References

Acemoglu, D. & Robinson, J. A. (2019). *The Narrow Corridor: States, Societies, and the Fate of Liberty*. New York: Penguin.

Albright, M. (2018). *Fascism: A Warning*. New York, NY: Harper Collins Press.

Al-Shawaf, L., Conroy-Beam, D., Asao, K. & Buss, D. (2015). Human Emotions: An Evolutionary Psychological Perspective. *Emotion Review*. 10.1177/1754073914565518.

Asch, S. E. (1951). Effects of Group Pressure Upon the Modification and Distortion of Judgment. In H. Guetzkow (ed.) *Groups, Leadership, and Men* (pp. 177–190). Pittsburgh, PA: Carnegie Press.

Atlas, S. W. (2022). *A Plague Upon Our House*. New York: Simon & Schuster.
Berkowitz, R. (2022). Lessons from Hannah Arendt on arresting our 'Flight from reality'. *Quillette*, 19th September 2022. https://quillette.com/2022/09/19/lessons-from-hannah-arendt-on-arresting-our-ideological/
Brennan, J. (2016). *Against Democracy*. Princeton, NJ: Princeton University Press.
Buss, D. (2019). *Evolutionary Psychology: The New Science of the Mind*. New York: Taylor & Francis.
Camus, A. (1955). *The Myth of Sisyphus and Other Essays*. New York: Alfred A. Knopf. ISBN 0-679-73373-6.
Camus, A. (1946), *The Stranger* (translated by Stuart Gilbert). New York: Alfred A. Knopf.
Caplan, B. (2008). *The Myth of the Rational Voter*. Princeton: University Press.
Dunbar, R. I. (1998). The Social Brain Hypothesis. *Evolutionary Anthropology: Issues, News, and Reviews: Issues, News, and Reviews*, 6(5), 178–190.
Durkheim, E. (1997). *The Division of Labour in Society*. New York: Free Press.
Ehrlich, P. (1968). *The Population Bomb*. The Sierra Club, New York: Ballantine.
Ehrlich, P. (1974). *The End of Affluence*. New York: Ballantine Books.
Fazio, R. H, Eiser, J. R. & Shook, N. J. (2004). Attitude Formation through Exploration: Valence Asymmetries. *Journal of Personality and Social Psychology*, 87, 293–311. https://doi.org/10.1037/0022-3514.87.3.293. PMID: 15382981.
Festinger, L. (1954). A Theory of Social Comparison Processes. *Human Relations*, 7(2), 117–140.
Fisher, M. (2002). *The Chaos Machine: The Inside Story of How Social Media Rewired Our Minds and our World*. New York: Little, Brown.
Forgas, J. P. (2019). Happy Believers and Sad Skeptics? Affective Influences on Gullibility. *Current Directions in Psychological Science*, 28(3), 306–313.
Forgas, J. P. (2021). *The Psychology of Populism: Tribal Challenges to Liberal Democracy*. Sydney: Centre for Independent Studies Occasional Paper. https://www.cis.org.au/publications/occasional-papers/the-psychology-of-populism-tribal-challenges-to-liberal-democracy/ See also video at https://www.youtube.com/watch?v=7XXWNz0Dkyo
Forgas, J. P. & Baumeister, R. F. (2019). (Eds.). *The Psychology of Gullibility: Fake News, Conspiracy Theories and Irrational Beliefs*. New York: Routledge.
Forgas, J. P., Crano, W. D. & Fiedler, K. (2021). *The Psychology of Populism: Tribal Challenges to Liberal Democracy*. New York: Routledge.
Forgas, J. P., Fiedler, K. & Crano, W. (2015). *Social Psychology and Politics*. New York: Routledge.
Frijters, P., Foster, G. & Baker, M. (2021). *The Great COVID Panic*. Austin, TX: Brownstone Institute.
Fukuyama, F. (2018). *Identity: The Demand for Dignity and the Politics of Resentment*. New York: Farrar, Straus, and Giroux.
Gigerenzer, G. (2015). *Simply Rational: Decision Making in the Real World*. New York: Oxford University Press.
Gilbert, D. (2006). Affective Forecasting: A User's Guide to Emotional Time Travel. In Forgas, J. P. (Ed.). *Affect in Social Thinking and Behavior*. New York: Psychology Press.
Harari, Y. N. (2014). *Sapiens: A Brief History of Humankind*. London: Random House.
Henrich, J. (2020). *The Weirdest People on Earth*. New York: Farrar, Straus, and Giroux.
Hobbes, T. (1651/2020). *Leviathan (Royal Classics Edition)*. London: Royal Classics.

Hoffmann, G.-L. (2003). *The Secret Dowry of Eve: Woman's Role in the Development of Consciousness*. Simon and Schuster. ISBN 978-1-59477-561-1

James, W. (1902/1985). *The Varieties of Religious Experience*. Boston: Harvard University Press.

Kahneman, D. (2013). *Thinking, Fast and Slow*. New York: Farrar Straus Giroux Inc.

Kant, I. (Ed.). (1784/2008). *Toward Perpetual Peace and Other Writings on Politics, Peace, and History* (D. L. Colclasure, Trans.). New Haven: Yale University Press.

Koestler, A. (1967). *The Ghost in the Machine* (1990 reprint ed.). Penguin Group. ISBN 0-14-019192-5.

Leary, M. (2007). *The Curse of the Self*. Oxford: University Press.

MacLean, P. D. (1990). *The Triune Brain in Evolution: Role in Paleocerebral Functions*. New York: Plenum Press. ISBN 0-306-43168-8. OCLC 20295730.

Mercier, H. & Sperber, D. (2017). *The Enigma of Reason*. Harvard University Press. https://doi.org/10.4159/9780674977860

Milgram, S. (1974). *Obedience to Authority: An Experimental View*. London: Tavistock Publications.

Mill, J. S. (1859/1982). *On Liberty*. Penguin: Harmondsworth.

Pinker, S. (2018). *Enlightenment Now: The Case for Reason, Science, Humanism, and Progress*. USA: Penguin Books.

Plato. (1943). *Plato's The Republic*. New York: Books, Inc.

Popper, K. (1945). *The Open Society and Its Enemies*. London: Routledge.

Sherif, M. (1936). *The Psychology of Social Norms*. New York: Harper & Brothers.

Tajfel, H. & Forgas, J. P. (2000). Social Categorization: Cognitions, Values, and Groups. In C. Stangor (Ed.), *Key Readings in Social Psychology. Stereotypes and Prejudice: Essential Readings* (pp. 49–63). New York: Psychology Press.

Tomasello, M. (1999). *The Cultural Origins of Human Cognition*. Boston, MA: Harvard University Press.

Tönnies, F. (2001). *Community and Civil Society*. Cambridge: Cambridge University Press.

Von Hippel, W. (2018). *The Social Leap: The New Evolutionary Science of Who We Are*. New York: Harper.

Walker, J. (2003). *The United States of Paranoia*. New York: Harper Collins.

Weber, M., 1864–1920. (1947). *Max Weber, the Theory of Social and Economic Organization*. New York: Free Press.

Zimbardo, P. (1989). *Loneliness: Theory, Research, and Applications*. Thousand Oaks, CA: Sage Publications.

2

THE EVOLUTION OF INSECURITY

William von Hippel and John Merakovsky

UNIVERSITY OF QUEENSLAND, AUSTRALIA

Abstract

To the best of our knowledge, *Homo sapiens* is the only species on the planet that can envision mutually contradictory futures, nested future scenarios, and unfelt needs. These capacities appear to have emerged more than 1.5 million years ago in our *Homo erectus* ancestors. These capacities played an outsized role in the human success story, as they enable us to simulate the future, thereby shaping our own destiny through the creation of complex plans. Nonetheless, these capacities have associated psychological costs, the most notable of which are anxiety and insecurity. An animal that cannot envision the future need not worry about it, but an animal that is aware of mutually contradictory future possibilities has a great deal to worry about. Nonetheless, the aversiveness of insecurity is highly adaptive, as it motivates people to shape the future in ways that enhance their security and safety. In this chapter, we trace the evolution of insecurity, the nature of its experience among hunter-gatherers, its role in the development of agriculture, and its manifestations in residents of modern nation-states.

The Evolution of Foresight

To some degree, the distinction between the immediate present and the immediate future is arbitrary, and particularly, within the bounds of that arbitrariness, it seems likely that all animals can envision the future. Indeed, animals tend to orient toward surprising events (McBride, 2012), providing clear evidence that they had expectations of the future based on their knowledge of the past (Suddendorf & Corballis, 1997). Thus, in a limited sense, all animals engage in prospection, envisioning a future that represents either a direct continuation

DOI: 10.4324/9781003317623-3

of the present or a future that is shifted by the application of reliable rules that result in changes to the present (e.g., sunrise and sunset).

At least 1.7 million years ago, our *Homo erectus* ancestors evolved the capacity to simulate a more complex version of the future than can be achieved via associative learning alone. The clearest evidence for that capacity can be found in the Acheulean tools that our ancestors invented at that point in time. In contrast to the simpler Oldowan tools they had inherited from their ancestors, which are simply slightly sharpened stones, the Acheulean tools of *Homo erectus* required advanced planning to make (see Figure 2.1). Not only does it take a fair bit of time to train people to create them, but when people who are trained to make them are asked what they would do next to finish a partially completed tool, fMRI reveals activation in frontal, planning regions of the brain and not just the motor cortex, which is activated when deciding how to finish making Oldowan tools (Stout et al., 2015).

This capacity to simulate the future and use that simulation to make decisions would have given our ancestors an enormous advantage over the other animals on the planet. Once our ancestors gained the capacity to envision future events, they would have evolved the ability to represent mutually contradictory futures, which, in turn, would have enabled them to prepare for various contingencies (Redshaw & Suddendorf, 2016). Because mutually contradictory futures often entail desired versus undesired outcomes, our ancestors' preparations would have been designed to increase the odds of the former and decrease the odds of the latter. Perhaps the simplest example of this process involves creating carry bags to bring tools with you that you do not need now but might need later (Langley & Suddendorf, 2020). But deliberative practice would likely have evolved soon after we gained the capacity to envision alternative futures. Practicing a skill requires more foresight and self-control than simply bringing along one's favorite tools, but practice would have been a likely consequence of the ability to envision the future and the desire to be prepared for it (Suddendorf, Brinums, & Imuta, 2016).

FIGURE 2.1 An Oldowan tool (Left Panel; Gallotti & Mussi, 2015) and an Acheulian tool (Right Panel; Diez-Martin et al., 2015), both from 1.7 million years ago.

Finally, once our ancestors had the ability to consider various futures and plan for them, they also gained the capacity to generate nested scenarios, envision how these scenarios might play out, and then make small (or large) tweaks to these simulations until they had developed a plan that seemed likely to succeed (Suddendorf, 2013). The power of these simulations becomes immediately apparent when we consider animals who cannot generate them. By way of example, consider an event that Jane Goodall (1986) witnessed when watching chimpanzees in the Gombe, Tanzania. The brief background to this story is that Melissa is a chimp who has just had a new baby, Passion is another chimp in Melissa's group, and Pom is Passion's adolescent daughter. Here is what Goodall wrote, in slightly abbreviated form:

> Melissa, with her three-week-old female infant, climbed to a low branch of a tree. Passion and her daughter Pom cooperated in the attack; as Passion held Melissa to the ground, biting at her face and hands, Pom tried to pull away the infant. Passion then grabbed one of Melissa's hands and bit the fingers repeatedly, chewing on them. Simultaneously Pom, reaching into Melissa's lap, managed to bite the head of the baby. Then, using one foot, Passion pushed at Melissa's chest while Pom pulled at her hands. Finally, Pom managed to run off with the infant and climb a tree. Melissa tried to climb also but fell back. She watched from the ground as Passion took the body and began to feed. Fifteen minutes after the loss of her infant, Melissa approached Passion. The two mothers stared at each other; then Melissa reached out and Passion touched her bleeding hand. An hour later Melissa again reached Passion, and the two females briefly held hands.

Hunter-gatherer human groups have often faced similar problems of cannibalism, murder, or excessive bullying, but all human groups solve them by making plans to attack the perpetrators while they sleep or to unite the many against the few (Boehm, 2009). In contrast, due to chimpanzees' inability to make complex plans, Melissa appeared to have little choice but to reconcile with her daughter's killers. Furthermore, this wasn't an isolated incident. Passion and Pom continued to kill and eat infants in their group for years. One mother lost three babies in a row, and it was then that Goodall realized that only one infant had survived its first month in the group in the last three years. Despite the simplicity and predictability of Passion and Pom's attacks, none of the mothers devised a successful strategy for dealing with this pair of cannibals, and the mother-daughter team devastated the reproductive potential of their group.

Chimpanzees can and do form coalitions, but cooperation is not their preferred or default mode of problem-solving (Bullinger et al., 2011; Mellis, Hare, & Tomasello, 2006). Chimps are also skilled manipulators, often using deception and other social strategies to gain status and avoid aggression

(Byrne & Whiten, 1988). Nonetheless, their social capacities are severely limited by their inability to envision a future that contains unfelt needs and their inability to simulate nested scenarios. As a consequence, there is no evidence that chimps ever use their partial theory of mind skills to enhance cooperative success, even though they occasionally use these skills in service of conflict and competition (Tomasello et al., 2005). In contrast, humans have combined their theory of mind with an inherently cooperative nature and a capacity to envision the future, and in so doing have created groups that are much more than the sum of their parts. Only in humans does theory of mind provide the basis for enhanced social cooperation, communication, and teaching (Krupenye & Call, 2019). In contrast to other animals, who are locked in the present, the combination of these abilities gives us enormous power to shape our own future.

The Psychological Costs of Foresight

Perhaps unsurprisingly, this capacity to envision the future and sometimes shape it comes with attendant costs. First, and perhaps most notably, because understanding the past helps us shape the future, humans are likely to spend a great deal of their mental lives reflecting on the past and envisioning possible futures (Suddendorf, Redshaw, & Bulley, 2022). This proclivity to live in a time other than the present is not necessarily a problem in and of itself, but it does introduce problems, most notably anxiety and insecurity (see also Hirschberger; Kruglanski & Ellenberg; Pyszczynski & Sundby, this volume). In contrast to fear, which is an intense emotion that dissipates quickly once the threat disappears, anxiety and insecurity are concerns about possible threats that may emerge in the future. Consequently, they are very difficult to ameliorate.

Anxiety and insecurity describe the subjective experience and emotional response to the mental construction of possible future threat scenarios (Miloyan et al., 2019). Anxiety manifests in different forms across the human lifespan, reflecting the specific vulnerabilities of each stage of life (Miloyan et al., 2019). During infancy, lack of proximity with the protective adult attachment figure results in separation anxiety and concomitant behavioral responses, such as crying, that draw the attention of caregivers (Bowlby, 1973; see also Mikulincer & Shaver, this volume). Once we enter childhood and adolescence, we become much more independent, but even in adulthood we still rely heavily on our social relationships to ensure our productivity and survival (see also Arriaga & Kumashiro; Murray & Lamarche, this volume). The ability to work toward group goals, engage in reciprocal relationships, and adhere to group norms and values contributes to an individual's social reputation, which was and is paramount for success in life (Romano et al., 2021; see also Crano & Hohman; Hogg & Gaffney, this volume). Indeed, the primary role of the self might be reputation maintenance (Baumeister, 2022).

Through this process of maturation from birth to adulthood, anxiety and insecurity shift from immediate survival concerns to broader concerns about reputation maintenance, threats to which lead to a form of insecurity known as social anxiety (Leary & Kowalski, 1997). Although social concerns might seem trivial in today's world, when physical survival is rarely a concern and hence coalition maintenance is no longer a matter of life and death, the processes that generate concern for these threats are the same today as they were in our ancestral past. For example, concerns with social reputation contribute to depression and suicidal ideation in adolescents (Wells et al., 2021).

Of course, thoughts of the future do not focus exclusively on factors that could go wrong, as people also spend a great deal of time thinking about how the future might pan out in desired ways and how to promote this possibility. This process evokes the emotion of hope, which is a uniquely human state that emerges from foresight and that motivates people to work optimistically toward a future goal (see also Fiedler & McCaughey, this volume). Hope represents an uncertain expectation of future goal achievement, often mediated through one's own agency. Given that humans traditionally achieved most of their goals through social cooperation (von Hippel, 2018), it is unsurprising that hope increases when people are in secure relationships (Moller et al., 2003).

Is Death Awareness An Inescapable Insecurity?

As should now be clear, once we evolved the capacity to envision the future, we gained the capacity to shape it as well. This capacity comes at a notable psychological cost in which humans spend a great deal of time worrying about unwanted future events. Nonetheless, from an evolutionary perspective, the aversiveness of worry is not a cost at all, as that aversiveness is what motivates and enables us to shape the future in ways that are beneficial to our survival and reproduction. Similarly, the more positive orientation toward the future – hope – also motivates and prepares us to be ready to capitalize on future opportunities. In combination, hope and worry lead us to form alliances, to practice, stockpile, and prepare, and more generally to engage in a wide variety of activities that might allow us to shape or at least exploit future events. Thus, from an evolutionary perspective, there seem to be no costs and only gains to the capacity to envision future events.

Unfortunately, there is an obvious exception to this rule. Our capacity to benefit by envisioning the future is limited to aspects of the future that can be exploited by our knowledge or shaped by our interventions. Knowledge of our eventual death is perhaps the most important aspect of the future that we cannot shape or easily exploit (beyond planning for its inevitability by leaving our assets to others). Although philosophers and eventually psychologists have studied how humans struggle with this knowledge for millennia, it is unclear why people respond to death as they do, and an evolutionarily informed theory remains elusive.

Building on Becker's *Denial of Death* (1973), Greenberg, Pyszczynski, and Solomon developed Terror Management Theory (1986), which outlines a variety of methods through which people attempt to cope with the inevitability of their own death (see also Pyszczynski & Sundby, this volume). The proposed psychological processes that provide protection against fear of one's own death range from the development of self-esteem (a sense of personal significance and value) to the construction of worldviews like religion that are not constrained by the physical self.

There is substantial empirical support for various aspects of Terror Management Theory (e.g., Pyszczynski, Solomon, & Greenberg, 2015), but the possibility that religion is a cultural adaption that circumvents death anxiety (through transcendence) is less clear (see also Fiske, this volume). For example, as Boyer (2001) notes, the explanations provided by different religions to accommodate unseen or unknown causation of physical phenomena in our world (particularly those that threaten our existence) are often more complicated than the mysteries they attempt to explain. Similarly, if religion were a cultural response that serves to allay anxiety about our inevitable death, one would expect to encounter less terrifying religious world constructs than our material one, which is not the case (Boyer, 2001). Furthermore, although more recent and highly successful religions like Islam and Christianity often provide clear explanations of what happens after death and how to maximize one's chances of a happy afterlife, the religions common in small-scale societies typically do not have these features.

Finally, it is worth noting that from an evolutionary perspective, survival is important only insofar as it enables successful reproduction. We can only experience our own death once, but we can experience the highly costly deaths of those close to us, particularly our offspring, many times – and indeed our ancestors lost nearly half of their offspring before they reached adulthood (Gurven & Kaplan, 2007). Thus, if there is a hierarchy of anxieties, and if that hierarchy is rooted in factors that threatened us in our ancestral past, then death anxiety would not sit at the top (Boyer, 2001). Rather, the anxiety of losing offspring might have been paramount. Or perhaps insecurity in the ability to attract a mate in the first place might have been more important, as a mate is a necessary precursor to worrying about one's children. Extending this line of thinking, perhaps the more general concern about one's social reputation might have been the greatest evolutionary threat that our ancestors faced, as mating and survival both depend on our social reputation. Dealing with these insecurities required specific strategies, whereas generalized death anxiety would have provided little to no value.

Fortunately, the attachment system provides some protection from these sources of insecurity. Attachment behaviors evolved to help human infants survive to adulthood (Simpson & Belsky, 2018) by obtaining proximity to their caregiver, which in turn provided them with food, safety, and the opportunity to learn the huge amount of information necessary to survive in ancestral environments (Mikulincer & Shaver, this volume). Unlike in other great apes, the attachment system does not become dormant in adolescence, as it appears to have been co-opted by natural selection in the form of pair-bonded romantic relationships that facilitate biparental care (Belsky, 2007; Fraley et al., 2005). But the attachment system goes well beyond romantic relationships, as humans rely on extensive friendship networks, ancestrally and today, to achieve their major life goals. Social and cultural learning, which lie at the foundation of human success, depend on attachment to other group members and to mentors who take a special interest in our success (Boyd, 2017; Laland, 2017).

Anxiety and Insecurity among Hunter-Gatherers

Food insecurity negatively impacts physical, social, emotional, and cognitive development in humans, and our survival as species has always depended on obtaining a sufficient quantity and quality of food to serve the caloric demands of our large brains. The high nutritional density of meat, compared to plants, thus played a critical role in our survival in the savannah, but even the best hunters are more likely to fail than succeed on any one hunt. As a consequence, our ancestors learned the value of obligatory sharing of the proceeds of the hunt. Additionally, given the propensity of meat to spoil quickly in low latitudes, the cost of sharing the proceeds of the hunt was comparatively low and easily offset by the future gains of reciprocal sharing from others. Thus, one of the universal dimensions of hunter-gatherer life in immediate return societies (those in which people have limited capacity to store food, and hence eat today what they killed today) is universal sharing of the proceeds of the hunt (Boehm, 2009). In this sense, cooperation became the primary behavioral strategy for offsetting the most important source of insecurity among hunter-gatherers.

Obligatory food sharing is a great equalizer in hunter-gatherer societies, but it does not completely level the playing field. Hunter-gatherer groups are nomadic, and they frequently split up and rejoin as they move from one location to another, following game and avoiding conflict with other groups. As a consequence of mandated sharing, the best hunters are not really better fed than other members of their group, but they do have the advantage that whatever group they are in will always have at least one good hunter. Thus, another universal strategy for avoiding insecurity about potential starvation is for humans to value hunters (particularly the best ones) in their group. Because hunting large game is also much easier for people who are not primary caregivers of small children, hunting is primarily a male activity and thus males are the ones who benefit from the esteem awarded to good hunters.

Perhaps the most notable evidence of the high esteem in which the best hunters are held is their greater reproductive success (Rueden & Jaeggi, 2016; Smith, 2004). Although widespread food sharing and the inability to accumulate material wealth are hallmarks of hunter-gatherer lifestyles, less successful hunters still have less access to mates due to their lower status. The end result of this combination of biological and cultural evolution is that planning enabled highly effective forms of hunting, which was then enhanced by obligatory sharing of the proceeds of the hunt. This universal strategy combatted food insecurity among our ancestors, but the varying abilities of individual hunters resulted in unequal social status and a resultant rise in status insecurity due to the strong relationship between status and mating success.

Lower-status males have limited options in dealing with this form of insecurity, which is based in an insecurity about their capacity to attract a mate. Fortunately, males can attract females through qualities other than their hunting prowess, as cooperativeness, kindness, and a willingness to care for others are nearly as important as hunting skills in ensuring the survival of one's family. Thus, an effective approach for attracting a partner among less skilled hunters is to increase the degree of provisioning and investment in biparental care (Kokko & Jennions, 2008). Due to the importance of such paternal care, females would have valued partners who would provision their young, thereby providing a mating strategy for lower-status males (Gavrilets, 2012). Nevertheless, lower-status men are less likely to marry and are at greater risk of cuckoldry. Given that conception is internal, and females are often motivated to seek additional partners for a host of reasons (Greiling & Buss, 2000), the risk of cuckoldry (and more broadly of relationship dissolution) introduces other forms of anxiety and insecurity, particularly for low-status individuals. The evolution of jealousy, mate guarding, and other strategies to minimize the loss of reproductive opportunities are a direct response to these forms of insecurity (Buss & Haselton, 2005).

In addition to these forms of insecurity, our hunter-gatherer ancestors also had to contend with the risks of pathogens, parasites, and predation. Although the latter are visible, successful predators are capable of remaining undetected until it is too late to avoid them, with the result that the three P's would have killed many of our would-be ancestors before they even knew they were at risk. These invisible (and nearly invisible) health and safety concerns would have been a major and ongoing source of insecurity, given the impossibility of avoiding them completely.

Perhaps unsurprisingly, these threats to our existence were more pressing prior to adulthood, by which point our immune system was stronger and our ability to protect ourselves from predators was at its maximum. Survival curves of hunter-gatherer societies show that humans were particularly likely to die in childhood, with an average of about 40% never making it to adulthood (Gurven & Kaplan, 2007). Thus, it comes as no surprise that anxiety and insecurity are common responses in children to the absence of their caregiver (Bowlby, 1973), as our ancestors were at the greatest risk when they were immature and relied on adult nurturance and support.

Insecurity and the Greatest Inflection Point in Human History

As discussed above, hunter-gatherers felt many of the same insecurities that we experience today – status insecurity, anxiety about being left out of the mating game, and worries about sickness. The latter concern has been mitigated (although not eliminated) by modern medicine, but the former concerns remain common. Nonetheless, hunter-gatherers' primary source of day-to-day insecurity, the worry about starvation, is no longer relevant for humans who have the good fortune to live in wealthy countries. The watershed event that eventually eliminated our worries about starvation was the development of agriculture, and indeed, people started planting seeds precisely to address the ever-present risk of starvation inherent in any immediate return society.

The invention of agriculture was clearly an idea whose time had come, given that societies in both the Middle East and China began agriculture at approximately the same time, 12,000 years ago. Some data raise the possibility that humans began planting food as soon as the climate was stable enough to enable farming (e.g., Feynman & Ruzmaikin, 2007), suggesting that it was the execution that was the primary challenge – not the idea itself. Although stuffing seeds into the ground so you know where the plants will grow may not be rocket science, the consequences were extraordinary. Agriculture had a far bigger impact on human lives than simply alleviating food insecurities, as it indirectly led to the development of cities, writing, science and technology, etc. Our ancestors who first started farming received none of these benefits, but the invention of agriculture is the single largest inflection point in the history of *Homo sapiens*.

As with other dramatic changes in lifestyle, the invention of agriculture solved one problem (food insecurity) but introduced many others. Our hunter-gatherer ancestors were probably largely indifferent to the weather (with the exception of long-term droughts and deadly floods), but the inevitable variability in rainfall in most temperate regions of the globe provided a new and important source of insecurity. Even when poor weather was not bad enough to cause everyone to starve, it could still easily ruin any one family of farmers. Similarly, pests like locusts and rats are of little concern to nomads, but they are deadly threats to farmers who store their food over the winter.

Animal husbandry, or the practice of farming animals rather than hunting them, was a similarly huge advance in stabilizing and expanding human food sources, but it too introduced a new set of problems. Most notably, living cheek by jowl with large numbers of domesticated animals introduced an apparently endless source of pandemics, as diseases that leap from animals to humans are often those against which we have limited or no immunities.

Despite these risks inherent in agriculture, the benefits clearly outweigh the costs, as the carrying capacity of the land (the number of humans who can live in a square kilometer) greatly increased with the advent of agriculture and the subsequent development of cities. Where once all humans were engaged in efforts to secure food for their livelihood, initially as hunter-gatherers and then as farmers, the increased efficiency of agriculture eventually released most people from the direct production of food so they could focus their efforts elsewhere. As a consequence, the nature of our insecurities has shifted rather dramatically, as having enough to eat is no longer a source of concern for most people.

Insecurity in Modern Nation-States

The creation of cities, and subsequently nation-states, played a central role in the development of writing, the arts and sciences, and almost all other modern human pleasures, but cities come with their own unique costs as well. Prior to the advent of cities, everyone spent their lives surrounded by people they knew well. With the move to cities, for the first time in human history, we began to spend our lives surrounded by strangers. Not only did this shift require a dramatic change in our psychology and cultural rules to accommodate our new existence (Henrich, 2020), but it also led to an important source of insecurity. People who are well known to us are understood risks and opportunities, but people who are unknown to us represent risks that are very difficult to calculate. Most are harmless, but assuredly all are not.

In response to this situation, humans invented a wide variety of new social mores as well as new laws and new enforcers of those laws. Hunter-gatherers had no police or formal government, in part because they spent their lives negotiating solutions to their problems directly with one another. Once we began to live in a world full of strangers, the development of impartial rules became much more important for the effective functioning of society, in large part because people are tempted to treat each other poorly if they are not known to each other and may never meet again. But trust is the basis of every well-functioning society, so a wide variety of rules and enforcers became necessary to enable sufficient trust for a functioning market economy (see also van den Bos; van Prooijen, this volume). Thus, to a large degree, the insecurities introduced by the ubiquity of strangers led to the creation of governance and law enforcement. The presence of laws and police might not eliminate the insecurities that humans feel, but they have made the world much safer than it used to be when we were all hunter-gatherers (Pinker, 2011), suggesting that felt insecurities and actual risks do not always track each other accurately.

Our increased safety is a very modern phenomenon; the immediate (and unfortunate) consequence of the development of cities was that humans began dominating one another in their efforts to be sure they benefitted from governance and law enforcement. Hunter-gatherers are fiercely egalitarian, and their nomadic way of life made it very difficult for any one person to enforce his will on anyone else. In contrast, once humans settled into farms and cities, our sedentary lifestyle and capacity to accumulate and store goods made it much easier for powerful people to exploit others. As a result, cities led to enormous inequality and despotism. It might seem that both would be rare, as the many who are poor and individually powerless could always take from the few who are rich and powerful. In practice, however, that has proven difficult, largely due to the networks of kin and allies that powerful leaders are able to foster and maintain through bribery and threat (Chagnon, 2013; see also Forgas, this volume). Furthermore, and importantly, because anarchy is often much worse for people's health and safety than despotic leaders, city dwellers eventually learned to trade their freedom and egalitarianism for inequality and despotism, even though no one wanted either (Hobbes, 1651).

The final source of insecurity that developed in modern nation-states was again an unintended consequence of the benefits wrought by science. Our hunter-gatherer ancestors worried every day about whether they would get enough to eat, but they never worried about their place in the universe. Their religions answered that question for them by linking them to family, to other life forms on earth, and most importantly, to their ancestors who were no longer living and their progeny who were not yet born. In this sense, their existence placed them in an unbroken chain from the past to the future, which connected them to a world that mattered. Scientific progress has broken that chain of meaningful connections and replaced it with a string of mindless processes that yielded our species and every other one via random chance, with our own particular existence being of negligible meaning or significance (see also Krueger & Gruening, this volume). In so doing, scientific advances also made religion less central in many of our lives, thereby replacing our previous understanding of our role in the universe and our importance and connection to a larger purpose with the uncomfortable reality that we are a trivial member of a trivial species living on one of many trillions of planets scattered among billions of galaxies.

Having indirectly demolished our understanding of our role in the universe via the creation of science, cities then began asking much more of us. People today are faced with an array of life choices never imagined by our ancestors, who never questioned what they would do for a living because everyone did much the same thing. Although our distant ancestors frequently faced uncertainty, they had little or no choice regarding their occupation, where they would live, the people with whom they would affiliate, etc. As a result, in our deep past, the struggle to find our place in society was limited to a few well-understood domains, such as whether a particular individual would try to be the best hunter, craftsman, or storyteller (keeping in mind that everyone engaged in all of these activities at least some of the time).

Modern living has introduced massive changes from one generation to the next, which in turn has put a much greater premium on individual life choices. For many people, the most important of those choices is finding and then pursuing their passion. The chance to pursue your passion in almost any direction is an enormous opportunity, but it is also a significant threat. People are expected to discover their calling among a seemingly endless array of opportunities – a process that can be both daunting and confusing. At the same time, our increased wealth, education, and urban living have reduced the depth of our connections to other members of our group.

As a result, modern humans experience a form of insecurity that our ancestors never encountered and that would probably make no sense to them. This insecurity is experienced as an almost overwhelming sense of choice in who we will become, which is then exacerbated by the fact that our diminished connections have left us bereft of the guidance that our ancestors could have sought had they been faced by these choices. For those of us who are lucky enough to discover our passion early in life, the freedom to go in any direction is an incredible gift. But for many of us, this freedom is a major source of insecurity, as every choice to pursue one possibility simultaneously precludes countless others.

References

Baumeister, R. F. (2022). *The self explained: Why and how we become who we are*. New York: Guilford Publications.

Becker, E. (1973). *The denial of death*. New York: Simon and Schuster.

Belsky, J. (2007). Childhood experiences and reproductive strategies. In R. Dunbar & L. Barrett (Eds.), *Oxford handbook of evolutionary psychology* (pp. 237–254). Oxford: Oxford University Press.

Boehm, C. (2009). *Hierarchy in the forest: The evolution of egalitarian behavior*. Cambridge, MA: Harvard University Press.

Bowlby, J. (1973). *Attachment and loss: Vol. 2. Separation: Anxiety and anger*. New York: Basic Books.

Boyd, R. (2017). *A different kind of animal: How culture transformed our species*. Princeton, NJ: Princeton University Press.

Boyer, P. (2001). *Religion explained: The evolutionary origins of religious thought*. New York: Basic Books.

Bullinger, A. F., Melis, A. P., & Tomasello, M. (2011). Chimpanzees, pan troglodytes, prefer individual over collaborative strategies towards goals. *Animal Behaviour, 82*(5), 1135–1141.

Buss, D. M., & Haselton, M. (2005). The evolution of jealousy. *Trends in Cognitive Sciences, 9*(11), 506–506.

Chagnon, N. A. (2013). *Noble savages: My life among two dangerous tribes--the Yanomamo and the anthropologists*. New York: Simon and Schuster.

Díez-Martín, F., Sánchez Yustos, P., Uribelarrea, D., Baquedano, E., Mark, D. F., Mabulla, A.,... & Domínguez-Rodrigo, M. (2015). The origin of the Acheulean:

The 1.7 million-year-old site of FLK West, Olduvai Gorge (Tanzania). *Scientific Reports, 5*(1), 1–9.

Feynman, J., & Ruzmaikin, A. (2007). Climate stability and the development of agricultural societies. *Climatic Change, 84*(3), 295–311.

Fraley, R. C., Brumbaugh, C. C., & Marks, M. J. (2005). The evolution and function of adult attachment. *Journal of Personality and Social Psychology, 89*(5), 731–746.

Gallotti, R., & Mussi, M. (2015). The unknown Oldowan: ~ 1.7-million-year-old standardized obsidian small tools from Garba IV, Melka Kunture, Ethiopia. *PLoS One, 10*(12), e0145101.

Gavrilets, S. (2012). Human origins and the transition from promiscuity to pair-bonding. *Proceedings of the National Academy of Sciences, 109*(25), 9923–9928.

Goodall, J. (1986). *The chimpanzees of Gombe: Patterns of Behavior.* Cambridge, MA: Harvard University Press.

Greenberg, J., Pyszczynski, T., & Solomon, S. (1986). The causes and consequences of a need for self-esteem: A terror management theory. In Baumeister, R. F. (Ed.), *Public self and private self* (pp. 189–212). New York: Springer.

Greiling, H., & Buss, D. M. (2000). Women's sexual strategies: The hidden dimension of extra-pair mating. *Personality and Individual Differences, 28*(5), 929–963.

Gurven, M., & Kaplan, H. (2007). Longevity among hunter- gatherers: A cross-cultural examination. *Population and Development Review, 33*(2), 321–365.

Henrich, J. (2020). *The WEIRDest people in the world: How the west became psychologically peculiar and particularly prosperous.* London: Penguin UK.

Hobbes, T. (1651). *Leviathan or the matter, Forme and power of a commonwealth Ecclesiasticall and civill.* London: Andrew Crooke.

Kokko, H., & Jennions, M. D. (2008). Parental investment, sexual selection and sex ratios. *Journal of Evolutionary Biology, 21*(4), 919–948.

Krupenye, C., & Call, J. (2019). Theory of mind in animals: Current and future directions. *Wiley Interdisciplinary Reviews. Cognitive Science, 10*(6), e1503–n/a.

Laland, K. N. (2017). *Darwin's unfinished symphony: How culture made the human mind.* Princeton, NJ: Princeton University Press.

Langley, M. C., & Suddendorf, T. (2020). Mobile containers in human cognitive evolution studies: Understudied and underrepresented. *Evolutionary Anthropology: Issues, News, and Reviews, 29*(6), 299–309.

Leary, M. R., & Kowalski, R. M. (1997). *Social anxiety.* New York: Guilford Press.

McBride, G. (2012). Ethology, evolution, mind & consciousness. *Journal of Consciousness Exploration & Research, 3*(7), 830–840.

Melis, A. P., Hare, B., & Tomasello, M. (2006). Engineering cooperation in chimpanzees: Tolerance constraints on cooperation. *Animal Behaviour, 72*(2), 275–286.

Miloyan, B., Bulley, A., & Suddendorf, T. (2019). Anxiety: Here and beyond. *Emotion Review, 11*(1), 39–49.

Moller, N. P., Fouladi, R. T., McCarthy, C. J., & Hatch, K. D. (2003). Relationship of attachment and social support to college students' adjustment following a relationship breakup. *Journal of Counseling and Development, 81*(3), 354–369.

Pinker, S. (2011). *The better angels of our nature: The decline of violence in history and its causes.* London: Penguin UK.

Pyszczynski, T., Solomon, S., & Greenberg, J. (2015). Thirty years of terror management theory: From genesis to revelation. In *Advances in experimental social psychology* (Vol. 52, pp. 1–70). Academic Press.

Redshaw, J., & Suddendorf, T. (2016). Children's and apes' preparatory responses to two mutually exclusive possibilities. *Current Biology, 26*(13), 1758–1762.

Romano, A., Giardini, F., Columbus, S., de Kwaadsteniet, E. W., Kisfalusi, D., Triki, Z.,... & Hagel, K. (2021). Reputation and socio-ecology in humans. *Philosophical Transactions of the Royal Society B, 376*(1838), 20200295.

Simpson, J. A., and Belsky, J. (2016). Attachment theory within a modern evolutionary framework. In Jude Cassidy and Phillip R. Shaver (Eds.), *Handbook of Attachment, Third Edition: Theory, Research, and Clinical Applications* (pp. 131–157). New York: Guilford Publications.

Smith. (2004). Why do good hunters have higher reproductive success? *Human Nature (Hawthorne, N.Y.), 15*(4), 343–364.

Stout, D., Hecht, E., Khreisheh, N., Bradley, B., & Chaminade, T. (2015). Cognitive demands of lower paleolithic toolmaking. *PLoS One, 10*(4), e0121804.

Suddendorf, T. (2013). *The gap: The science of what separates us from other animals.* New York: Basic Books.

Suddendorf, T., Brinums, M., & Imuta, K. (2016). *Shaping one's future self: The development of deliberate practice.* Oxford: Oxford University Press.

Suddendorf, T., & Corballis, M. C. (1997). Mental time travel and the evolution of the human mind. *Genetic, Social, and General Psychology Monographs, 123*(2), 133–167.

Suddendorf, T., Redshaw, J., & Bulley, A. (2022). *The invention of tomorrow: Foresight and the human quest to control the future.* NY: Basic Books.

Tomasello, M., Carpenter, M., Call, J., Behne, T., & Moll, H. (2005). Understanding and sharing intentions: The origins of cultural cognition. *Behavioral and Brain Sciences, 28*(5), 675–691.

von Hippel, W. (2018). *The Social Leap.* New York: HarperCollins.

Von Rueden, C. R., & Jaeggi, A. V. (2016). Men's status and reproductive success in 33 nonindustrial societies. *Proceedings of the National Academy of Sciences, 113*(39), 10824–10829.

Welch, R. D., & Houser, M. E. (2010). Extending the four-category model of adult attachment: An interpersonal model of friendship attachment. *Journal of Social and Personal Relationships, 27*(3), 351–366.

Wells, G., Horwitz, J., & Seetharaman, D. (2021). The Facebook files. *Wall Street Journal*, Sep 14, https://www.wsj.com/articles/the-facebook-files-11631713039.

3
THE INTERACTIVE ROLE OF DEATH, UNCERTAINTY, AND THE LOSS OF SHARED REALITY ON SOCIETAL AND INDIVIDUAL INSECURITY

Tom Pyszczynsky and Jonathan Sundby

UNIVERSITY OF COLORADO, COLORADO SPRINGS, USA

Abstract

Terror management theory (TMT) posits that human awareness of the inevitability of death gives rise to the potential for existential terror, which is managed by an anxiety-buffering system consisting of cultural worldviews, self-esteem, and close relationships. This chapter uses TMT as a point of departure for an analysis of the vicious cycle of psychological distress and maladaptive responses set in motion by turbulent world events. While many turbulent world events involve direct threats to continued existence (wars, terrorism, and pandemics), others undermine the psychological structures that provide protection from death anxiety (radicalization, extremism, and political divisiveness). People typically attempt to manage the anxiety instigated by troublesome world events by clinging to their cultural worldviews, which leads to suboptimal problem-solving and derogation of or fighting against those with different worldviews. These tendencies often increase existential distress and exacerbate the problems that set the cycle in motion in the first place. The possibility of more constructive responses to these threats is discussed.

What's Going On? A Terror Management Perspective on the Current Age of Anxiety

What lies at the root of the feelings of anxiety, insecurity, and hopelessness that many people are grappling with in recent years? The litany of problems facing today's world has become depressingly familiar. Unabetted environmental degradation could radically change the habitability of our planet in the near future. Wars, terrorism, and other forms of ethnic, religious, and political violence have taken the lives of multitudes of people and led to large-scale migrations

DOI: 10.4324/9781003317623-4

that the world struggles to accommodate. Political extremism and divisiveness are increasing, characterized by the rise of demagogic populist leaders that eschew norms of civility and democracy while eroding basic human rights (Forgas, Crano & Fiedler, 2021). A side effect of this divisiveness is the splintering of families and friendships and a reduced capacity to openly communicate with others for fear of arousing anger. Racial injustice and income inequality undermine the ability of many people to meet their basic needs and undermine the sense of justice needed for psychological equanimity. Rapid changes in cultural beliefs, values, and social norms regarding sex, gender, and other issues have led to confusion and resentment among many who feel their understanding of some of the most basic aspects of life – such as the biological distinction between two sexes – has been called into question. Random mass murders in schools, supermarkets, places of worship, and other instances of violent crime make even the most mundane daily activities potentially dangerous. And, of course, the global COVID-19 pandemic has, as of this writing, taken the lives of an estimated 15 million people worldwide (Adams, 2022), left countless families grieving, and disrupted virtually all aspects of ordinary life while causing economic and interpersonal hardship and divisive debates about how to best manage the threat it poses (see also Arriaga & Kumashiro; Kreko, this volume). Of course, this is just a highly selective listing of some of the "highlights" of the catalog of recent threats to psychological equanimity. What's going on?

It has often been argued that we are living in a uniquely threatening era, an "age of anxiety," which is taking a toll on the psychological well-being of much of the world's population. Consistent with this notion, research has shown that rates of anxiety (Goodwin et al., 2020), depression (Ettman et al., 2020), substance abuse (Grucza et al., 2018; Haight et al., 2018; Stringfellow et al., 2022), domestic violence (Boserup et al., 2020), suicide (Hedegaard et al., 2018), violent crime (Abt et al., 2022; Lopez, 2022), and other forms of distress and psychopathology have increased over recent years. Of course, the threats facing today's world have many precedents over the course of human history, and a case could be made that today's problems are no worse or even less severe (Pinker, 2011) than those faced in previous historical eras. W. H. Auden coined the phrase, "age of anxiety," in an epic poem written 75 years ago in the wake of World War II that won him a Pulitzer Prize that inspired numerous commentaries focused on how *most eras since then* were particularly disturbing.

Regardless of whether the current era is or is not more troublesome than previous ones, it is clear that the world is currently facing a panoply of problems that are taking their toll on psychological well-being (see also Hirschberger; van den Bos; van Prooijen, this volume). In this chapter, we address the psychological processes set in motion by these stressors and how human responses to them exacerbate these problems in a vicious cycle of distress and social pathology. Toward this end, we use terror management theory (TMT; Greenberg et al., 1985; Solomon et al., 1991, 2015) and other ideas from recent experimental existential psychology to shed light on this vicious cycle of troublesome world events and individual and group responses to them that often exacerbate problems.

Terror Management Theory

TMT was inspired by cultural anthropologist Ernest Becker's (1971, 1973, 1975) attempts to synthesize and integrate a wide range of ideas from diverse scholarly disciplines to yield what he hoped would become a "general science of man." When we proposed TMT in 1986, we viewed it as a conceptual vehicle to integrate diverse theories and findings in psychology, especially within our own discipline of social psychology. Thirty-six years later, we think the theory has been at least moderately successful in this regard, in that it has been used to shed light on diverse aspects of human behavior and domains of social psychological research, including attitudes, attachment, interpersonal relationships, romantic love, conformity, intergroup conflict, aggression, prejudice, stereotyping, social cognition, disgust, health, and sexual ambivalence. It has also led to a variety of theoretical offshoots, extensions, and alternative perspectives (e.g., Goldenberg & Arndt, 2008; Jonas et al., 2013). On the other hand, TMT has also been a divisive perspective, leading to expressions of both strong endorsements and vociferous condemnations from other psychologists (see Pyszczynski et al., 2015, for an overview of theoretical extensions and criticisms, along with our responses to them).

Building on Becker's seminal ideas, TMT argues that the fear of death plays an important role in diverse aspects of human behavior. It focuses on the conflict between two highly adaptive consequences of human evolution: diverse biological and psychological systems that function to keep organisms alive long enough to reproduce and care for their offspring, thus passing on their genes; and sophisticated human cognitive abilities that facilitate gene perpetuation in many ways, including increasing the flexibility and adaptability of the human behavioral repertoire.

But there is a downside to sophisticated human intellect (see also von Hippel & Merakovsky, this volume). It leads to an awareness of the inevitability of death, which gives rise to the potential for *existential terror* in an animal highly motivated to stay alive. Because the experience of terror would undermine adaptive behavior necessary for survival, as well as being extremely unpleasant, it put a press on the ideas our ancestors were developing with their emerging sophisticated cognitive capacities. This created a preference for ideas that were helpful in managing terror by detoxifying death. People were more likely to generate such ideas to understand themselves and the world in which they lived and were more likely to share these ideas with others. Consequently, death-denying ideas became major parts of the conceptions of reality that people abstracted from their experiences and became part of cultural knowledge passed down across generations (see also Fiske, this volume). The ever-present potential for terror influenced the ideas that people generated over the millennia and continues to influence the ideas they generate and use to manage existential anxiety to this day.

TMT posits that the potential for terror engendered by awareness of death is managed by a tripartite anxiety-buffering system, consisting of cultural worldviews, self-esteem, and close interpersonal relationships. *Cultural worldviews* are shared beliefs that consist of three basic components: (1) a theory of reality that provides answers to basic questions about life and one's place in the universe while imbuing existence with meaning, structure, significance, and permanence; (2) standards of value that serve as guides for behavior and make it possible to attain self-esteem; and (3) the promise of literal or symbolic immortality to those who believe in their worldview and live up to its standards. *Literal immortality* entails beliefs that life will continue after physical death, as exemplified by concepts of heaven, reincarnation, or joining with ancestral spirits. *Symbolic immortality* comes from being part of and contributing to something greater than oneself that will continue long after one has died, such as a family, community, nation, or the memories of others.

Self-esteem is the sense that one is a valuable contributor to a meaningful universe that is attained by believing that one is living up to the standards of one's cultural worldview. Self-esteem is a sense of personal or collective value (Turner & Tajfel, 1980) that depends heavily on the cultural worldview to which one subscribes. What provides value within the context of one worldview (e.g., attaining massive wealth at the expense of others; killing members of rival groups) might undermine self-esteem in another. The standards of value through which one attains self-esteem are initially acquired through interaction with one's parents and other significant others and are augmented by diverse experiences and information over the course of one's life.

Mikulincer, Florian, and Hirschberger (2003) convincingly argued that *attachments and close interpersonal relationships* should be included as a third component of the anxiety-buffering system, in addition to the two discussed above and specified in initial presentations of TMT (Greeberg et al., 1986). Attachment to one's parents or primary caregivers is the infant's initial source of security and distress management. Cultural worldviews and self-esteem develop and acquire their ability to buffer anxiety through a developmental process that builds on the attachment system (for a thorough discussion of this process, see Pyszczynski et al., 2015). A large literature has shown that attachment relationships continue to manage anxiety and distress throughout the lifespan (for reviews, see Mikulincer et al., 2003; Mikulincer & Shaver; Arriaga & Kumashiro; Murray & Lamarche, this volume). Romantic relationships appear to be especially important sources of emotional security for adolescents and adults and have much in common with early childhood attachments.

Effective functioning of all three anxiety-buffer components requires *consensual validation* from others. As Festinger (1954) argued, people rely on social reality for validation of their beliefs and perceptions when there is no way of objectively determining their veracity, as is the case with most aspects of the worlds in which we live. Faith in, or certainty regarding, one's worldview, self-esteem, and close relationships increase when others share one's beliefs about these things and decrease when others view them differently (see also Cooper & Pearce, this volume). This is one reason why people are bothered when others have worldviews different from their own or view them negatively and are attracted to those who share their own worldviews or view them positively.

Empirical Evidence for TMT

As of this writing, over 1200 studies have tested and supported hypotheses derived from TMT. This research has been conducted in at least 35 different countries, representing diverse cultures. As would be expected, given that the theory posits that the specific ways in which people derive meaning and self-esteem depend on their cultural worldviews, there is cultural variability in how people cope with the problem of death (for a review, see Park & Pyszczynski, 2016). As with most findings in psychology, there have also been some failures to replicate (Rodríguez-Ferreiro et al., 2019; Sætrevik & Sjåstad, 2022; Schnidler et al., 2021), though some, but not all, of these studies suffered from major methodological problems (Klein et al., 2022; see Chatard et al., 2020).

Research has supported a network of converging hypotheses derived from TMT. This research shows that: (1) reminders of death (*mortality salience;* MS) increase commitment to one's worldview, self-esteem, and relationships, and increase defense of these entities when threatened; (2) bolstering self-esteem, worldview, or relationships makes one less prone to anxiety and anxiety-related behavior in response to threats; (3) threats to worldview, self-esteem, and relationships increase the accessibility of death-related thoughts; and (4) self-esteem striving, cultural worldview defense, and affirming close relationships reduce death thought accessibility (DTA) and the need for further terror management defenses in response to MS; this suggests that the three components of the anxiety buffer are psychologically interchangeable; Hart, Shaver, & Goldenberg, 2005). Meta-analyses have found strong evidence that reminders of death increase commitment to one's worldview (Burke et al., 2010) and that threats to one's worldview increase the accessibility of death-related thoughts (Steinman & Updegraff, 2015). We will discuss tests of TMT hypotheses, especially relevant for understanding current world events and their relationship to psychological insecurity, in later sections of this chapter.

Alternative explanations have been offered for some specific findings in this literature – for example, it has been debated whether the effects of MS are driven by the problem of death, per se, or by the uncertainty, incongruity, meaning threat, uncontrollability, or other uncomfortable experiences that covary with death. Despite the debate about precisely what it is about death that affects people's behavior, there appears to be a general consensus that thoughts of death affect behavior in ways consistent with TMT. For a discussion of these debates and a broad review of the TMT literature, see Pyszczynski et al. (2015). Despite the possibility of alternative explanations for specific findings, we have yet to see an alternative analysis that can, or has attempted to, account for the breadth of converging findings from tests of the logically distinct hypotheses that have been used to empirically assess TMT.

Different Defenses for Conscious and Unconscious Death-related Thoughts

As research on TMT unfolded, inconsistent findings led us to refine the theory to distinguish between psychological defenses used to manage conscious and non-conscious death-related ideation. We noticed that whereas studies that (quite inadvertently) included a delay between the MS induction and assessment of worldview defense dependent measures consistently yielded theory-consistent effects, studies that did not include such delays tended not to show effects. We also noticed that a few studies that employed more intensive and impactful MS manipulations than we were using did not yield significant effects. We then conducted studies comparing the effect of MS with and without delay/distraction and the effects of subtle and more potent MS inductions that confirmed these patterns: MS increased worldview defense when there was a delay between the induction and worldview defense assessment and when the MS induction was relatively subtle but not when it was more potent (Greenberg et al., 1994). This led us to rethink the theory and realize that worldview defense and the pursuit of self-esteem are not the only ways people manage death anxiety; they also do so in more direct ways by denying their vulnerability, engaging in health-promoting behavior, and simply suppressing death-related thoughts.

These findings and considerations led us to refine the theory to posit that people manage death anxiety with two distinct systems, referred to as *proximal and distal defenses* (Pyszczynski et al., 1999). When death-related thoughts are in conscious awareness or focal attention, proximal defenses are deployed to suppress such thoughts or push death into the distant future by denying one's vulnerability to things that could end one's life, planning a healthier lifestyle, or actually engaging in healthier behaviors. However, when death-related thoughts are no longer in the focal attention but are still highly accessible, people activate distal defenses, which are symbolic in nature and focused on one's worldview and self-esteem.

Conscious awareness of death requires defenses that "make sense" and deal directly with the problem of death, but they do little to quell the anxiety stemming from the ultimate inevitability of death. These concerns are assuaged by distal defenses that are not logically related to death but imbue one's life with meaning and value. Research testing this model showed that proximal defenses emerge shortly after reminders of death, but distal defenses emerge only when the death reminder is followed by a delay and distraction (for a review, see Arndt et al., 2004). Also consistent with the distinction between conscious and non-conscious processes, distal defenses emerge immediately after MS when death primes are presented subliminally and thus bypass conscious attention. Distal defenses have been shown to reduce the accessibility of death-related thoughts, which is presumably how they manage anxiety (see Arndt et al., 2002).

The Vicious Cycle of a Turbulent World and the Undermining of Psychological Equanimity

Applying TMT to the problems of today's turbulent world (and other troublesome eras over the course of history) suggests a vicious cycle of stressful world events increasing DTA and the potential for existential anxiety, which leads to attitudes and behaviors that often exacerbate the problems and undermine the security of those who do not share one's cultural worldview, leading to further DTA and distress, more maladaptive responses, and so on. Despite people's best efforts to use their worldviews, self-esteem, and close relationships to manage anxiety and maintain equanimity, these psychological entities are frequently assaulted by events occurring both within the confines of their individual lives and those occurring within the broader world. This dialectic interplays between world events and people's anxiety-buffering systems likely characterize the current age of anxiety, as well as those of previous and probably future historical eras. From the perspective of TMT, world events undermine security in two general ways: they either directly remind people of their vulnerability and mortality or they threaten the integrity of their worldviews, self-esteem, and close relationships, thereby undermining the protection from anxiety these structures provide (see also van den Bos, this volume).

Turbulent world events increase DTA. Most of the challenges facing today's world provide direct reminders of one's vulnerability and mortality. Indeed, many of them entail the possibility of dying, often in terrible ways. Wars, terrorism, and ethno-religious-political violence are obvious threats to life and limb, as are violent crime and seemingly random mass murders that have become a near-daily presence on our TV screens and news feeds. Disease and pandemics continue to take the lives of people everywhere, with the ongoing COVID-19 pandemic estimated as having claimed the lives of over 15 million people worldwide (Adams, 2022) being particularly salient at this juncture in history. Climate change, once a distant threat, has suddenly become present through more extreme weather patterns and natural disasters. It is estimated that in the last half-century, two million people have died from natural disasters, and scientists estimate that similar events will only become more frequent and intense as the effects of climate change compound (World Meteorological Organization, 2021). Above and beyond the objective threat posed by such calamities, constant access to news of such events on the internet and 24-hour news networks increases their accessibility and subjective likelihood.

Furthermore, prejudice and violence toward both minority and majority groups are frequent threats to the anxiety-buffering function of group membership. This can lead to fear and thoughts of death, which typically leads to deeper ties to one's in-group. Disdain toward groups of various types often erupts in violence, sometimes making one's social identity a literal threat to continued existence. Political divisiveness and associated protests can also pose a direct threat to life. Though most political movements and protests are peaceful, many are not. Protests and rallies on both the right and left have sometimes turned into riots in which innocent people were killed. These events – whether experienced in person or through the media – likely increase perceptions of danger, leading to thoughts of death and greater in-group identification.

Turbulent world events threaten the anxiety-buffering system. These same events also threaten the psychological structures that manage anxiety and provide security. Some world events threaten basic assumptions about the world that help people feel safe and secure. For example, many of the events noted in the previous paragraph entail seemingly random death or misfortune that likely undermine belief in a just world (Lerner, 1980). The fact that people are killed by strangers while shopping at a local grocery store, a terrorist while attending a concert, or a natural or manmade disaster virtually anywhere undermines the sense that one can be spared such outcomes by being a good person and "doing the right things." Such events also challenge beliefs that other people are generally good and decent, that a caring god will protect us from horrible outcomes, or that medicine and science can protect us from disease and disaster. When people perceive that violence is directed toward people similar to themselves because of the ethnic, racial, religious, national, or ideological social categories to which they belong, the threat to security conveyed by such animosity may be even greater.

Many disturbing world events are perpetrated by people whose behavior is guided by cultural worldviews very different from one's own. TMT suggests that the mere existence of people with worldviews different from one's own is threatening because it undermines the social consensus needed to maintain faith and certainty regarding one's own worldviews. In an increasingly divided world, in which those with the most extreme positions on both the left and right sides of the political spectrum tend to get the most attention in both traditional and social media, people of all beliefs and persuasions experience an ongoing onslaught of beliefs and values different from their own. These views are often expressed with extreme rhetoric and disdain for those with different worldviews. As many have suggested (e.g., Haidt, 2022), the 24-hour news cycle and ubiquity of social media and the internet have exponentially increased exposure to threats to one's own beliefs and values. This trend overlaps with the increasing polarization of the media (Jurkowitz et al., 2020), which likely leads to more extreme attitudes and cements political and cultural positions (Brauer et al., 1995).

Threats increase DTA. Research has shown that threats to one's worldview, self-esteem, or close relationships increase DTA (Hayes et al., 2010; Steinman & Updegraff, 2015). Research has also demonstrated that reminders of major world events, such as terrorism (Das et al., 2009; Landau et al., 2004) and COVID-19 (Fairlamb, 2021), lead to higher DTA. Fairlamb (2021) documented the real-world association between external events and DTA by measuring DTA at multiple times during the pandemic and finding that it varied directly with COVID-19 death toll numbers: as COVID-19 death rates increased DTA levels were elevated. Since the pandemic took away many sources of security (such as relationships and opportunities for self-esteem), it's conceivable that people had even less anxiety-buffering resources against these threats at the time when reminders of death in the form of lockdowns, masks, news reports, and casual conversations with friends were ubiquitous (Pyszczynski et al., 2021).

Turbulent world events lead to derogation of those different from oneself. In classic TMT studies, researchers have found that MS leads to derogation and prejudice toward outgroup members (Greenberg et al., 1994; see also Hogg & Gaffney, this volume). Recent work has built on this oft-documented finding to demonstrate its external validity beyond the confines of the lab. It is unfortunate that there have been many opportunities to test the link between threat and violence in recent years.

In particular, the onset of the COVID-19 pandemic enabled TMT researchers to collect evidence of TMT defenses playing out in the real world. Multiple studies and theoretical papers have outlined the existence of proximal defenses and distal defenses that have cropped up in the wake of the pandemic (Courtney et al., 2020; Pyszczynski et al., 2021). Social media studies have been especially helpful for identifying TMT defenses during the COVID-19 pandemic and the subsequent lockdowns. As universities and their laboratories shut down, relatively new big data techniques were used to fill the void. These studies assessed the internet habits of people around the world during the pandemic and identified several TMT defenses (see also Kreko, this volume). Evers et al. (2021) and Chew (2022) found an increase in death-related internet searches in the United States and Singapore, respectively. Other researchers used similar techniques to find that people were engaging in classic TMT defenses through the internet. Kwon and Park (2022) found that some of the defenses people engaged in were culturally dependent, with American Twitter users engaging in political polarization, racial divisiveness, and critiques of government incompetence, whereas Indian users more often shared pertinent information and emphasized close relationships. Li et al. (2020) found elevated levels of social media posts expressing support for family and religion in China during a strict lockdown. Tellingly, a common phrase on Weibo, a Chinese social media site, during the pandemic was "God Bless China" – an interesting phrase for a country where more than half of the population identifies as an atheist (Li et al., 2020; US Department of State, 2021).

Following the leader. Throughout the course of history, those with power and wealth have had a disproportionate impact on the worldviews that are accepted by the masses, either through their greater capacity for persuasion that power and wealth provide or their capacity for coercion and proclaiming by decree what their followers must believe in and abide by. Just as kings, clerics, shamans, and tribal leaders determined the official belief systems in past eras, contemporary leaders continue to dictate the official belief systems of most nations, religions, and ideological movements. On the other hand, as Enlightenment ideas have spread, values of self-determination and freedom of choice have become increasingly impactful and central to many cultural worldviews (see also Krueger & Gruening, this volume). Nonetheless, many contemporary people's belief systems continue to fall in line with those promoted by charismatic leaders. Consistent with this notion, research has shown that after the 9/11 terrorist attacks in the US, reminders of either death or terrorism increased support for then President George W. Bush and his policies related to the war in Iraq and curtailing domestic freedoms to combat terrorism (Landau et al., 2004). More recently, Cohen et al. (2017) found that MS increased support for Donald Trump, both when he was a candidate and after he was elected president.

This tendency to adopt the worldviews of charismatic leaders, even when such adherence is not imposed by force, is likely due to many psychological factors. It seems likely that part of the appeal of such leaders is that they tap into the psychological needs of potential followers, express their fears and frustrations, and promote ideas and policies that seem likely to remedy these problems (see also Forgas, this volume). It is also likely that powerful and wealthy people have financial and other resources that make it possible to get their message out in more compelling ways and thus reach larger numbers of people. Conformity to one's in-group also likely plays a powerful role. As members of one's in-group buy into a leader's worldview, this can promote both informational and normative social influence, leading other group members to fall in line. As noted above, research showing that MS increases admiration for charismatic leaders and conformity to their worldviews suggests that aligning one's own worldview with those of powerful leaders serves a terror management function.

Feeding the cycle by derogating those with different worldviews. When death-related thoughts are highly accessible, people often cling to their in-groups. This can manifest as closer personal relationships and increased identification with one's community (Vail III et al., 2012). However, there is a darker, often symbiotic, form of the same phenomenon – derogation of out-groups. Becker (1973) hypothesized that the mere presence of groups with diverging worldviews threatens one's anxiety buffer. The finding that MS increases derogation of those who criticize or deviate from one's own worldview is one of the most frequently replicated findings in the TMT literature. For example, Greenberg et al. (1990) found that MS increased Americans' derogation of a supposed foreign student who criticized the US and, among Christian participants, increased negative evaluations of a Jewish student. Such disdain is likely to lead the targets of such negativity to respond in kind and exacerbate the conflict.

TMT researchers have also documented this process outside the laboratory. For instance, researchers were conducting a longitudinal study in Poland when the COVID-19 pandemic hit. As the virus took hold, they subsequently documented a rise in authoritarianism, as well as more hostile attitudes toward sexual minorities and non-gender conforming women (Golec de Zavala et al., 2020). A similar phenomenon has been found in response to terrorism, with Europeans becoming more hostile toward people of Middle Eastern descent in response to either a reminder of terrorism or an MS induction (Das et al., 2009). Studies in the aftermath of the Charlie Hebdo terrorist attack in France showed more negative evaluations of people of North African descent in the weeks after the attack, but this did not occur among participants who placed a high value on the French value of colorblind equality (Cohu, Maisonneuve, & Teste, 2016; Nugler et al., 2016). This latter finding is consistent with previous research showing that responses to MS depend on a person's cultural worldview (Park & Pyszczynski, 2015).

Support for violence. In its most extreme form, derogation and prejudice toward out-groups can become violent. Beyond the terrible human price that this violence often enacts, it also ignites a cycle of fear and retaliation that deepens these patterns and makes peace less attainable. Becker (1973) and many TMT theorists have hypothesized that our fear of death can drive this violence as a way to defend our own cultural worldviews against the existential threat of the outgroup. Physical fear in the context of strife and violence often makes one's death more salient, thus further driving defensiveness of one's worldview and hostility toward those with different worldviews. This pattern cements a vicious cycle of fear, death thoughts, and violence that becomes harder to dislodge the longer it continues.

Support for this link between death anxiety and violence is provided by research demonstrating that, beyond simple derogation in the previously discussed studies, MS also increases support for violence against outgroup members and the desire to punish those who violate in-group norms. In one of the earliest TMT studies, Rosenblatt et al. (1989) found that judges who were reminded of their deaths before sentencing of a prostitute set bail nearly nine times higher on average than judges in a control condition. The researchers theorized that this was due to the increased need to uphold social values – in this case, by incarcerating those who violate them. Physical aggression and support for political violence have also been found to follow MS. McGregor et al. (1998) found that MS increased physical aggression toward a person who criticized their political ideology, and Pyszczynski et al. (2006) found that MS increased Iranians' support for a person who advocated terrorist violence against Americans and conservative Americans' support for extreme violent tactics to counter terrorism, including the use of nuclear and chemical weapons.

Pyszczynski and colleagues (2021) theorized that this same process was likely at play during the COVID-19 pandemic. Protests on both sides of the political spectrum in America were a common occurrence during the height of the lockdowns, which sometimes spilled over into violence (Caputo et al., 2020; Dress, 2022). From the perspective of TMT, the turn toward mass protest was an attempt for people to cement their value as meaningful actors within their world at a time when death was constantly in the background and many other sources of self-esteem – such as relationships and work – were taken away (Pyszczynski et al., 2021). It's also likely that the turmoil that people perceived in society, in addition to the distress brought on by the virus, exacerbated this sense of uneasiness and led to increased death thoughts. Confronted with both the deaths caused by the virus and the perceived violence of their dissimilar neighbors, people were primed to support harming those they perceived as "others." Indeed, some research before the 2020 presidential election found that nearly 1 in 5 Americans with strong political beliefs were willing to endorse political violence if their candidate lost (Diamond et al., 2020) – something that unfortunately came to fruition with the Capitol Insurrection on January 6, 2020. Hindsight seems to have only exacerbated these divisions. Almost two years after the storming of the Capitol, a poll found that 33% of Republicans still supported the actions of the insurrectionists (Dress, 2022).

Existential threat undermines rational decision-making. Though one would hope that world crises would motivate people and leaders to marshal their resources to think more carefully and rationally to understand the causes of problems in ways that point to useful solutions, powerful psychological forces work against rationality in difficult times. Kruglanski's (2010, see also Kruglanski and Ellenberg, this volume) *lay epistemology theory* sheds light on how threatening world events and the existential threats they entail can disrupt rational thinking. From this perspective, human thought is driven by varying levels of two broad epistemic motives. The *need for specific closure* entails a preference for specific conclusions that meet one's psychological needs. As we've been suggesting throughout this chapter, threatening world events activate existential fears and thus increase people's need for conclusions consistent with their pre-existing worldviews and self-concepts. On the other hand, the *need for general closure* entails a need for conclusions that provide structure and quick answers, even inadequate ones, to pressing questions. Challenging world events typically entail a great deal of uncertainty and ambiguity that increases the need for general closure, which motivates people to freeze their cognitive activities without optimal information gathering and analysis, thereby seizing on conclusions that provide simple, easily digestible understandings. Both of these motives can readily overwhelm the need for accurate conclusions or the fear of invalidity that orients people to thorough, unbiased information gathering and careful analysis of information in the service of conclusions that best fit available information. These forces are likely to lead to faulty conclusions, that in turn, lead to policies oriented more toward meeting psychological needs than providing optimal solutions to pressing problems.

Impact on out-groups. All of these maladaptive responses to the existential threats inherent in turbulent world events are likely to threaten the worldviews and self-esteem, and perhaps the lives, of those with worldviews and social identities different from one's own or different from those of powerful and influential groups within a society. As noted above, more powerful groups have greater influence on policies and often derogate and retaliate against those who don't go along. This often leads to parallel responses against the mainstream from less powerful groups, in the form of digging in on their own beliefs, derogation of mainstream views, protests, and sometimes violence. Just as existential threat exacerbates the tendency of mainstream groups to derogate and silence outsiders, these negative responses, especially when combined with the many other death-related threats impinging on awareness, undermine the anxiety buffers of out-groups. This is likely to motivate resistance, ultimately further inflaming arguments, conflict, protests, and sometimes violence. Again, the essence of our analysis is that turbulent world events instigate psychological processes and responses that lead to vicious cycles of increasingly maladaptive responses to crises that make matters worse.

Can the Cycle Be Broken?

Our analysis paints a rather bleak and pessimistic picture of the crises currently facing humankind. This analysis is based on several decades of research on how people most commonly cope with existential threats. However, TMT does *not* imply that hostility, disrupted rationality, and disdain for those who are different are the *only* ways of managing death anxiety. How people respond to existential threat is thought to depend on the specific beliefs and values of their cultural worldviews that are influential and salient at the time. When life-affirming values oriented toward optimal decision-making and the welfare of all people are dominant, these values are expected to lead people toward more constructive, open-minded, compassionate behavior.

Research has shown that activating values of tolerance (Greenberg et al., 1994), compassion (Rothschild et al., 2009), close relationships (Weiss et al., 2008), common humanity (Motyl et al., 2011), or shared fate (Pyszczynski et al., 2012) eliminates the hostile, closed-minded responses that MS often produces. Indeed, in some of these studies, activating pro-social values in combination with MS actually led to reduced hostility toward long-standing enemies. Other research has shown that mindful meditation (Park & Pyszczynski, 2015) or reminding intrinsically religious people of their faith (Jonas & Fisher, 2006) can counter the effects of MS. These findings suggest that it is far from inevitable that people will respond to existential threats in ways that exacerbate problems.

Unfortunately, these more encouraging findings all come from laboratory studies in which the informational environment was carefully controlled. Translating these findings into practical strategies for influencing large groups of people fed a steady diet of information designed to promote particular ideological agendas and inflame their passions against those with different worldviews poses a daunting challenge. We suspect it's far easier to blame others for world problems than to come together with people with divergent perspectives to find common ground and make concessions and sacrifices needed to solve difficult problems. Perhaps crises will reach a level of severity at which putting aside differences and cooperating toward common goals becomes more appealing than the polarization that has currently overwhelmed public discourse.

Conclusion

The philosopher William James once said that death is "the worm at the core" of the human condition. Over the past three decades, TMT researchers have explored how this "worm" influences our attitudes and behaviors. In this chapter, we have outlined how death may also be playing a role in our current "Age of Anxiety." Martin Luther King (1967) conceptualized violence as a "cycle" that perpetuates itself through hatred and retribution. We expand on this idea and decades of TMT research to suggest that thoughts of death may be part of the engine driving this cycle. As we move deeper into the 21st century, insight into how our fear of death influences these human patterns and drives human behavior could be critical in helping lessen the power of these cycles and in building a healthier, more open, and more tolerant society.

References

Abt, T., Bates-Chamberlain, C., Bocanegra, E., Carillo, P., ... & Webster, D. (2022). (rep.). *Violent Crime Working Group*. Council on Criminal Justice. Retrieved July 1, 2022, from https://counciloncj.org/violent-crime-working-group/.

Adam, D. (2022, January 18). *The pandemic's true death toll: Millions more than official counts*. Nature News. Retrieved July 13, 2022, from https://www.nature.com/articles/d41586-022-00104-8.

Arndt, J., Greenberg, J., & Cook, A. (2002). Mortality salience and the spreading activation of worldview-relevant constructs: Exploring the cognitive architecture of terror management. *Journal of Experimental Psychology: General, 131*(3), 307.

Arndt, J., Cook, A., & Routledge, C. (2004). The blueprint of terror management. In J. Greenberg, S. L. Koole, & T. Pyszczynski (Eds.), *Handbook of experimental existential psychology* (pp. 35–53). New York: The Guilford Press.

Becker, E. (1971). *Birth and death of meaning* (2nd ed.). Free Press.

Becker, E. (1973). *The denial of death*. Free Press.

Becker, E. (1975). *Escape from evil*. Free Press.

Boserup, B., McKenney, M., & Elkbuli, A. (2020). Alarming trends in US domestic violence during the COVID-19 pandemic. *The American Journal of Emergency Medicine, 38*(12), 2753–2755.

Brauer, M., Judd, C. M., & Gliner, M. D. (1995). The effects of repeated expressions on attitude polarization during group discussions. *Journal of Personality and Social Psychology, 68*(6), 1014.

Burke, B. L., Martens, A., & Faucher, E. H. (2010). Two decades of terror management theory: A meta-analysis of mortality salience research. *Personality and Social Psychology Review, 14*(2), 155–195.

Caputo, A., Craft, W., & Gilbert, C. (2020, June 30). *What happened at Minneapolis' 3rd Precinct – and what it means*. APM Reports. Retrieved August 18, 2022, from https://www.apmreports.org/story/2020/06/30/what-happened-at-minneapolis-3rd-precinct

Chatard, A., Hirschberger, G., & Pyszczynski, T. (2020). A word of caution about Many Labs 4: If you fail to follow your preregistered plan, you may fail to find a real effect. *PsyArXiv,* https://doi.org/10.31234/osf.io/ejubn.

Chew, P. K. (2022). Big data analysis of terror management theory's predictions in the COVID-19 pandemic. *OMEGA-Journal of Death and Dying*. https://doi.org/10.1177/00302228221092583

Cohen, F., Solomon, S., & Kaplin, D. (2017). You're hired! Mortality salience increases Americans' support for Donald Trump. *Analyses of Social Issues and Public Policy, 17*(1), 339–357.

Courtney, E. P., Goldenberg, J. L., & Boyd, P. (2020). The contagion of mortality: A terror management health model for pandemics. *British Journal of Social Psychology, 59*(3), 607–617.

Das, E., Bushman, B. J., Bezemer, M. D., Kerkhof, P., & Vermeulen, I. E. (2009). How terrorism news reports increase prejudice against outgroups: A terror management account. *Journal of Experimental Social Psychology, 45*(3), 453–459.

Diamond, L., Drutman, L., Lindberg, T., Kalmoe, N., & Mason, L. (2020, Oct 1). *Americans increasingly believe violence is justified if the other side wins*. POLITICO. Retrieved August 18, 2022, from https://www.politico.com/news/magazine/2020/10/01/political-violence-424157

Dress, B. (2022, June 21). *New poll finds 33 percent of GOP support actions of Jan. 6 rioters*. The Hill. Retrieved August 18, 2022, from https://thehill.com/homenews/state-watch/3526469-new-poll-finds-33-percent-of-gop-support-actions-of-jan-6-rioters/

Du, H., Jonas, E., Klackl, J., Agroskin, D., Hui, E. K., & Ma, L. (2013). Cultural influences on terror management: Independent and interdependent self-esteem as anxiety buffers. *Journal of Experimental Social Psychology, 49*(6), 1002–1011.

Ettman, C. K., Abdalla, S. M., Cohen, G. H., Sampson, L., Vivier, P. M., & Galea, S. (2020). Prevalence of depression symptoms in US adults before and during the COVID-19 pandemic. *JAMA Network Open*, *3*(9), e2019686-e2019686.

Evers, N. F., Greenfield, P. M., & Evers, G. W. (2021). COVID-19 shifts mortality salience, activities, and values in the United States: Big data analysis of online adaptation. *Human Behavior and Emerging Technologies*, *3*(1), 107–126.

Fairlamb, S. (2021). The relationship between COVID-19-induced death thoughts and depression during a national lockdown. *Journal of Health Psychology*, 13591053211067102.

Festinger, L. (1954). A theory of social comparison processes. *Human Relations*, *7*(2), 117–140.

Forgas, J. P., Crano, W. D. & Fiedler, K. (2021). *The psychology of populism: Tribal challenges to liberal democracy*. New York: Routledge / Psychology Press.

Goldenberg, J. L., & Arndt, J. (2008). The implications of death for health: A terror management health model for behavioral health promotion. *Psychological Review*, *115*(4), 1032.

Golec de Zavala, A., Bierwiaczonek, K., Baran, T., Keenan, O., & Hase, A. (2021). The COVID-19 pandemic, authoritarianism, and rejection of sexual dissenters in Poland. *Psychology of Sexual Orientation and Gender Diversity*, *8*(2), 250.

Goodwin, R. D., Weinberger, A. H., Kim, J. H., Wu, M., & Galea, S. (2020). Trends in anxiety among adults in the United States, 2008–2018: Rapid increases among young adults. *Journal of Psychiatric Research*, *130*, 441–446.

Greenberg, J., Pyszczynski, T., & Solomon, S. (1986). The causes and consequences of a need for self-esteem: A terror management theory. In R. Baumeister (Ed.), *Public self and private self* (pp. 189–212). New York: Springer.

Greenberg, J., Pyszczynski, T., Solomon, S., Rosenblatt, A., Veeder, M., Kirkland, S., & Lyon, D. (1990). Evidence for terror management theory II: The effects of mortality salience on reactions to those who threaten or bolster the cultural worldview. *Journal of Personality and Social Psychology*, *58*(2), 308–318. https://doi.org/10.1037/0022-3514.58.2.308

Greenberg, J., Simon, L., Pyszczynski, T., Solomon, S., & Chatel, D. (1992). Terror management and tolerance: Does mortality salience always intensify negative reactions to others who threaten one's worldview? *Journal of Personality and Social Psychology*, *63*(2), 212.

Greenberg, J., Pyszczynski, T., Solomon, S., Simon, L., & Breus, M. (1994). Role of consciousness and accessibility of death-related thoughts in mortality salience effects. *Journal of Personality and Social Psychology*, *67*(4), 627–637.

Grucza, R. A., Sher, K. J., Kerr, W. C., Krauss, M. J., Lui, C. K., McDowell, Y. E.,... & Bierut, L. J. (2018). Trends in adult alcohol use and binge drinking in the early 21st-Century United States: A meta-analysis of 6 national survey series. *Alcoholism: Clinical and Experimental Research*, *42*(10), 1939–1950.

Haight, S. C., Ko, J. Y., Tong, V. T., Bohm, M. K., & Callaghan, W. M. (2018). Opioid use disorder documented at delivery hospitalization—United States, 1999–2014. *Morbidity and Mortality Weekly Report*, *67*(31), 845.

Hart, J., Shaver, P. R., & Goldenberg, J. L. (2005). Attachment, self-esteem, worldviews, and terror management: Evidence for a tripartite security system. *Journal of Personality and Social Psychology*, *88*(6), 999.

Hayes, J., Schimel, J., Arndt, J., & Faucher, E. H. (2010). A theoretical and empirical review of the death-thought accessibility concept in terror management research. *Psychological Bulletin*, *136*(5), 699.

Hedegaard, H., & Warner, M. (2021). *Suicide mortality in the United States, 1999–2019*. National Center for Health Statistics.

Hirschberger, G., & Ein-Dor, T. (2006). Defenders of a lost cause: Terror management and violent resistance to the disengagement plan. *Personality and Social Psychology Bulletin, 32*(6), 761–769.

Jonas, E., & Fischer, P. (2006). Terror management and religion: Evidence that intrinsic religiousness mitigates worldview defense following mortality salience. *Journal of Personality and Social Psychology, 9*, 553–560.

Jurkowitz, M., Mitchell, A., Shearer, E., & Walker, M. (2022, March 28). *U.S. media polarization and the 2020 election: A nation divided*. Pew Research Center's Journalism Project.

King Jr, M. L. (2010). *Where do we go from here: Chaos or community?* (Vol. 2). Beacon Press.

Klein, R. A., Cook, C. L., Ebersole, C. R., Vitiello, C., Nosek, B. A., Hilgard, J.,... & Ratliff, K. A. (2022). Many Labs 4: Failure to replicate mortality salience effect with and without original author involvement. *Collabra: Psychology, 8*(1), 35271.

Kwon, S., & Park, A. (2022). Understanding user responses to the COVID-19 pandemic on Twitter from a terror management theory perspective: Cultural differences among the US, UK and India. *Computers in Human Behavior, 128*, 107087.

Landau, M. J., Solomon, S., Greenberg, J., Cohen, F., Pyszczynski, T., Arndt, J.,... & Cook, A. (2004). Deliver us from evil: The effects of mortality salience and reminders of 9/11 on support for President George W. Bush. *Personality and Social Psychology Bulletin, 30*(9), 1136–1150.

Lerner, M. J. (1980). The belief in a just world. *The Belief in a Just World: A Fundamental Delusion*. Springer, Boston, MA.

Li, S., Wang, Y., Xue, J., Zhao, N., & Zhu, T. (2020). The impact of COVID-19 epidemic declaration on psychological consequences: A study on active Weibo users. *International Journal of Environmental Research and Public Health, 17*(6), 2032.

Lopez, G. (2022, April 17). A violent crisis. *The New York Times*. Retrieved July 1, 2022, from https://www.nytimes.com/2022/04/17/briefing/violent-crime-ukraine-war-week-ahead.html

McGregor, H. A., Lieberman, J. D., Greenberg, J., Solomon, S., Arndt, J., Simon, L., & Pyszczynski, T. (1998). Terror management and aggression: Evidence that mortality salience motivates aggression against worldview-threatening others. *Journal of Personality and Social Psychology, 74*(3), 590.

Mikulincer, M., Florian, V., & Hirschberger, G. (2003). The existential function of close relationships: Introducing death into the science of love. *Personality and Social Psychology Review, 7*(1), 20–40.

Mikulincer, M., & Shaver, P. R. (this volume). Attachment security and coping with existential concerns. In J. Forgas, W. Crano, & K. Fiedler (Eds), *The psychology of insecurity: Seeking certainty where none can be had*. New York: Routledge.

Motyl, M., Hart, J., Pyszczynski, T., Weise, D., Maxfield, M. & Seidel, S. (2011). Subtle priming of group similarity eliminates threat-induced implicit and explicit hostility toward immigrants and Arabs. *Journal of Experimental Social Psychology, 47*, 1179–1184.

Park, Y. C., & Pyszczynski, T. (2016). Cultural universals and differences in dealing with death. In L A. Harvell & G. S. Nisbett (Eds), *Denying death: An interdisciplinary approach to terror management theory* (pp. 193–213). New York: Taylor & Francis.

Park, Y. C., & Pyszczynski, T. (2019). Reducing defensive responses to thoughts of death: Meditation, mindfulness, and Buddhism. *Journal of Personality and Social Psychology, 116*(1), 101.

Pinker, S. (2011). *The better angels of our nature: Why violence has declined.* Penguin Books.

Pyszczynski, T., Greenberg, J., & Solomon, S. (1999). A dual-process model of defense against conscious and unconscious death-related thoughts: An extension of terror management theory. *Psychological Review, 106*(4), 835.

Pyszczynski, T., Motyl, M. S., Vail, K. E., Hirschberger, G., Arndt, J., & Kesebir, P. (2012). A collateral advantage of drawing attention to the problem of global warming: Increased support for peacemaking and decreased support for war. *Peace and Conflict: Journal of Peace Psychology, 18*, 354–368.

Pyszczynski, T., Solomon, S., & Greenberg, J. (2015). Thirty years of terror management theory: From genesis to revelation. In M. Zanna (Ed.), *Advances in experimental social psychology* (Vol. 52, pp. 1–70). Academic Press.

Pyszczynski, T., Lockett, M., Greenberg, J., & Solomon, S. (2021). Terror management theory and the COVID-19 pandemic. *Journal of Humanistic Psychology, 61*(2), 173–189.

Rosenblatt, A., Greenberg, J., Solomon, S., Pyszczynski, T., & Lyon, D. (1989). Evidence for terror management theory: I. The effects of mortality salience on reactions to those who violate or uphold cultural values. *Journal of Personality and Social Psychology, 57*(4), 681.

Rothschild, Z. K., Abdollahi, A., & Pyszczynski, T. (2009). Does peace have a prayer? The effect of mortality salience, compassionate values, and religious fundamentalism on hostility toward ourgroups. *Journal of Experimental Social Psychology, 45*, 816–827.

Sætrevik, B., & Sjåstad, H. (2022). Mortality salience effects fail to replicate in traditional and novel measures. *Meta-Psychology, 6*, 1–20. https://doi.org/10.31234/osf.io/dkg53.

Schindler, S., Reinhardt, N., & Reinhard, M. A. (2021). Defending one's worldview under mortality salience: Testing the validity of an established idea. *Journal of Experimental Social Psychology, 93*, 104087.

Solomon, S., Greenberg, J., & Pyszczynski, T. (1991). A terror management theory of social behavior: The psychological functions of self-esteem and cultural worldviews. In M. Zanna (Ed.): *Advances in experimental social psychology* (Vol. 24, pp. 93–159). Cambridge, MA: Academic Press.

Solomon, S., Greenberg, J., & Pyszczynski, T. (2015). *The worm at the core: On the role of death in life.* Random House.

Steinman, C. T., & Updegraff, J. A. (2015). Delay and death-thought accessibility: A meta-analysis. *Personality and Social Psychology Bulletin, 41*(12), 1682–1696.

Stringfellow, E. J., Lim, T. Y., Humphreys, K., DiGennaro, C., Stafford, C., Beaulieu, E.,... & Jalali, M. S. (2022). Reducing opioid use disorder and overdose deaths in the United States: A dynamic modeling analysis. *Science Advances, 8*(25), eabm8147.

Tajfel, H., Turner, J. C. (1979). An integrative theory of intergroup conflict. *Organizational Identity: A Reader, 56*(65), 9780203505984-16.

U.S. Department of State. (2021, May 12). China (Includes Tibet, Xinjiang, Hong Kong, and Macau) 2020 International Religious Freedom Report. U.S. Department of State. Retrieved July 11, 2022, from https://www.state.gov/reports/2020-report-on-international-religious-freedom/china/

Vail III, K. E., Juhl, J., Arndt, J., Vess, M., Routledge, C., & Rutjens, B. T. (2012). When death is good for life: Considering the positive trajectories of terror management. *Personality and Social Psychology Review, 16*, 303–329.

Weise, D., Pyszczynski, T., Cox, C., Arndt, J., Greenberg, J., Solomon, S., & Kosloff, S. (2008). Interpersonal politics: The role of terror management and attachment processes in political preferences. *Psychological Science, 19*, 448–455.

World Meteorological Organization. (2021, September 9). *Weather-related disasters increase over past 50 years, causing more damage but fewer deaths*. Retrieved August 18, 2022, from https://public.wmo.int/en/media/press-release/weather-related-disasters-increase-over-past-50-years-causing-more-damage-fewer-deaths

4

THE UNCERTAINTY CHALLENGE

Escape It, Embrace It

Arie W. Kruglanski and Molly Ellenberg

UNIVERSITY OF MARYLAND, USA

Abstract

Whereas in both lay psychology and scientific analysis, uncertainty is typically associated with negative emotional reactions (as in "fear of the unknown"), we offer a differentiated perspective whereby the response to uncertainty depends on individuals' mindset and focus and can range over a large spectrum of affective reactions, from anxiety to exuberance. In our framework, we distinguish between "objective" uncertainty, the actual likelihood of something happening, which is largely unknowable, and subjective uncertainty, the believed likelihood and nature of it happening. Moreover, responses to subjective uncertainty can vary as a function of past conditioning of such uncertainty to positive and negative experiences, as well as prior proximal such experiences. We cite empirical evidence for our theoretical analysis, including our own studies of reactions to uncertainty.

The Myth of the Dark Unknown

To many people, the notion of uncertainty has a negative connotation, and the idea that people generally desire certainty seems like an obvious truth. This belief is commonplace among laypeople and scientists alike.

"It is the unknown people fear when they look upon death and darkness," thought Albus Dumbledore, the headmaster of Hogwarts School of Witchcraft and Wizardry (in the Harry Potter legend). "We fear that which we cannot be seen," remarked Tite Kubo, the Japanese Manga artist. The author H. P. Lovecraft (1927) thought that the oldest and strongest kind of fear is people's fear of the unknown, and the psychologist N. R. Carleton also speculated that it is the fear of the unknown that is possibly the most fundamental fear

DOI: 10.4324/9781003317623-5

(Carleton, 2016, p. 5). Systematic research in psychology similarly emphasized the negative reactions to uncertainty encapsulated in the concept of the *intolerance* of uncertainty (IU) or ambiguity (e.g., Frenkel-Brunswik, 1948; Grenier, Barrette & Ladouceur, 2005).

There is evidence that fear of the unknown has an evolutionary basis (Brosschot, Verkuil, & Thayer, 2016; see also von Hippel, this volume). In environments replete with dangers, some degree of fear toward the unknown should be adaptive, as long as it didn't compromise useful exploratory activities (for food, shelter, and mates, c.f. Carleton, 2016). Some authors, therefore, have proposed that fear is the "default" reaction to uncertainty and novelty that can be overridden by learning (Thayer et al., 2012). It represents an evolved, cautious approach that prepares the organism for danger but is readily suspended given evidence that the situation is safe. The evolutionary argument suggests the existence of inborn individual differences in the IU. This notion is supported by evidence about the heritability of anxiety in response to novel, unknown situations (Sanchez, Kendall, & Comer 2016).

During the COVID-19 pandemic, the view that uncertainty is inherently frightening and threatening has been often repeated (see also Arriaga & Kumashiro; Kreko, this volume). Millions around the globe have been experiencing severe angst about the unprecedented (and hence uncertain) situation that the pandemic created. National data from across the world attested to a significant spike in reported distress tied to the pandemic. According to the World Health Organization, there has been a 35% increase in the prevalence of distress in China, 60% in Iran, and 45% in the US. And in the Amhara region of Ethiopia, there has been a threefold increase in the frequency of depression and anxiety relative to the pre-pandemic rate, although the region has also been mired in civil war during the same time period (United Nations, 2020); figures from all the different corners of the world attest to the same thing; wherever you look, people are deeply troubled and disconcerted.

The fact that during the pandemic people experience angst and uncertainty need not mean that the two are causally related. The pandemic generates both uncertainty and angst. But the angst might not come from the uncertainty as such. It might come from the negative outcomes – bad things that could happen to people because of the pandemic – negative events that are linked to the uncertainty: increased likelihood of severe illness, death, loss of one's loved ones, loss of one's job, restrictions on freedom, etc. Consider that trivial, irrelevant, uncertainties do not evoke distress at all. Uncertainty about the weather (50% chance of rain) produces nary a noticeable change in our emotions, nor does uncertainty about the outcome of an athletic competition between teams we do not care about or about political elections in a faraway land. In fact, research has also found that, whereas already insecure romantic relationships became weaker during the pandemic, secure romantic relationships actually became stronger during COVID-19 (see Arriaga & Kumashiro;

Mikulincer & Shaver; Murray & Lamarche, this volume), perhaps because these individuals viewed the uncertainty as an opportunity rather than a threat.

Just think! Would you rather have a 90% chance of contracting the virus or a 50% chance? Most people would likely opt for the 50% even though it connotes *maximal uncertainty*, yet the 90%, while much more certain, is intuitively much less preferable and more anxiety provoking. So, perhaps it is not the uncertainty as such then that is so troubling, but rather the adverse consequences that the pandemic is associated with and hence makes it more likely.

In the present paper, we submit this possibility to careful scrutiny. To do so, we first define what we mean by uncertainty, and in particular, by *affective uncertainty* of major present interest. We also discuss the overlapping features of uncertainty and ambiguity and consider work in the area of social cognition that examined the effects of prior knowledge (i.e., momentary and chronic accessibility of constructs) on the interpretation of ambiguous stimuli (see also Krueger & Gruening, this volume). We then extrapolate the interpretative effects of prior knowledge and apply them to people's affective reactions to uncertain situations. Finally, we examine the consequences of different reactions to uncertain situations, that is, "escaping" and "embracing" uncertainty, and consider what these mean for psychological phenomena on intra, inter, and group levels of analysis.

Defining Uncertainty

We define uncertainty as ignorance about the likelihoods of different imagined possibilities. Uncertainty in this sense is a matter of degree. For instance, having not heard the weather forecast and thus having no idea whether it will rain tomorrow is more uncertain than knowing that the probability (e.g., as estimated by a meteorological service) is 50%, which in turn is less certain than knowing that the probability is 90%, etc. Additionally, one can extrapolate that maximal uncertainty also covers a complete lack of control over the situation, whereas situations over which one has more control are definitionally less uncertain. Here, our interest is in the likelihood of affectively relevant outcomes varying in the degree of their "goodness" or "badness," or the degree to which they satisfy or frustrate individuals' motives.

Ambiguity and Uncertainty

All ambiguous situations contain an element of uncertainty in the sense of unclarity as to the likelihoods of different possibilities. Typically, however, the term "ambiguity" has referred to the uncertain description of a *present stimulus* (e.g., how likely it is that a given person is reckless or adventurous, how likely they are to be a lawyer or an engineer). In contrast, "uncertainty" has generally referred to a future event (e.g., whether Putin will invade Poland, whether it will rain tomorrow). For present purposes, however, what matters is the ignorance (i.e., unknown likelihood) aspect that is common to situations classified as uncertain or ambiguous. From this perspective, factors that shape the experience of ambiguity should also affect the experience of uncertainty.

Dispelling Ambiguity

There is ample evidence that concepts of which the individual is currently conscious can affect the interpretation of ambiguous stimulus information. Much of this work has been carried out in the realm of social cognition and has specifically to do with the factor of construct accessibility. Higgins (1996; see also Higgins & King, 1981) defined accessible constructs as constructs that are stored in memory and "are readily used in information processing" (Higgins, 1996, p. 133). Furthermore, construct accessibility is determined by momentary exposure and chronic activation, that is, "recent and frequent priming of the construct" as well as "the construct's interconnectedness with other stored constructs" (ibid.). Here, "priming" refers to (natural or manipulated) events that stimulate or activate some stored constructs.

Interpreting Ambiguous Stimuli

1 **Effects of momentary accessibility**. In a social cognition classic, Higgins and colleagues (1977) semantically primed constructs that were subsequently used (without participants' awareness) to interpret an ambiguous stimulus, a description of a target person. Higgins et al. (1977) found that participants tended to use the primed constructs in characterizing the target person in the reading comprehension task. Moreover, participants exposed earlier to the negatively toned prime "reckless" evaluated the target more negatively overall than participants exposed to the positive prime "persistent." This study and similar others offer ample evidence that momentarily, that is, recently primed, as well as frequently primed concepts (e.g., Srull & Wyer, 1979) can influence individuals' subsequent evaluation of ambiguous stimuli.

Other types of research support this conclusion. In the clinical realm, Eysenck and colleagues (1991) showed that currently anxious participants interpreted ambiguous sentences in more threatening ways than recovered anxious participants or non-anxious participants. Dearing and Gotlib (2009) induced negative mood in daughters of depressed mothers, and those of never-disordered mothers. They found that the former more than the latter interpreted ambiguous words more negatively and less positively, and ambiguous stories more negatively, suggesting that the momentary negative mood activated available negative ideation in depressed individuals' memory.

2 **Effects of chronic accessibility.** The idea that contents of individuals' personalities affect their interpretation of ambiguous stimuli has long served as the cornerstone of various projective techniques like the Thematic Apperception Test or the Rorschach, for example. In social cognition research, Higgins and Brendl (1995) and Bargh and Pratto (1986) demonstrated that chronically accessible constructs have the similar effect on interpretation of ambiguous stimuli as momentarily primed constructs.

Researchers also juxtaposed recent and frequent priming in a test of a "synapse model" of chronic and momentary accessibility effects (Higgins et al., 1985). According to this model, stimulation of a construct through priming increases its action potential to a fixed, maximum level, which then dissipates over time. Moreover, the more frequently a construct has been stimulated, the slower the pace of the dissipation, and the longer will its action potential remain sufficiently high to be applied and override other sources of activation. In support of the "synapse model," Higgins et al. (1985) demonstrated that *recent* priming predominated over *frequent* priming in the interpretation of ambiguous stimuli when the delay between the priming and the judgment was short (15 seconds); in contrast, frequent priming predominated when the delay was long (120 seconds). It thus appears that frequent priming builds a kind of resilience that allows it to withstand the impact of recent priming over the long term, whereas recent priming has momentary effects of brief duration.

Consistent with these findings is evidence that chronic affective negativity prompts the negative interpretation of ambiguous stimuli. Constans and colleagues (1999) showed that socially anxious subjects show more threatening interpretations of ambiguous interpersonal events than do low-anxiety participants (see also Mikulincer & Shaver; Murray & Lamarche; Arriaga & Kumashiro, this volume). Other lines of research converge on the same conclusion (see Schoth & Liossi, 2017, for a review).

Confronting Uncertain Situations: The Dynamics of Hope and Fear

Meeting a new person, contemplating a new business venture, making a proposal of some sort or applying for a new job are everyday instances of uncertain situations people often encounter. Some individuals confront those with hope, others with fear and apprehension. What factors underlie those divergent reactions? Drawing on results in the realm of ambiguity, we suggest that people's affective reactions to uncertainty should depend on their chronic or momentary accessibility of positive and negative experiences (whether in a given domain or globally). In the same way that construct accessibility determines our interpretations of present ambiguous stimuli, our accessibility of good and bad outcomes should affect our expectations of what the future will bring (see also Fiedler & McCaughey; Krueger & Gruening, this volume).

Chronic Accessibility Effects

According to the present theory then, chronic preponderance of positive outcomes should make positive expectations chronically accessible. This should induce an optimistic orientation and make people more hopeful that the uncertain situation will be resolved in a desired way. In fact, several items of the widely used optimism scale (Life Orientation Test – Revised (LOT-R; Glaesmer et al., 2012) directly imply positive expectancies for future (hence, uncertain) outcomes, for instance: "In uncertain times, I usually expect the best," "I'm always optimistic about my future" and "Overall, I expect more good things to happen to me than bad." Consistent with such positive expectancies is evidence that optimism and *hope* are positively related (e.g., Alarcon, Bowling, & Khazon, 2013).

Consistent with the present theory, it is of interest also that past outcomes (suggesting a chronic accessibility of positive or negative experiences) appear to determine the degree of people's optimism or pessimism. In this vein, Ek and colleagues (2004) carried out a population-based longitudinal study of young adults ($n = 8673$) that examined early childhood predictors of adult (at age 31) optimism. Consistent with the present theory, it was found that the degree to which the child was wanted by its parents, the father's initial socioeconomic status and subsequent changes in it, school achievement, vocational education, and work history significantly predicted optimism at adulthood, as did a good financial situation and being married. If optimism and hope indicate a positive affective reaction to uncertainty, as we have suggested, this means that the history of positive outcomes predicts such a reaction (see also Mikulincer & Shaver, this volume).

In a recent study, we found convergent support for this hypothesis. In our study of 495 American adults (48.2% male), we found that participants with more positive childhood perceptions of their parents (including the parents' perceived warmth, involvement, and autonomy support) were higher in optimism, which in turn negatively predicted IU. IU subsequently negatively predicted perceptions of uncertain events, comprising: a blind date, the first day of school, the birth of a new sibling, a job interview, a pop quiz, a game of Bingo, a swim in the ocean, a haunted house, the COVID-19 pandemic, a terrorist attack, a trip to a foreign country for the first time, an economic recession, a war and an undiagnosed illness.

Our theory implies also the "other side of the coin", namely that a history of *negative* experiences should induce a perception of those experiences being more pervasive, inducing in turn an IU and a negative perception of uncertain events. Consistent with this implication, previous research has found that people with negative life experiences demonstrate a high IU (Mittal & Griskevicius, 2014). And in the above-described study, we find that people who had experienced more adverse childhood experiences [ACES; Felitti et al., 1997] were lower in optimism, which in turn negatively predicted IU. IU subsequently negatively predicted perceptions of the uncertain events referred to above. Thus, it seems that it is not uncertainty as such that leads people to feel anxious and threatened. Rather, people's expectations of negative outcomes (partially determined by their life histories) are the source of their IU.

Other research is consistent with this conclusion. For example, Gold-Spink and colleagues (2000) found that in couples in which one partner was diagnosed with multiple sclerosis, optimism was negatively related to uncertainty, suggesting that optimistic individuals dispelled the inherent uncertainty that MS induces by interpreting it in a positive way.

The chronic effects of past outcomes on attitudes toward uncertainty replicate on the cultural level in determining national mentality. Specifically, history of threats that a society has experienced was shown to shape members' attitudes toward uncertainty (see also van den Bos; Forgas; Kreko; van Prooijen, this volume). Gelfand and colleagues (2011) carried out a 33-nation study in which people's IU (measured via the need for structure scale) was positively predicted by the countries' ecological and historical threats (e.g., high population density, resource scarcity, a history of territorial conflict and disease and environmental threats). This result was replicated for differences between the states in the US (Harrington & Gelfand, 2014).

Orehek et al. (2010, Study 1) found that percentages of Muslims in Dutch neighborhoods, perceived as a source of threat, significantly predicted Dutch persons' need for cognitive closure. Though these results are correlational, they seem to attest to the idea that persistent (chronic) perception of threat elevates individuals' aversion to uncertainty (i.e., need for closure), more than they attest to the opposite causal direction, of Dutch persons' need for closure affecting the percentage of Muslims in Dutch neighborhoods.

Momentary Accessibility Effects

Drawing on accessibility theory (see Higgins, 1996), we have suggested that affective reactions to uncertain situations should also be affected by immediate, momentarily accessible outcomes. A variety of evidence supports that hypothesis. Webber et al. (2018, Studies 4–8) manipulated negative outcomes (recall of humiliation and loss of significance) and found that this elevated participants' need for cognitive closure, representing an aversion to uncertainty. In turn, the elevated need for closure mediated greater commitment to extreme ideological positions (Obaidi et al., 2022).

A series of experiments by Erev and his colleagues (2020) yielded results supportive of the notion that the recent activation of affectively valenced outcomes influences individuals' responses to uncertainty. In these studies, participants chose between different options with unknown payoffs (i.e., making decisions under uncertainty), after which they received full feedback (varying in desirability), allowing them to make future decisions based on immediate past experiences. The results suggested that by manipulating the feedback, participants under- or over-weighted the likelihood of rare positive and negative events. Participants disregarded the objective probabilities of wins and losses, which they learned from all of the previous trials, and instead based their decisions on the immediate past five outcomes that were highly accessible in memory: if those choices yielded a positive gain, participants overestimated the likelihood of a gain on the next trial, and if those recent prior choices resulted in a loss – participants overestimated the likelihood of a loss.

In a landmark study also relevant to this issue, Johnson and Tversky (1983) asked participants to read newspaper articles designed to induce either negative or positive moods. Participants were then asked to estimate the frequency of annual fatalities in different types of risk categories, including various diseases, natural hazards and homicide. It was found that negative mood induction increased participants' estimates of the likelihood of fatalities, and positive mood induction decreased those risk estimates. Subsequent research attested to more nuanced effects, specifically that negative emotions such as fear, which signal weakness and vulnerability, were associated with risk avoidance, whereas negative emotions such as anger, which signal empowerment, were associated with risk-seeking (Lerner & Keltner, 2001; Raghunathan & Pham, 1999). Similarly, recent work by Forgas (2022) found that inducing a negative mood increased negative cognitions, including increased vigilance and suspicion.

Work by Schwarz et al. (1987) suggests that the effects of momentary affective states, or the use of the *affect-as-information heuristic*, primarily affect judgments of uncertain, unfamiliar situations (see also Forgas, 1995), whereas judgments about familiar topics are unaffected by mood. Consistent with this notion, Strack et al. (1988) induced affective states through unobtrusively manipulated facial expressions and found that these had a mood-congruent influence on judgments of unfamiliar stimuli. And Otto and Eichstaedt (2018) used a large data set of social media sentiments to find that on days when city residents expressed more positive moods on Twitter, residents of these cities engaged in more risk-taking behavior, implying greater optimism or hope, that is, a positive affective reaction to uncertainty.

Relevant to the notion that negative outcomes induce a tendency to feel negatively about uncertainty is the work on "self-uncertainty." For instance, in one of van den Bos's (2001) studies, the experimenter asks the participant: "Please briefly describe the emotions that the thought of your being uncertain arouses in you" and "Please describe as accurately as you can what you think physically will happen to you as you feel uncertain." Uncertainty about the self implies something undesirable that evokes negative emotions. Van den Bos and his colleagues indeed find that in those conditions, participants for whom self-uncertainty was aroused affirmed their commitment to the value of fairness and desired greater control over decisions. In other words, they escaped the negatively tinted uncertainty into positive images of self (as committed to values and having control; see also van den Bos, this volume).

In another study, McGregor and Marigold (2003) exposed some participants to uncertainty threats by having them think of an unresolved personal dilemma. Specifically, participants were instructed to select a dilemma "that made them feel very uncertain, one that they had not already solved, and that took the form of 'should I ... or not?'" (p. 840). It was found that participants responded to this manipulation with what the authors label *compensatory conviction*, expressed as clarity about the self that "helps keep unwanted thoughts out of awareness." These results too suggest that, when coupled with unpleasant possibilities, uncertainty prompts the tendency to escape it, by embracing more appealing self-views.

Sources of Affect Interacting: The Affective Synapse Model

According to the synapse model of priming effects (Higgins et al., 1985) discussed earlier, recent priming has a strong momentary effect that overrides the effects of frequent priming. Its effect dissipates more quickly than those of frequent priming so that the latter predominates after a long enough delay. These effects pertain to the use of primed constructs toward the interpretation of ambiguous stimuli. But what might these processes imply as far as the affective responses to uncertainty are concerned? We assume that the logic of the synapse model applies here as well, with one addition, concerning the differential response to a recent affective outcome by individuals with a history of positive versus negative outcomes.

Specifically, we assume that the response to momentary/recent outcomes will be stronger for individuals whose history suggested the opposite outcomes. In other words, we expect that individuals whose prior history was affectively positive overall would react more strongly to a negative recent outcome (e.g., will be more upset); similarly, individuals whose prior history was affectively negative should react more strongly to a positive recent outcome (e.g., be more excited). In the long run, and consistent with the synapse model – those with a positive affective history would regain their optimism and positive expectancies for the future, whereas those with a negative affective history would regain their pessimism *vis-à-vis* uncertain situations.

We know of no specific evidence that tests our "affective synapse model" in its entirety. Yet there exist suggestive findings that fit some of its implications. Consider the hypothesis that prior affective history will augment individuals' reactions to an oppositely valenced recent outcome. In Ainsworth's (e.g., 1971, 1978) "strange situation" research, it was found that securely attached children, that is, those with a more positive affective history, reacted more strongly to the negative event (the mother's departure) than infants with avoidant attachment styles, though they were not more distressed than those with resistant attachment styles. In an entirely different domain, McGregor & Marigold (2003) found that participants with high self-esteem (presumably associated with a history of positive outcomes) reacted more strongly to personal uncertainty threats (by having stronger compensatory convictions) than their low-esteem counterparts.

As for the long term, there is consistent evidence that individuals with high esteem (i.e., persons presumed to have had a positive outcome history) are more resilient, that is, they retain their optimism despite momentary adversities more than individuals with low esteem (e.g., Dumont, & Provost, 1999). Admittedly, these findings are merely suggestive and partial. The full implications of the "affective synapse model" merit more specific probing in subsequent research.

Interim Summary

Drawing on the common elements of ambiguity and uncertainty, we have proposed that uncertainty isn't experienced as inherently pleasant or unpleasant. Rather, affective reactions to uncertainty depend on affectively valenced categories that are active in an individual's mind when confronting a given uncertain situation. In turn, the degree to which a given cognitive category is active is a function of its frequent priming in the individual's past experience, resulting in its *chronic accessibility*, as well as its recent priming, determining its *momentary accessibility*.

Thus, building off the work of Higgins and colleagues (1985), we proposed an *affective synapse model*, whereby immediately after encountering a given valenced outcome, individuals whose chronic accessibility of affect is of *the same sign as that* outcome will be less affected by it than individuals whose chronic accessibility of affect is of the opposite sign to the outcome. This difference, however, would dissipate after a longer delay between encountering the outcome and judgment, and come to reflect the chronically accessible, affectively valenced category, rather than the affective valence of the recent outcome.

Escaping or Embracing Uncertainty

Affective reactions to uncertainty have downstream consequences. A negative reaction should prompt an escape from the uncertainty, whereas a positive reaction involves embracing and exploring it (see also Fiedler & McCaughey, this volume).

Escaping Uncertainty

Escaping into rigidity. To the extent that uncertainty has a negative connotation, one should be motivated to escape it, if the opportunity arises. At a minimum, this would mean the embracement of any certainty, even a negatively tinged one, to the extent that its negativity was less than that evoked by the original uncertainty. In our work on the need for cognitive closure (e.g., Kruglanski & Webster, 1996; Roets et al., 2015), we cited considerable evidence for individuals' "seizing and freezing" on any knowledge affording information, that is, forming a quick closure (e.g., in impression formation or stereotyping) and holding on to it, thus becoming relatively impervious (closed-minded) to subsequent, potentially inconsistent, information.

In turn, these cognitive effects were found to affect various social psychological phenomena on the interpersonal, group and intergroup levels of analysis. At the interpersonal level, people with a high (vs. low) need for closure were found to exhibit lesser empathy (Nelson, Klein, & Irvin, 2003) and lesser communicative ability (Richter & Kruglanski, 1999), due to their tendency to freeze on their own perspective and a reluctance to adjust it by taking into account the other's viewpoint. In the realm of interpersonal negotiations, DeDreu and colleagues (1998) found that participants with a high (vs. low) need for closure are more affected by "focal points" (such as the initial strategy of reaching an agreement or a norm based on prior negotiation outcomes). Too, negotiators with a high (vs. low) need for closure tended to make smaller concessions, particularly in the presence of a high focal point (a higher expected negotiation outcome) a stereotypically competitive opponent (a business student) (De Dreu et al., 1998). Finally, research by Rubini and Kruglanski (1997) suggests that the language of people with a high need for closure tends toward greater abstraction than that of individuals with a low need for closure.

At the group level of analysis, individuals with a high (vs. low) need for closure (whether manipulated or measured) were found to exhibit a tendency to be task-oriented rather than socio-emotionally oriented during group activities (De Grada et al., 1999). Moreover, groups composed of individuals with a high (vs. low) need for closure were found to exhibit a stronger preference for ingroup consensus (Kruglanski, Webster, & Klem, 1993) and pressures toward uniformity (De Grada et al., 1999). Groups composed of individuals with a high (vs. low) need for closure also displayed a greater tendency to focus on commonly available (vs. unique) information (Webster, 1993), a stronger tendency to reject opinions deviating (Kruglanski & Webster, 1991), and a stronger proclivity toward autocracy (De Grada et al., 1999; Pierro et al., 2003). Finally, research by Livi (2003) established that groups composed of individuals with a high (vs. low) need for closure exhibit greater norm stability across changes in membership.

On the intergroup level of analysis, it was found that individuals with a high (vs. low) need for closure exhibit a stronger ingroup bias (Webster & Kruglanski, 1994; Kruglanski et al., 2002), particularly when the group is perceived as homogeneous rather than diverse in terms of members' attributes (see also Hirschberger; Hogg & Gaffney, this volume). Further, individuals with a high need for closure were found to exhibit a pronounced linguistic intergroup bias (Maas & Arcuri, 1992), a tendency by group members to describe positive ingroup and negative outgroup behaviors in abstract terms, suggesting their temporal and trans-situational stability, and to describe negative ingroup and positive outgroup behaviors in concrete terms, suggesting their transiency (Webster, Kruglanski, & Pattison, 1997).

Escaping into affirmation. Escaping the negative implication of uncertainty, which is unpleasant for one's self-concept, may be served by embracing assured opinions with neutral or even slightly negative implications. Yet, escaping unpleasant uncertainty is served even better by embracing a downright positive view, that is, seeking specifically positive, affirming knowledge. Indeed, considerable research demonstrates that whenever possible, people react to aversive uncertainty by seeking a positive type of certainty that replaces the aversive feeling with a positively tinged emotion. Specifically, research on personal or self-uncertainty by Hogg, van den Bos, McGregor and their colleagues compellingly demonstrates that when uncertainty is aversive because it implies one's own frailty and lack of confidence, individuals seek self-affirmation either by endorsing cherished values or by membership in extreme groups characterized by devotion to an empowering ideal of some sort (van den Bos, this volume).

For instance, in van den Bos's study (2001) on personal uncertainty described previously, participants' reactions to scenarios in which they were allowed or not allowed to voice their opinion about a vignette depicting a fair or unfair situation were markedly affected by their experimentally induced uncertainty. Participants whose sense of empowerment and significance was undermined by the uncertainty manipulation acted in ways aimed to restore their positive sense of self, driving away the negative feeling induced by the experimental manipulation of personal uncertainty.

Similarly, McGregor's identity consolidation theory (McGregor, 1998, 2003) views personal uncertainty as a particularly aversive self-threat; research guided by this theory finds that experimental inductions of such uncertainty via having participants ruminate about difficult personal dilemmas led them to exaggerate their convictions about social issues, values, goals and identifications (McGregor et al., 2001), representing what the authors call "compensatory convictions," whose function is to distract individuals' attention from unpleasant uncertainties. McGregor and Marigold (2003) further demonstrated that the tendency to exhibit compensatory convictions is particularly pronounced among individuals with high self-esteem, who are assumed to be particularly defensive after self-worth threats.

Hogg (2012) highlighted people's motivation to feel certain about themselves and the aversive nature of self-uncertainty (Hogg & Gaffney, this volume). He theorized and found empirical support for the proposition that, to reduce the negative affect produced by the experience of self-uncertainty, people will be motivated to join extreme groups because these offer members a strong sense of social identity and hence certainty. Whereas Hogg (2012) emphasized the entitative nature of extreme groups, their internal homogeneity, common goals and common fate, we would add another reason consistent with his hypothesis: extreme groups are typically strongly committed to their central goal. Extremism has been portrayed as the process of concentrating on one goal only and the readiness to sacrifice others (Kruglanski et al., 2021 a, b). To do so means to endow the central goal with particular importance. Serving it, therefore, lends one a strong sense of self-worth, which is of particular value to individuals for whom it was undermined by the experience of self-uncertainty.

Therefore, research on people's reactions to personal or self-uncertainty highlights the motivation to remove the negative implications of uncertainty for one's sense of personal significance and self-worth (Kruglanski et al., 2022) via various ways of self-affirmation, either by endorsing cherished values, demanding a say in important issues, or endorsing and approving of extreme groups committed to a significance-lending ideal.

Embracing Uncertainty

The thesis of this chapter has been that uncertainty as such is neither necessarily aversive nor necessarily pleasurable. Instead, it can be imbued with positive or negative valence to different degrees depending on the valence of outcomes that are associated with it. And reactions to uncertainty are uniquely determined by such valence. In the preceding sections of this paper, we discussed ways of *escaping* or avoiding uncertainty, whether by developing a cognitive rigidity or by seeking self-affirmation. Next, we consider the approach mode of responding to uncertainty, prompted by its degree of positive valence.

Creativity. We assume that when uncertainty has a positive valence, the individual will likely want to prolong their stay in the uncertain situation. We also assume that this desire is likely to be served by exploring the possibilities that this situation offers. The notion that a positive orientation toward uncertainty promotes exploration has received indirect support from several research domains in psychology. In the area of personality, McCrae (1987) reported a positive correlation between the openness to experience factor of the Big Five and creativity. Subsequent research replicated this finding (e.g., Li et al., 2015).

In social psychology, Chirumbolo and colleagues (2004) found a negative correlation between the need for cognitive closure (whether manipulated or measured) and creativity. Individuals high on the need for closure were less creative during group discussion than those who were low on this need. Furthermore, groups composed of high need for closure individuals were less creative than groups composed of low need for closure participants. This tendency was mediated by greater conformity pressures in the high need for closure groups. The negative correlation between the need for closure and creativity was also obtained in subsequent studies (e.g., Chirumbolo et al., 2005). Finally, in the domain of organizational psychology, Kirrane and colleagues (2019) found a negative relation between insecure attachment, connoting a need for security and certainty, and creativity at work.

Exploration and initiation. In Ainsworth's (e.g., Ainsworth & Bell, 1970) classic "strange situation" research, securely attached kids, that is, those whose history of prior experiences had been largely positive and reassuring, were found to recover quickly from the stress occasioned by their mother's brief departure, whereas anxiously and ambivalently attached children were found to take a long time to get back to normal. After some brief clinging, the securely attached children left the mother and excitedly resumed their exploration of the laboratory environment. For securely attached babies, then, the supportive mother is a safe haven to whom they can turn when in distress; she serves as a secure base from which they can launch their exploration of the unknown (Mikulincer & Shaver, this volume).

The positive relation between secure attachment and exploration has received robust support in subsequent research. Studies by Grossmann et al. (2008) find a strong link between secure child–parent relationships and the eagerness to engage in autonomous exploration. Rothbaum and colleagues (2000) note the replication of the attachment-exploration relation across different cultures and species (e.g., Harlow & Zimmerman, 1959) and suggest that it may reflect a biological basis, albeit moderated by culture. Thus, in several studies, Japanese babies tended less to explore the lab environment than US babies when left alone in the strange situation and manipulated toys less often when their mothers returned (e.g., Caudill & Weinstein, 1969; Takahashi, 1990).

The intimate connection between attachment and exploration is hardly limited to young children. For instance, Beck (2006) found in an interesting study that religious students displaying a secure attachment to God (as measured by the Attachment to God Inventory) were more engaged in the theological exploration were more tolerant of Christian faiths different from their own and reported more peace and less distress during their exploratory investigations. Indeed, God and other spiritual or religious figures can be attachment figures to whom one can turn in moments of uncertainty; a secure attachment to those figures can serve as a secure base from which to explore that uncertainty (see Mikulincer & Shaver, this volume).

Similarly, Elliot and Reis (2003) report several studies attesting to the idea that "secure attachment in adulthood affords unimpeded, appetitive exploration in achievement setting and that insecure attachment in adulthood interferes with exploration in achievement settings by evoking avoidance motivation." (p. 328). And in two studies whose approaches closely anticipate the present analysis, both chronic and momentary (i.e., primed) accessibility of secure attachment constructs were related to participants' motivation to explore and their stated interest in novel stimuli.

Recapitulation and Conclusion

Beyond their hardwired, genetic predisposition, individuals' reactions to uncertainty are determined by both chronic and momentary accessibility of positive and negative thoughts weighted by their magnitude. The greater the accessibility/magnitude of negative thoughts, the more negative will be the individual's reaction to uncertainty; likewise, the greater the accessibility/magnitude of positive thoughts, the more positive will be the individual's reaction to uncertainty. Notably, this could also include positive thoughts reflecting self-efficacy and resilience; even when uncertainty momentarily activates negative feelings, an individual with a long-term history of positive outcomes may eventually come to accept the uncertainty with the understanding that they can effectively handle the situation. These reactions to uncertainty define its valence, which in turn determines the value of avoiding/escaping the uncertainty or approaching/embracing it. In the presence of a sufficiently pronounced expectancy of successfully carrying out the approach or the avoidance, it will combine with their value to form an approach or avoidance goal; under the appropriate circumstances (of this goal's dominance over other salient objectives), such a goal will result in the corresponding behavior (i.e., escaping or embracing the uncertainty in actual behavior (cf. Kruglanski et al., 2018; Ajzen & Kruglanski, 2020)).

Escaping uncertainty can take the form of *cognitive rigidity*. Such escape represents a recoil, or push away from uncertainty as such, because of the inchoate negative valence it has acquired for the individual. There is, however, also the pull of positive experiences, and given the choice, individuals would prefer highly positive, rather than merely neutral or slightly positive self-affirmations. Escaping uncertainty was found to have significant downstream consequences, including the tendency toward stereotyping, striving for consensus, autocracy and intolerance of diversity (Kruglanski et al., 2006). It was also found to promote the readiness to embrace conspiracy theories as well as support for violence against outgroups. Embracing positively valenced uncertainty was found to express itself in creativity as well as in extensive exploration of the environment.

By identifying the factors that determine people's affective responses to uncertainty, the present analysis points the way to controlling those responses for the benefit of the individuals involved and others in their social networks. This knowledge could be of particular utility in today's quickly changing world and the attendant uncertainties that such change inevitably ushers in. Particularly in instances where "dark clouds on the horizon" seem to accompany the uncertainty, it may be possible to refocus individuals' attention on possible rewarding opportunities that the uncertain situation affords. This means that the best strategy is trying to accentuate the positive, eliminate the negative and not worry too much about what lies in between, as suggested in Johnny Mercer's hit song of the 1940s.

References

Ainsworth, M. D. S., & Bell, S. M. V. (1970). Attachment, exploration, and separation: Illustrated by the behavior of one-year-olds in a strange situation. *Child Development, 41*, 49–67.

Ainsworth, M. D. S., Bell, S. M., & Stayton, D. J. (1971) Individual differences in strange situation behavior of one-year-olds. In H. R. Schaffer (Ed.) *The origins of human social relations.* London and New York: Academic Press. Pp. 17–58.

Ainsworth, M. D. S., Blehar, M. C., Waters, E., & Wall, S. (1978). *Patterns of attachment: A psychological study of the strange situation.* Hillsdale, NJ: Erlbaum.

Ajzen, I., & Kruglanski, A. W. (2020). Reasoned action in the service of goal pursuit. *Psychological Review, 126*(5), 774.

Alarcon, G. M., Bowling, N. A., & Khazon, S. (2013). Great expectations: A meta-analytic examination of optimism and hope. *Personality and Individual Differences, 54*(7), 821–827.

Bargh, J. A., & Pratto, F. (1986). Individual construct accessibility and perceptual selection. *Journal of Experimental Social Psychology, 22*(4), 293–311.

Beck, R. (2006). God as a secure base: Attachment to God and theological exploration. *Journal of Psychology and Theology, 34*(2), 125–132.

Brosschot, J. F., Verkuil, B., & Thayer, J. F. (2016). The default response to uncertainty and the importance of perceived safety in anxiety and stress: An evolution-theoretical perspective. *Journal of Anxiety Disorders, 41*, 22–34.

Carleton, R. N. (2016). Fear of the unknown: One fear to rule them all? *Journal of Anxiety Disorders, 41*, 5–21.

Caudill, W., & Weinstein, H. (1969). Maternal care and infant behavior in Japan and America. *Psychiatry, 32*(1), 12–43.

Chirumbolo, A., Livi, S., Mannetti, L., Pierro, A., & Kruglanski, A. W. (2004). Effects of need for closure on creativity in small group interactions. *European Journal of Personality, 18*(4), 265–278.

Chirumbolo, A., Mannetti, L., Pierro, A., Areni, A., & Kruglanski, A. W. (2005). Motivated closed-mindedness and creativity in small groups. *Small Group Research, 36*(1), 59–82.

Constans, J. I., Penn, D. L., Ihen, G. H., & Hope, D. A. (1999). Interpretive biases for ambiguous stimuli in social anxiety. *Behaviour Research and Therapy, 37*(7), 643–651.

Dearing, K. F., & Gotlib, I. H. (2009). Interpretation of ambiguous information in girls at risk for depression. *Journal of Abnormal Child Psychology, 37*(1), 79–91.

De Dreu, C. K., Koole, S. L., & Oldersma, F. L. (1999). On the seizing and freezing of negotiator inferences: Need for cognitive closure moderates the use of heuristics in negotiation. *Personality and Social Psychology Bulletin, 25*(3), 348–362.

De Grada, E., Kruglanski, A. W., Mannetti, L., & Pierro, A. (1999). Motivated cognition and group interaction: Need for closure affects the contents and processes of collective negotiations. *Journal of Experimental Social Psychology, 35*(4), 346–365.

Dumont, M., & Provost, M. A. (1999). Resilience in adolescents: Protective role of social support, coping strategies, self-esteem, and social activities on experience of stress and depression. *Journal of Youth and Adolescence, 28*(3), 343–363.

Ek, E., Remes, J., & Sovio, U. (2004). Social and developmental predictors of optimism from infancy to early adulthood. *Social Indicators Research, 69*(2), 219–242.

Elliot, A. J., & Reis, H. T. (2003). Attachment and exploration in adulthood. *Journal of Personality and Social Psychology, 85*(2), 317.

Erev, I., Ert, E., & Plonsky, O. (2020). Six Contradicting Deviations From Rational Choice, and the Possibility of Aggregation Gain.

Eysenck, M. W., Mogg, K., May, J., Richards, A., & Mathews, A. (1991). Bias in interpretation of ambiguous sentences related to threat in anxiety. *Journal of Abnormal Psychology, 100*(2), 144.

Felitti, S. A., Chan, R. L., Gago, G., Valle, E. M., & Gonzalez, D. H. (1997). Expression of sunflower cytochrome c mRNA is tissue-specific and controlled by nitrate and light. *Physiologia Plantarum, 99*(2), 342–347.

Forgas, J. P. (1995). Mood and judgment: the affect infusion model (AIM). *Psychological Bulletin, 117*(1), 39.

Forgas, J. P. (in press), The evolutionary functions of sadness: The cognitive and social benefits of negative affect. In Al Shawaf, L. (Ed.). *The Oxford Handbook of Evolution and Emotions.* Oxford: University Press.

Frenkel-Brunswik, E. L. S. E. (1948). Intolerance of ambiguity as emotional perceptual. *Personality of Assessment, 40,* 67–72.

Gelfand, M. J., Raver, J. L., Nishii, L., Leslie, L. M., Lun, J., Lim, B. C.,... & Yamaguchi, S. (2011). Differences between tight and loose cultures: A 33-nation study. *Science, 332*(6033), 1100–1104.

Glaesmer, H., Rief, W., Martin, A., Mewes, R., Brähler, E., Zenger, M., & Hinz, A. (2012). Psychometric properties and population-based norms of the life orientation test revised (LOT-R). *British Journal of Health Psychology, 17*(2), 432–445.

Gold-Spink, E., Sher, T. G., & Theodos, V. (2000). Uncertainty in illness and optimism in couples with multiple sclerosis. *International Journal of Rehabilitation and Health, 5*(3), 157–164.

Grenier, S., Barrette, A. M., & Ladouceur, R. (2005). Intolerance of uncertainty and intolerance of ambiguity: Similarities and differences. *Personality and Individual Differences, 39*(3), 593–600.

Grossmann, K., Grossmann, K. E., Kindler, H., & Zimmermann, P. (2008). A wider view of attachment and exploration: The influence of mothers and fathers on the development of psychological security from infancy to young adulthood. In J. Cassidy & P. R. Shaver (Eds.), *Handbook of attachment: Theory, research, and clinical applications* (pp. 857–879). The Guilford Press.

Harlow, H. F., & Zimmermann, R. R. (1959). Affectional response in the infant monkey: Orphaned baby monkeys develop a strong and persistent attachment to inanimate surrogate mothers. *Science, 130*(3373), 421–432.

Harrington, J. R., & Gelfand, M. J. (2014). Tightness–looseness across the 50 united states. *Proceedings of the National Academy of Sciences, 111*(22), 7990–7995.

Higgins, E. T. (1996). Activation: Accessibility, and salience. In E. T. Higgins & A. W. Kruglanski (Eds.), *Social psychology: Handbook of basic principles* (pp. 133–168). The Guilford Press.

Higgins, E. T., Bargh, J. A., & Lombardi, W. J. (1985). Nature of priming effects on categorization. *Journal of Experimental Psychology: Learning, Memory, and Cognition, 11*(1), 59.

Higgins, E. T., & Brendl, C. M. (1995). Accessibility and applicability: Some "activation rules" influencing judgment. *Journal of Experimental Social Psychology, 31*(3), 218–243.

Higgins, E. T., & King, G. (1981). Accessibility of social constructs: Information processing consequences if individual and contextual variability. In N. Cantor & J. F. Kihlstrom (Eds.), *Personality, cognition, and social interaction* (pp. 69–121). Erlbaum.

Higgins, E. T., Rholes, W. S., & Jones, C. R. (1977). Category accessibility and impression formation. *Journal of Experimental Social Psychology, 13*(2), 141–154.

Hogg, M. A. (2012). Self-uncertainty, social identity, and the solace of extremism. In M. A. Hogg & D. L. Blaylock (Eds.), *Extremism and the psychology of uncertainty* (pp. 19–35). Wiley-Blackwell.

Johnson, E. J., & Tversky, A. (1983). Affect, generalization, and the perception of risk. *Journal of Personality and Social Psychology, 45*(1), 20.

Kirrane, M., Kilroy, S., Kidney, R., Flood, P. C., & Bauwens, R. (2019). The relationship between attachment style and creativity: The mediating roles of LMX and TMX. *European Journal of Work and Organizational Psychology, 28*(6), 784–799.

Kruglanski, A. W., Molinario, E., Jasko, K., Webber, D., Leander, N. P., & Pierro, A. (2022). Significance-quest theory. *Perspectives on Psychological Science*, 17456916211034825.

Kruglanski, A. W., Shah, J. Y., Fishbach, A., Friedman, R., Chun, W. Y., & Sleeth-Keppler, D. (2018). A theory of goal systems. In A. W. Kruglanski (Ed.), *The motivated mind* (pp. 207–250). London: Routledge.

Kruglanski, A. W., Shah, J. Y., Pierro, A., & Mannetti, L. (2002). When similarity breeds content: Need for closure and the allure of homogeneous and self-resembling groups. *Journal of Personality and Social Psychology, 83*(3), 648.

Kruglanski, A. W., Szumowska, E., & Kopetz, C. (2021). The call of the wild: How extremism happens. *Current Directions in Psychological Science, 30*(2), 181–185.

Kruglanski, A. W., Szumowska, E., Kopetz, C. H., Vallerand, R. J., & Pierro, A. (2021). On the psychology of extremism: How motivational imbalance breeds intemperance. *Psychological Review, 128*(2), 264.

Kruglanski, A. W., & Webster, D. M. (1991). Group members' reactions to opinion deviates and conformists at varying degrees of proximity to decision deadline and of environmental noise. *Journal of Personality and Social Psychology, 61*(2), 212.

Kruglanski, A. W., & Webster, D. M. (1996). Motivated closing of the mind: Its cognitive and social effects. *Psychological Review, 103*(2), 263–283.

Kruglanski, A. W., Webster, D. M., & Klem, A. (1993). Motivated resistance and openness to persuasion in the presence or absence of prior information. *Journal of Personality and Social Psychology, 65*(5), 861.

Lerner, J. S., & Keltner, D. (2001). Fear, anger, and risk. *Journal of Personality and Social Psychology, 81*(1), 146.

Li, W., Li, X., Huang, L., Kong, X., Yang, W., Wei, D., ... & Liu, J. (2015). Brain structure links trait creativity to openness to experience. *Social Cognitive and Affective Neuroscience, 10*(2), 191–198.

Livi, S. (2003). Il bisogno di chiusura cognitiva e la trasmissione delle norme nei piccoli gruppi [The need for cognitive closure and norm transmission in small groups]. *Unpublished doctoral dissertation, University of Rome "La Sapienza," Rome, Italy*.

Lovecraft, H. P. (1927). *The Colour Out of Space*. Penguin UK.

Maass, A., & Arcuri, L. (1992). The role of language in the persistence of stereotypes. In G. R. Semin & K. Fiedler (Eds.), *Language, interaction and social cognition* (pp. 129–143). Sage Publications, Inc.

McCrae, R. R. (1987). Creativity, divergent thinking, and openness to experience. *Journal of Personality and Social Psychology, 52*(6), 1258.

McGregor, H. A., Lieberman, J. D., Greenberg, J., Solomon, S., Arndt, J., Simon, L., & Pyszczynski, T. (1998). Terror management and aggression: evidence that mortality salience motivates aggression against worldview-threatening others. *Journal of Personality and Social Psychology, 74*(3), 590.

McGregor, I., & Marigold, D. C. (2003). Defensive zeal and the uncertain self: What makes you so sure? *Journal of Personality and Social Psychology, 85*(5), 838.

McGregor, I., Zanna, M. P., Holmes, J. G., & Spencer, S. J. (2001). Compensatory conviction in the face of personal uncertainty: going to extremes and being oneself. *Journal of Personality and Social Psychology, 80*(3), 472.

Mittal, C., & Griskevicius, V. (2014). Sense of control under uncertainty depends on people's childhood environment: A life history theory approach. *Journal of Personality and Social Psychology, 107*(4), 621.

Nelson, D. W., Klein, C. T., & Irvin, J. E. (2003). Motivational antecedents of empathy: Inhibiting effects of fatigue. *Basic and Applied Social Psychology, 25*(1), 37–50.

Obaidi, M., Skaar, S. W., Ozer, S., & Kunst, J. R. (2022). Measuring extremist archetypes: Scale development and validation. *PloS One, 17*(7), e0270225.

Orehek, E., Fishman, S., Dechesne, M., Doosje, B., Kruglanski, A. W., Cole, A. P.,... & Jackson, T. (2010). Need for closure and the social response to terrorism. *Basic and Applied Social Psychology, 32*(4), 279–290.

Otto, A. R., & Eichstaedt, J. C. (2018). Real-world unexpected outcomes predict city-level mood states and risk-taking behavior. *PloS One, 13*(11), e0206923.

Pierro, A., Mannetti, L., De Grada, E., Livi, S., & Kruglanski, A. W. (2003). Autocracy bias in informal groups under need for closure. *Personality and Social Psychology Bulletin, 29*(3), 405–417.

Raghunathan, R., & Pham, M. T. (1999). All negative moods are not equal: Motivational influences of anxiety and sadness on decision making. *Organizational Behavior and Human Decision Processes, 79*(1), 56–77.

Richter, L., & Kruglanski, A. W. (1999). Motivated search for common ground: Need for closure effects on audience design in interpersonal communication. *Personality and Social Psychology Bulletin, 25*(9), 1101–1114.

Roets, A., Kruglanski, A. W., Kossowska, M., Pierro, A., & Hong, Y. Y. (2015). The motivated gatekeeper of our minds: New directions in need for closure theory and research. In J. M. Olson & M. P. Zanna (Eds.), *Advances in experimental social psychology* (Vol. 52, pp. 221–283). Academic Press.

Rothbaum, F., Weisz, J., Pott, M., Miyake, K., & Morelli, G. (2000). Attachment and culture: Security in the United States and Japan. *American Psychologist, 55*(10), 1093.

Rubini, M., & Kruglanski, A. W. (1997). Brief encounters ending in estrangement: Motivated language use and interpersonal rapport in the question–answer paradigm. *Journal of Personality and Social Psychology, 72*(5), 1047.

Sanchez, A. L., Kendall, P. C., & Comer, J. S. (2016). Evaluating the intergenerational link between maternal and child intolerance of uncertainty: a preliminary cross-sectional examination. *Cognitive Therapy and Research, 40*(4), 532–539.

Schoth, D. E., & Liossi, C. (2017). A systematic review of experimental paradigms for exploring biased interpretation of ambiguous information with emotional and neutral associations. *Frontiers in Psychology, 8*, 171.

Schwarz, N., Strack, F., Kommer, D., & Wagner, D. (1987). Soccer, rooms, and the quality of your life: Mood effects on judgments of satisfaction with life in general and with specific domains. *European Journal of Social Psychology, 17*(1), 69–79.

Srull, T. K., & Wyer, R. S. (1979). The role of category accessibility in the interpretation of information about persons: Some determinants and implications. *Journal of Personality and Social Psychology, 37*(10), 1660.

Strack, F., Martin, L. L., & Stepper, S. (1988). Inhibiting and facilitating conditions of the human smile: A nonobtrusive test of the facial feedback hypothesis. *Journal of Personality and Social Psychology, 54*(5), 768.

Takahashi, K. (1990). Are the key assumptions of the 'Strange Situation' procedure universal? A view from Japanese research. *Human Development, 33*(1), 23–30.

Thayer, J. F., Åhs, F., Fredrikson, M., Sollers III, J. J., & Wager, T. D. (2012). A meta-analysis of heart rate variability and neuroimaging studies: implications for heart rate variability as a marker of stress and health. *Neuroscience & Biobehavioral Reviews, 36*(2), 747–756.

United Nations (2020, May 14). UN leads call to protect most vulnerable from mental health crisis during and after COVID-19. *UN News*. Retrieved from https://news.un.org/en/story/2020/05/1063882

Van den Bos, K. (2001). Uncertainty management: The influence of uncertainty salience on reactions to perceived procedural fairness. *Journal of Personality and Social Psychology, 80*(6), 931.

Webber, D., Babush, M., Schori-Eyal, N., Vazeou-Nieuwenhuis, A., Hettiarachchi, M., Bélanger, J. J.,... & Gelfand, M. J. (2018). The road to extremism: Field and experimental evidence that significance loss-induced need for closure fosters radicalization. *Journal of Personality and Social Psychology, 114*(2), 270.

Webster, D. M. (1993). Motivated augmentation and reduction of the over attribution bias. *Journal of Personality and Social Psychology, 65*(2), 261.

Webster, D. M., & Kruglanski, A. W. (1994). Individual differences in need for cognitive closure. *Journal of Personality and Social Psychology, 67*(6), 1049.

Webster, D. M., Kruglanski, A. W., & Pattison, D. A. (1997). Motivated language use in intergroup contexts: Need-for-closure effects on the linguistic intergroup bias. *Journal of Personality and Social Psychology, 72*(5), 1122.

5
INSECURITY CAN BE BENEFICIAL

Reflections on Adaptive Strategies for Diverse Trade-Off Settings

Klaus Fiedler and Linda McCaughey

HEIDELBERG UNIVERSITY, GERMANY

Abstract

While human (and animal) behavior is doubtlessly striving for security as an ideal end of all important behaviors (regarding health, mating, achievement, survival, etc.), many means toward this end are energized by insecurity. From Bjork and colleagues, we have learned that good learning is effortful, not easy, or fluent. Partial reinforcement schedules lead to better performance than secure reinforcement, as impressively shown by Lawrence and Festinger. Inferiority and minority status trigger better argumentation than a superior majority status (Moscovici). Bischof's Zürich model of social motivation offers a refined ethological and evolutionary account for the twofold need for both familiarity and unfamiliarity, security and novelty. We presume that the malleability of trade-off problems allows ordinary people and researchers to misperceive the asymmetry of security versus insecurity strategies, favoring the former and neglecting the latter. A taxonomy of established trade-off paradigms demonstrates that (a) there seem to be more normative reasons for insecurity strategies, although (b) the pertinent literature renders security strategies more prominent and intuitively plausible.

Both Security and Insecurity Can Be Beneficial

Two seemingly contradictory truisms refer to benefits of both security and insecurity. On the one hand, human (and animal) behavior doubtlessly strives for security as an ideal end of virtually all important behaviors. Subjective well-being and life satisfaction, an edifying marriage, intellectual and physical achievement, and of course survival profit from security and high confidence. On the other hand, the means toward these ends are often

DOI: 10.4324/9781003317623-6

energized by insecurity (see also von Hippel & Merakovsky; Kruglanski & Ellenberg, this volume). For a satisfying life event – a happy marriage, an academic degree, or an Olympic medal in sports – to be most rewarding, the positive outcomes must not be self-evident and fully expected. The higher the prior insecurity, due to deprivation and scarcity of reward, the lower the aspiration level, and the normative expectations, the stronger the satisfaction gained from attaining these goals. Under the latter conditions, the resulting performance and the long-term learning effect are often stronger.

To be sure, the apparent contradiction need not be framed as a paradox but can be easily translated into a meaningful and logically sound psychological message. One way to make sense of the seeming paradox is to assume that reward value is not a zero-order function of absolute outcome values X but a derivative function of the experienced change in outcome values ΔX (i.e., gains and losses). Defining ΔX as the difference $X_{posterior} - X_{prior}$ between posterior and prior states implies that to maximize reward value, $X_{posterior}$ must not only be high; ΔX also increases when the prior value X_{prior} is low. Another possibility is that the ultimate reward is a dialectic function of contrasting experiences of positive and negative reinforcement, success, and disappointment. In any case, the aim of this chapter is to provide a meaningful psychological interpretation of a seeming paradox, the benefits of insecurity.

More than a paradox. In what situations should one expect to find any evidence for real benefits of insecurity? What conditions must be met for insecurity to trigger adaptive behavior and a positive balance of benefits and costs? Before we address these crucial questions, we must clarify what we mean by "insecurity", that is, how we define insecurity. It is essential to disentangle the term "insecurity" from such dreadful causes or concomitants as war, object loss, threatening disease, unemployment, or serious failure experiences that are often confounded with miserable feelings of insecurity but that do not belong to the psychological core meaning of the term (see also Arriaga & Kumashiro; Kreko, this volume). To understand its beneficial potential, we must separate "insecurity" – which is simply the opposite of "security" and refers to uncertain, unfamiliar, and risky situations – from those aversive eliciting conditions, which are responsible for the extremely negative connotations surrounding salient insecure episodes. Abstracting from these dreadful sources of negative affect, in the present chapter, we rather define the core meaning of insecurity as internal states paired with ecological settings entailing non-secure outcomes, reflecting risk, novelty, alien character, or imperfect control. We believe that for insecurity to be present it is sufficient that only a subset of these defining criteria is met. Thus, the present chapter is concerned with potential benefits of risky, unfamiliar, and hard-to-control situations and strategies embracing and often actively seeking insecurity, as contrasted with secure, familiar, and perfectly controllable tasks.

To anticipate an intriguing insight of the present chapter, a review of the theoretical literature will show that well-established theoretical constraints will almost always highlight adaptive advantages of insecurity, although fulfillment of security is presupposed to be a primary goal of behavior. As we shall see, there are various theoretical reasons why insecurity must be beneficial in certain situations, but hardly any sound reason why security should foster achievement, well-being, or adaptive behavior.

Insecurity and Creative Self-Regulation

Basic trade-off structures. A variety of trade-offs that one faces every day afford one of the fundamental reasons why insecurity, specifically a strategy that involves insecurity, is sometimes beneficial or even necessary. To understand the relative benefits and losses of secure and insecure strategies, respectively, it is essential to consider the trade-off structure that underlies adaptive cognition and behavior. A trade-off can be characterized as a stable conflict between competing strategies that draw on the same resource, such that increasing or improving in one respect leads to a loss in the other, and vice versa. One would not refer to a trade-off if either of the two respects could be improved without impacting the other, without there being a conflict that needed to be resolved between the two strategies. The disadvantages that one must accept in exchange for reaping the benefits of improving one aspect can vary. Hence, trade-offs can be renegotiated flexibly depending on the underlying utility function, which is malleable, allowing adaptive agents to construe trade-offs in different ways. This means that the optimum can fluctuate rather than being invariant. Because of this very malleability or flexibility in the construal of the trade-off utility function, it is possible that people develop preferred strategies that work against an obviously superior normative strategy (see also Hirt, Eyink & Heyman; Cooper & Pearce; van Prooijen, this volume). Specifically, with reference to a variety of trade-offs, normative principles often highlight the superiority of insecurity-oriented strategies, although most people develop an affective, quasi-morally motivated preference for security-oriented behavior. It is interesting to note that the same preference for security-oriented behavior and the same neglect of normatively superior insecurity strategies can be found in behavioral science (see also Jussim et al., this volume). As we shall see in the following review of empirical results, the normative superiority of insecurity-based strategies is often sorely neglected.

Speed-accuracy trade-off. However, let us first illustrate how insecurity can be beneficial at all, using as a very common example the trade-off between speed and accuracy, before we delineate a more comprehensive overview of other universal trade-offs that engulf our everyday behavior. In a speed-accuracy trade-off, an individual can increase the total payoff either quantitatively by increasing the number of decisions completed in a given time period or qualitatively by improving the quality of a few carefully completed decisions. For instance, a therapist may be speed-oriented and treat many patients, or she may be accuracy-oriented and use her capacity to treat a few patients at the highest level of care. Similarly, a winemaker can produce a large quantity of rather inexpensive wine by letting his vines grow many shoots with many grapes or reduce the amount of fruit formed by pruning the vines, which produces a much smaller number of higher-quality grapes that can be turned into more expensive high-quality bottles.

A useful way to implement speed-accuracy trade-offs experimentally is by linking speed and accuracy to the amount of information sampled by decision-makers before they make a choice between two options. Thus, on every trial of a sequential sample-based choice task, Fiedler, McCaughey, Prager, Eichberger, and Schnell (2021) allowed their participants to sample binary observations about two investment funds' previous success or failure until they believed to have sufficient information to choose the better fund. With an increasing sample size, the rate of correct choices (accuracy) increased, but the total number of choices completed in a fixed time period (speed) decreased. However, crucially, because the number of completed tasks decreased linearly with growing sample size whereas the accuracy increase was clearly sublinear (i.e., negatively accelerated), speed (or the number of completed choices) was a much stronger determinant of their cumulated payoff (+1 ECU = economic currency unit for all correct and −1 ECU for all incorrect choices) than accuracy.

From a normative point of view, a strategy based on little information that enabled fast decisions, though accompanied by a feeling of insecurity, would have been far superior to one that focused more on accuracy (see also Krueger & Gruening, this volume). This is a paradigmatic example of an insecure strategy (which in this case means one based on little information and hence prone to errors) being by far superior to a more secure strategy. Embracing this insecurity would have allowed participants to achieve very high payoffs. However, people may not always be able or willing to take advantage of an insecure strategy. Before we elaborate on other examples of insecurity's benefits, we use this example to discuss the reasons why many people may fail to recognize the advantage of a courageous insecurity strategy and instead often prefer a security strategy, facilitated by the malleable utility function of the speed-accuracy trade-off.

Evidence for the neglect of insecurity benefits of speed-accuracy trade-offs. Despite the clear-cut speed advantage built into the depicted task, almost all participants exhibited a pronounced accuracy bias, reducing their final payoff by several hundred percent. They obviously struggled to accept the insecurity implied by a fast strategy and preferred to focus and increase accuracy. Even when explicit feedback and manipulated upper limits of sample size encouraged participants to recognize the speed dominance and even when participants could only win from correct but never lose from incorrect choices, they continued to oversample and to disregard the advantage of speed over accuracy. Indeed, the oversampling tendency was so strong and the insensitivity to the speed advantage was so common that most participants wasted around 15 Euro of possible payoff they could have gained in the sequential choice game.

A major factor that seemed to have played a great role in participants' failure to maximize their payoff is their fondness for high accuracy. Apparently, the accuracy bias was hardly experienced as a source of weakness or as a deficit in rationality. It rather appeared to reflect a quasi-moral strength, as if a careful, accuracy-oriented investment strategy was particularly responsible and superior to a "reckless" speed strategy. Drawing large samples and refraining from premature choices could be apparently framed as a desirable, responsible, reliable behavior, reflecting a utility function that was not only sensitive to a trade-off between factors for maximizing payoff but also exhibiting the quasi-moral advantage of accuracy. We dare to assume that most participants were happy with, or proud of, their accuracy bias, notwithstanding the waste of possible payoff (which they may, however, not have been fully aware of), even though a small minority of speed-oriented players were happy with their maximal payoff, which is the usual standard by which high performance is measured.

Both security and insecurity can be beneficial. Developing a conservative preference for these security-oriented values and neglecting the obvious advantages of insecurity is only possible because the malleability of trade-offs renders both strategies justifiable. Depending on how one's investment trade-off is framed, either as responsible action that satisfies one's super-ego or the leading norms of one's social environment or as a chance to win a payoff competition, one can be proud of or happy with either strategy (see also Krueger & Gruening, this volume). This win-win structure of the speed-accuracy trade-off may prevent people from using competitive feedback learning as a premise for maximizing their total payoff.

In this chapter, we argue that the same message as the introductory example of a speed-accuracy trade-off holds for a variety of other prominent trade-off paradigms that have been the focus of intensive research in behavioral science. Even when insecurity is normatively superior, one must consider that many people have acquired a disposition to seek security. The structure of malleable trade-offs with flexible utility functions facilitates this seemingly irrational preference for security, despite all the benefits of insecurity.

While any particular psychological explanation for this discrepancy would presumably constitute an unwarranted simplification, it seems easy to understand why the social-learning environment facilitates the development of secure and cautious strategies as opposed to insecure and risk-abiding strategies. On one hand, secure behaviors – involving social approval, conformity, familiar people and tasks, and insurance – are intrinsically comfortable and less demanding and challenging than insecure behaviors – involving autonomy, non-conformist action, disobedience, foreign people and unfamiliar tasks, and risky prospects (see also Mikulincer & Shaver, this volume). On the other hand, the action outcomes triggered by secure strategies are more immediate and more evaluable than outcomes triggered by insecure strategies (Hsee & Zhang, 2010). That is, the dangers, threats, accidents, violent attacks, frustrations, and failure experiences that can be avoided by secure strategies are incontestable and experienced directly, with temporal and spatial contiguity to the security-oriented decisions. In contrast, the enjoyment of novelty, curiosity, risk-taking, new horizons, sensation seeking, and other insecurity strategies is often less direct and detached from immediate outcomes (see Crano & Hohman, this volume). Just as speed is less evaluable than accuracy, enjoying new horizons and friends and unfamiliar hobbies takes more time and patience than exploiting immediate rewards for secure activities.

Most importantly, perhaps, attributional mechanisms can facilitate the development of security-focused social norms. Memorable research on victim blame have illustrated this intriguing point (Bohner, Siebler & Raaijmakers, 1999; Viki & Abrams, 2002). The acceptance of sexist and victim-blaming rape myths, such as the belief that women's behavior or appearance provokes rape, can bring an illusion of security for other women. They can maintain the belief that similar crimes will not happen to them as long as they do not engage in any allegedly provocative, insecure behavior.

Analogical discrepancy in real-life monetary investment. The same discrepancy between widespread preferences for security despite an objective advantage of insecurity can be found in real-life behavioral finance. Granting that a timely task for virtually everybody in the 21st century is to care for one's own retirement fund, a notable discrepancy exists between the general reluctance to accept a reasonable amount of risk and the uncontested advantage of risk-oriented investment funds in the long run. In an enlightening piece of translational research, Kaufmann, Weber, and Ainsley (2013) provided intriguing evidence on real investment strategies. When investing money into their own retirement funds, people ought to overcome the typical conservative bias that characterizes most people's strategies and understand that risky investment strategies are met with more success than too conservative strategies. A well-established finding in behavioral finance is that people tend to be too conservative to maximize their retirement funding. Consistent with this premise, Kaufmann et al. (2013) found that banking customers who underwent systematic feedback training to better understand the impact of risk in investment markets actually learned, to some degree, to overcome their restrictive conservative attitude. They learned to increase their risk preference and thereby to benefit significantly from financial risk taking.

Ecological basis of insecurity benefits. A plausible explanation for the superiority of risky investments can be found in an ecological analysis of life's gambles conducted by Pleskac and Hertwig (2014). Based on an analysis of roulette gambling, insurance, or trading, these authors found evidence for a strong and regular negative relationship between value and frequency of occurrence. Just as in the economy, high prices are reflective of the scarcity of goods and rare objects are generally considered more precious than common objects. The following quotation in Pleskac and Hertwig (2014, p. 2002) illustrates the wisdom of this negative correlation (between value and probability): "One doesn't discover new lands without consenting to lose sight of the shore for a very long time" (Gide, 1973, p. 309). Analogous to the business rule "nothing ventured, nothing gained", the same wisdom holds for social action and leisure-time behavior: "No risk, no fun".

Pleskac and Hertwig (2014) present several explanations for why rare and scarce objects are more precious than common objects, referring to a fair bet (i.e., balanced winning and losing chances) or open markets (where the most sought-after goods are most expensive). In any case, risk taking and exploration of insecure ground constitute preconditions for the attainment and exploitation of many attractive goals.

A Taxonomy of Trade-Offs Reflecting the Dialectics of (In)security

So far, we have discussed the advantage of risky speed strategies in situations where a prevailing norm (i.e., payoff structure) calls for risk-abiding behavior. In the speed-accuracy trade-off setting underlying the sample-based choice task used by Fiedler et al. (2021), the normative rule of the game is to maximize the total payoff. To reach this goal, participants have to make "quick and dirty" choices based on small samples. This task-specific norm follows logically from the fact that, with increasing sample sizes, the number of completed choices decreases faster than the accuracy of choices increases. Yet, most individual participants' preferred strategies diverge from this norm. Only a few participants manage to keep the sample sizes very low (and thereby attain by far the highest payoff); most participants oversample dramatically (foregoing the chance to gain a much higher monetary payoff). Apparently, such a strong oversampling tendency is reflective of a personal utility function that is sensitive not only to payoff but also to the idealist goal of solving choice problems carefully and responsibly (see also Krueger & Gruening, this volume).

We now want to turn to the broader discussion of a variety of distinct trade-offs that have become the focus of prominent research paradigms, which highlight the same general message. Many participants are reluctant to recognize and exploit the advantage of insecurity strategies, and the research literature seems to reflect a distinct bias that favors security. The prominent paradigms listed in the taxonomy of Table 5.1 all share a common feature – a trade-off between insecure strategies (like speedy choices based on small samples) and secure strategies (like cautious decisions after extended information searches). The taxonomy highlights the dependence of adaptive behavior on two factors, the normative advantage of paradigm-specific security and insecurity advantages (see rows of the taxonomy table) and the tendency in behavioral science, and in personal dispositions, to favor security or insecurity (columns of the table). As is evident from the distribution of trade-off tasks in Table 5.1, the majority of paradigms confirm the same pattern as the trade-offs depicted at the outset. Neither participants nor researchers appear to savor the typical normative superiority of insecurity-based strategies.

Two regularities of the taxonomy deserve to be emphasized. First, it seems noteworthy that for so many prominent trade-off paradigms (concentrated in the upper part of the taxonomy), normative considerations call for

TABLE 5.1 Taxonomy of trade-off paradigms focusing on security (middle column) or insecurity (right column) as prominent social motives. For most paradigms (in the upper part of the table), existing theoretical norms call for insecurity rather than security

	Prominent research on security. Appreciated motive	Neglected research on insecurity. Depreciated motive
Normative insecurity dominance	Speed-accuracy trade-off: Oversampling	Speed-accuracy trade-off: Under-sampling
	Risk/uncertainty aversion (Ellsberg, 1961)	Risky vs. cautious investment (Kaufmann et al., 2013)
	Familiarity advantage (Zajonc, 1980)	Incest barriers (Bischof, 1975)
	Fluency; flow experience Reber et al. (1998)	Disfluency (Alter et al., 2007; Fiedler, 2013)
	Averaging/wisdom of crowd (Surowiecki, 2004)	Minority vs. majority information (Moscovici & Personnaz, 1980; Gruenfeld, 1995)
	Nudging and rich reward (Thaler & Sunstein, 2008)	Difficult learning more effective (Bjork & Bjork, 1994; Koriat, 2008)
	Poor vs. rich reward. Law of effect (Thorndike, 1927)	Partial vs. full reinforcement (Lawrence & Festinger, 1962)
No clear-cut normative dominance	Delay of gratification (Metcalfe & Mischel, 1999)	Ability to exploit the present
	Moral intuition (Haidt & Graham, 2007)	Moral mastery (Kelly, 1971)
Normative security dominance	————	————

insecurity-driven strategies. Second, although opposite perspectives allow participants and researchers to consider both the virtues and vices of security and insecurity, there is a conspicuous asymmetry of prevalent dispositions and prominent research biases toward security (left column) and much less prevalent dispositions and neglected evidence on insecurity benefits (gray-shaded right column).

Review of Psychological Evidence on Prominent Trade-Offs

Let us now deal in the remainder of this chapter with neglected benefits of insecurity, following the entries of Table 5.1. We already have discussed the first two entries rows of the taxonomy, concerning the curious accuracy bias in the speed-accuracy trade-off and the asymmetry between the common risk and uncertainty aversion and the risk-abiding strategies as keys to the successful investment decisions. Let us next discuss the regulation of behavior as a function of familiarity versus novelty, fluency versus disfluency, and frequency of occurrence versus scarcity.

(In)security of fluency(disfluency). In a seminal article titled "preferences need no inferences", Zajonc (1980) established the automaticity of mere exposure, that is, the enhanced attractiveness of a repeatedly presented stimulus (hologram or polygons), even when recognition memory is not aware of the repetition. Supporting this fundamental bias in favor of familiarity, the attractiveness advantage of repeated stimuli was reported by researchers of memory (Whittlesea & LeBoe, 2003), social psychology (Reber, Schwarz & Winkielman, 1998), and evolution science (Gigerenzer & Goldstein, 2002) pointing to the survival advantage of familiar, repeatedly experienced stimuli. Moreover, the so-called fluency heuristic in decision research (Hertwig, Herzog, Schooler & Reimer, 2008) generalizes the enhanced attractiveness of a single prior exposure and propagates a preference for the more fluent (i.e., more frequently experienced) choice option. The so-called truth bias complements the enhanced attractiveness of familiar stimuli, showing that repeatedly presented propositions are also more likely to be judged as true than non-familiar propositions.

Indeed, a huge literature is concerned with the reward value of familiarity, and the evidence for the familiarity-type of security seems overwhelming. Nevertheless, counterevidence has been offered for decades, reflecting a disfluency advantage in decisions and problem-solving (Alter, Oppenheimer, Epley & Eyre, 2007; Fiedler, 2013) and sometimes an attractiveness advantage for unexpected, deviant, or slightly irregular stimuli (Berlyne, 1966; Pocheptsova, Labroo & Dhar, 2010). Moreover, a general preference law in favor of frequent or most familiar stimuli is incompatible with the aforementioned ecological correlation between scarcity (likelihood of occurrence) and preciousness (value) demonstrated by Plekac and Hertwig (2014). Yet, this counterevidence against familiarity and in favor of disfluency figures much less prominently in the literature; it is hardly included in textbooks on social psychology and emotion research.

The neglect of insecurity benefits becomes particularly obvious in the practical ignorance of Bischof's (1975) Zürich model of social psychology. Although the cybernetic model of familiarity-driven behavior regulation (Gubler & Bischof, 1991) is clearly more sophisticated than the popular exposure notion and deeply anchored in ethological research on incest barriers in biology, it has been neglected in the literature on adaptive social behavior. According to this sophisticated approach, there are limits to the exploitation of familiarity, and switching between familiar and unfamiliar strategies is essential for personal growth (Pocheptsova et al., 2010; Woolley & Fishbach, 2022). The notion of incest barriers affords a biological model that helps psychologists understand the dialectic interplay of security and insecurity. Because unrestricted familiarity would foster incest, with all its degenerative consequences, nature has evolved an instinctive behavioral mechanism (e.g., the migration of young deer reaching sexual maturity) that functions as an incest barrier, rendering novelty and unfamiliarity attractive under distinct conditions (Bischoff, 1975; Gubler & Bischof, 1991). Young, promiscuous animals find unfamiliar, exotic stimuli more attractive than familiar stimuli, based on a long history of repeated exposure. By analogy, Bischof's model assumes that moderated by a general feeling of self-confidence, social support, and ecological safety, organisms switch in an adaptive manner between preferences for well-known versus exotic stimuli, familiar versus unknown task settings, or the exploitation of approved options versus the exploration of novel options (see also Mikulincer & Shaver, this volume). Considering only the familiarity branch in this dialectic pair would convey an impoverished picture of adaptive learning and growth.

Minority versus majority influence, driven by conversion versus compliance. Analogous to the dialectics of novelty-based insecurity and familiarity-based security, it is worthwhile to compare the differential means of minority and majority influence. According to Moscovici and Personnaz's (1980) famous writings, there is a fundamental difference in the way majorities and minorities exert social influence. Majority members are in a secure position. To exert strong influence on weaker minorities, they need not mobilize good arguments; they can merely rely on a compliance mechanism. That is, their best "argument" is the high majority consensus. Minority members, in contrast, are in a clearly more insecure position. They can only persuade majorities with good and striking arguments, through a mechanism called conversion. The crucial difference between compliance and conversion is that, in the latter case, persuasion is more than conformity service exhibited in the public. Getting the receiver of a persuasive communication to become a "convert" means that he or she is genuinely convinced, in private as in public settings (Maass & Clark, 1984).

Yet, despite the insecurity associated with the minority role, their conversion influence is more long-lasting and less contingent on public conformity pressure. Because minority arguments are also more intellectually convincing than mere majority compliance, it seems justified to conclude that the underlying process renders minority influence more effective and durable than majority influence (Crano, 2001; Crano & Seyranian, 2007; Maass & Clark, 1984; see also Crano & Hohman, this volume). Striking support for this conclusion comes from Gruenfeld's (1995) content analysis of the argument quality of Supreme Court members who belong to the minority versus majority party.[1] This repeatedly demonstrated superiority of minority arguments reflects the same pattern we have already observed with other trade-offs. Despite the insecurity of minorities, who are by definition weaker than majorities, the minorities' conversion strategy is superior according to normative-intellectual standards to the majorities' compliance strategy. The long-term benefits of insecurity are strikingly evident in the historical fact (noted by Moscovici & Personnaz, 1980) that virtually all innovations and enduring cultural and technical changes have originally started as deeply insecure, exotic, and non-conformist minority ideas (see also Hirschberger, this volume).

Should effective learning be easy or difficult? The same trade-off structure that characterizes the acquisition (through repeated exposure) and persuasive communication of knowledge (social influence) can be found in direct and indirect memory. (Direct memory refers to the performance in recall or recognition tasks, whereas indirect memory is manifested in memory-based inferences, judgments, and decisions). Although learning and memory under secure task conditions (i.e., optimal presentation, complete stimuli, certain feedback, and low difficulty) are less error-prone and more likely to lead to reward than insecure task conditions (blurred presentation, degraded stimuli, uncertain feedback, high difficulty), the so-called generation effect testifies to a strong advantage for insecure memory. One of the most striking findings from decades of research on effective learning and memory under experimental control, or on effective transfer from learning to practical conditions, is that difficult learning is more likely to foster good memory than a feeling of easiness and flow during learning. As summarized in a compelling article by Bjork and Bjork (2011), when exercising under unfamiliar and uncomfortable conditions (e.g., when tennis players exercise with a ball machine operating in random mode), the resulting learning effect and the transfer to real-life settings (e.g., real tennis matches) are stronger than when learning environments try to match the transfer settings.

Yet, despite this advantage of difficult and insecure training techniques in sports, many athletes and coaches resort to cautious, low-risk strategies. Even when a prevailing (scientific) norm says that ability will, in the long run, profit from difficult and uncomfortable learning conditions, the malleable trade-off structure allows agents to stick to conservative and more comfortable strategies, reflecting an illusion of security in sports, in politics, economy, and academic settings. Interleaving training sessions, for example, to learn to execute a few different movement patterns practiced in random order leaves participants less confident about their performance than participants who had blocked training sessions (during which each pattern was trained multiple times before moving on to the next). However, insecure participants actually performed the patterns better later on (Bjork & Bjork, 2011).

In contrast, when learning is experienced as smooth and easily accomplished, an illusion of learning misleads learners into a deceptive overestimation of their achievement and a premature truncation of further learning efforts. As Bjork and Bjork (2011) have pointed out, learning and performance are two separate things, and feeling confident while training because practice performance is good does not mean that actual progress leading to enduring competence has been achieved.

Hundreds of experiments conducted in the so-called generation-effect paradigm highlight the memory advantage of learning under insecure conditions. In a typical experiment of this paradigm, participants are presented with a mixed list of words that are either presented completely or in a degraded format (with some letters missing or obscured) so that participants themselves have to generate the word rather than simply read a word. The typical finding is enhanced memory of generated as compared to read words (e.g., Fiedler, Nickel, Asbeck & Pagel, 2002; Slamecka & Graf, 1978). While the comparison of read and generated stimuli confounds several differences, Fiedler, Lachnit, and Fay (1992) held the generation task constant but manipulated the difficulty or the degree of degradation. For instance, in a counterbalanced design, participants had to generate the missing word (e.g., "wheels") in a sentence frame that rendered the generation difficulty low ("the car rolls on four w-e—s"), intermediate ("the car rolls on w-e—s"), or high ("car w-e—s"). The higher the difficulty of generative distance between the degraded stimulus frame and the complete stimulus, the higher the rate of correctly reproduced target words in a subsequent surprise recall test.

The scope and empirical support for the generation effect are immense, not just in orthodox learning and memory research but also long-established in the realm of social cognition. Quite in line with the depth of processing idea, it has been shown repeatedly that when an increasing amount of cognitive work is needed to make sense of unexpected and hard-to-understand information, the resulting memory improves. For instance, the seminal research by Hastie and Kumar (1979) has shown that unexpected and stereotype-inconsistent behaviors may be more likely to be remembered than expected behaviors that fit the secure frame of stereotypes or scripted expectations.

While the work of Hastie and Kumar is four decades old, Lawrence and Festinger's (1962) eye-opening monograph *Deterrents and Reinforcements* is as old as six decades. It highlights the applicability of dissonance theory to animal learning, analogous to the notion of underjustification (see also Cooper & Pearce, this volume). Just as an extra reward can undermine the intrinsic motivation to exhibit behavior unconditionally, lowering the threshold for reinforcement can be counter-productive for animal learning. Every trainer of dogs, horses, or other animals would subscribe to the superiority of partial reinforcement as a means of inducing stable and enduring animal learning, consistent with a long tradition of behaviorist research. In other words, training animals in an insecure environment, in which performance is not rewarded on every trial but only on every second or third trial, will make learning more stable and resistant to extinction than learning in a secure schedule that warrants predictable reinforcement on every trial.

When animals experience the reinforcement of a behavior (e.g., running down a runway toward a goalbox as a rat) with a positive stimulus (e.g., finding food in the goalbox) in a partial instead of a complete way (i.e., the goalbox does not contain food every time), the behavior is exhibited longer once the reinforcement has stopped altogether. The extinction of learned behavior takes longer under a partial reinforcement schedule as compared to a complete reinforcement schedule.

The interpretation of the partial reinforcement effect as an insecurity benefit seems obvious. The extinction resistance had long been attributed to the insecurity (uncertainty) of the reward structure. Both Amsel (1967) and Capaldi (1967) extended this idea with their own theories beginning in the 1960s, Amsel with his Frustration Theory and Capaldi with the Sequential Theory. According to Mellgren (2012), what both theories continued to have in common with each other and the original idea was the crucial role assigned to new stimuli or experiences introduced in the conditioning, in the rat's case, the new experience of finding the goalbox empty. Whether this leads to frustration or triggers memory processes, it is the indeterminacy, that is, the uncertainty introduced by the new event that plays the crucial role of a catalyst increasing the resistance to extinction. The resistance to extinction also increases when reward is delayed, likewise introducing uncertainty that seems central to more sustainable learning.

In any case, Lawrence and Festinger's (1962) analogy to dissonance theory highlights the insight that partial reinforcement is by no means peculiar to animal learning. It also applies to human behavior in manifold contexts, such as consumer choices, healthy nutrition, and work-related achievement (Frey, 1993), and it is at the heart of the strong ecological correlation of scarceness and subjective value. The more consumers must pay for a bottle of wine, the longer students have to wait for access to a graduate place in a prestigious university, the stronger will be the resulting attitude and motivation. In other words, deprivation and insecurity are the key to maximal enjoyment and motivated achievement.

(In)security of (dis)fluency: Insecurity benefits priming (Fiedler et al., 2005). Why should this be the case? Why should task features that entail insecurity (such as difficulty, novelty, risk, uncertainty, deprivation, and degradation) foster rather than hinder adaptive learning? A plausible answer is actually not too hard to find. The more challenging a problem or learning task to be solved, the stronger and more sustainable is the resulting learning effect. A controlled vaccination infection challenges and thereby fosters the immune system. A student's strongest learning effect is achieved through a demanding research project. The ultimate rivalry of an Olympic competition pushes an athlete's further development. Or, exposure to a multi-lingual environment can create optimal conditions for a young child's intellectual growth.

A major experimental model of adaptive behavior is the priming paradigm. Priming means to prepare an organism for fast and appropriate responses to prime stimuli that signal dangers, opportunities, or significant events in a complex and uncertain environment. A plethora of pertinent experiments converge in demonstrating stronger priming effects under insecure conditions. Priming effects are enhanced when prime words are infrequent rather than frequent (Chan, Ybarra & Schwarz, 2006), when the stimulus-onset asynchrony (i.e., the time between prime and target onset) is short (Hermans, Spruyt & Eelen, 2003), when primes are blurred or degraded (Alexopoulos, Fiedler & Freytag, 2012), and when the prime insecurity is not reduced in a kind of pre-priming.

For illustration, in a series of experiments, Fiedler, Schenck, Menges & Watling (2005) assessed the accuracy and latency required to recognize a trait word (e.g., "hostile") that gradually appeared behind a mask, preceded by a pictorial prime that either matched (a silhouette showing an aggressor stabbing a victim) or mismatched the target trait. When a preceding assessment task asked participants to confirm the hostility of the prime picture, the facilitation of the subsequent target trait recognition was eliminated. Many experiments have demonstrated that prior sensitization to the prime contents – a procedure that Sparrow and Wegner (2006) called "unpriming" – eliminates or strongly reduces the priming effect. Thus, analogous to the learning and memory advantage of incomplete or degraded stimuli in the generation effect, preparatory processing also benefits from insecurity and uncertainty.

(In)security of time discounting: Delay of gratification. It is not by coincidence that normative principles can be found for benefits of insecurity for the vast majority of trade-offs. These benefits are not only overlooked by many experimental participants, whose preferred strategies exhibit a distinct security bias. We hasten to add that the same bias in favor of security that characterizes naïve participants is also present in scientific research. Bischof's (1975) seminal work on incest barriers is clearly less prominent in the literature than Zajonc's (1980) work on familiarity; a disfluency advantage is far less well known than a fluency effect (Reber et al., 1998).

Lawrence and Festinger's (1962) ground-breaking work on dissonance in animal learning is fully ignored in the fashionable literature on nudging (Thaler & Sunstein, 2008), and Bjork and Bjork's (1994) lesson on sustainable learning did not play the slightest role in the recent debate on public learning during the pandemic.

Table 5.1 reveals that while we could not find trade-offs with a clear-cut normative advantage for security-based strategies, there are at least a couple of trade-offs that render the choice between security and insecurity strategies an open empirical question. To understand this point, consider the trade-off inherent in the famous delay-of-gratification task (Mischel & Metcalfe, 1999). In many important self-regulation encounters related to healthy nutrition, impulse control, partner choice, or economic action, individuals must choose between an immediate outcome and a more delayed, and ideally more sustainable, gratification at a later time. The ability to wait for delayed gratification and to refrain from the impulsive consumption of immediate gratification is generally considered a sign of prudence and maturity. However, from a normative standpoint, apparently self-evident superiority of a delay-of-gratification strategy is hardly justified. Specifying a normatively appropriate strategy depends on a closer analysis of the agent's subjective utility function. The economic notion of "time discounting" reminds us of the truism that subjective value or utility may not be the same if we cannot enjoy a holiday trip, a dinner, or a romantic encounter right now. It is hard to say on *a priori* normative grounds what inability constitutes the greater obstacle on the way to happiness: the inability to wait for delayed gratification in the future or the inability to enjoy one's life in the here and now.

Assuming that the secure strategy in a delay-of-gratification paradigm is to exploit the immediately present gratification, whereas waiting for a somewhat uncertain future gratification entails more insecurity, it seems once more obvious that benefits of insecurity are not too far-fetched a possibility in this paradigm either.

Concluding Remarks

Throughout this chapter, we have gathered evidence for benefits of insecurity, correcting for a prevailing bias in favor of security. Frankly speaking, this was not quite intended from the beginning. Although we did start from the insight that malleable trade-offs facilitate the framing of both security and insecurity tasks as advantageous, we originally wanted to come up with a mix of trade-offs that could account for the development of both security and insecurity strategies. Likewise, the taxonomy of Table 5.1 was originally intended to include a reasonable sample of trade-offs of both types, with norms favoring both kinds of strategies. At the end, we had to give up this plan. We were unable to find trade-off paradigms for which normative accounts or well-established empirical laws called for security. Rather, as it turned out, there is a conspicuous asymmetry such that available norms, if they exist, often prioritize risky, insecurity-oriented strategies, even though prevalent dispositions and research paradigms render security goals more prominent.

In the absence of a patent explanation for this asymmetry, we are not overly surprised by the imbalanced outcome of our inquiry. Just as people value a good meal higher than food deprivation, a victory in sports higher than a defeat, or a certain gain more than a lottery with the same expected value (Ellsberg, 1961), they doubtlessly prefer security to insecurity. Nevertheless – and this is no contradiction at all – people probably learn from deprivation, from a defeat, and from a gamble under uncertainty; a good meal is even better after a period of deprivation, a victory is most enjoyable after the experience of a defeat, and winning an uncertain lottery is more rewarding than receiving a fixed outcome (Mellers, Schwartz, Ho & Ritov, 1997). At the end, the take-home message is, of course, not that insecurity should become an ironic goal. Rather, as already anticipated at the outset, although "behavior is doubtlessly striving for security as an ideal end of all important behavior …, many means toward this end are energized by insecurity."

Author Note: The work underlying this article was supported by grants provided by the Deutsche Forschungsgemeinschaft to the first author (FI 294/29-1; FI 294/30–1).

Note

1 In this context, the minority or majority status depends on what political party, Democrats or Republicans, is more frequently represented in the Supreme Court.

References

Alexopoulos, T., Fiedler, K., & Freytag, P. (2012). The impact of open and closed mindsets on evaluative priming. *Cognition & Emotion, 26*(6), 978–994.

Alter, A. L., Oppenheimer, D. M., Epley, N., & Eyre, R. N. (2007). Overcoming intuition: Metacognitive difficulty activates analytic reasoning. *Journal of Experimental Psychology: General, 136*(4), 569–576.

Amsel, A. (1967). Partial reinforcement effects on vigor and persistence. In K. W. Spence & J. T. Spence (Eds.), *The psychology of learning and motivation* (pp. 1–65). New York: Academic.

Berlyne, D. E. (1966). Curiosity and exploration: Animals spend much of their time seeking stimuli whose significance raises problems for psychology. *Science, 153*(3731), 25–33.

Bischof, N. (1975). A systems approach toward the functional connections of attachment and fear. *Child Development*, 801–817.

Bjork, E. L., & Bjork, R. A. (2011). Making things hard on yourself, but in a good way: Creating desirable difficulties to enhance learning. *Psychology and the Real World: Essays Illustrating Fundamental Contributions to Society*, 2, 59–68.

Bohner, G., Siebler, F., & Raaijmakers, Y. (1999). Salience of rape affects self-esteem: Individual versus collective self-aspects. *Group Processes and Intergroup Relations*, 2, 191–199.

Capaldi, E. J. (1967). A sequential hypothesis of instrumental learning. In K. W. Spence & J. T. Spence (Eds.), *Psychology of learning and motivation* (pp. 67–156). New York: Academic.

Chan, E., Ybarra, O., & Schwarz, N. (2006). Reversing the affective congruency effect: The role of target word frequency of occurrence. *Journal of Experimental Social Psychology*, 42(3), 365–372.

Crano, W. D. (2001). Social influence, social identity, and ingroup leniency. In C. K. W. De Dreu & N. K. De Vries (Eds.), *Group consensus and minority influence: Implications for innovation* (pp. 122–143). Malden, MA: Blackwell.

Crano, W. D., & Seyranian, V. (2007). Majority and minority influence. *Social and Personality Psychology Compass*, 1(1), 572–589.

Ellsberg, D. (1961), Risk, ambiguity and the savage axioms. *Quarterly Journal of Economics* 75, 643–669.

Fiedler, K. (2013). Fluency and behavior regulation: Adaptive and maladaptive consequences of a good feeling. In C. Unkelbach & R. Greifender (Eds.), *The experience of thinking: How the fluency of mental processes influences cognition and behaviour* (pp. 234–254). New York: Psychology Press.

Fiedler, K., Lachnit, H., Fay, D., & Krug, C. (1992). Mobilization of cognitive resources and the generation effect. *The Quarterly Journal of Experimental Psychology Section A*, 45(1), 149–171.

Fiedler, K., McCaughey, L., Prager, J., Eichberger, J., & Schnell, K. (2021). Speed-accuracy trade-offs in sample-based decisions. *Journal of Experimental Psychology: General*, 150(6), 1203–1224.

Fiedler, K., Nickel, S., Asbeck, J., & Pagel, U. (2003). Mood and the generation effect. *Cognition and Emotion*, 17(4), 585–608.

Fiedler, K., Schenck, W., Watling, M., & Menges, J. I. (2005). Priming trait inferences through pictures and moving pictures: The impact of open and closed mindsets. *Journal of Personality and Social Psychology*, 88(2), 229–244.

Frey, B. S. (1993). Shirking or work morale?: The impact of regulating. *European Economic Review*, 37(8), 1523–1532.

Goldstein, D. G., & Gigerenzer, G. (2002). Models of ecological rationality: The recognition heuristic. *Psychological Review*, 109, 75–90.

Gruenfeld, D. H. (1995). Status, ideology, and integrative complexity on the US Supreme Court: Rethinking the politics of political decision making. *Journal of Personality and Social Psychology*, 68(1), 5–20.

Gubler, H., & Bischof, N. (1991). A systems theory perspective. In M. E. Lamb & H. Keller (Eds.), *Infant development: Perspectives from German-speaking countries* (pp. 35–66). Lawrence Erlbaum Associates, Inc.

Haidt, J., & Graham, J. (2007). When morality opposes justice: Conservatives have moral intuitions that liberals may not recognize. *Social Justice Research*, 20(1), 98–116.

Hastie, R., & Kumar, P. A. (1979). Person memory: Personality traits as organizing principles in memory for behaviors. *Journal of Personality and Social Psychology*, 37(1), 25.

Hermans, D., Spruyt, A., & Eelen, P. (2003). Automatic affective priming of recently acquired stimulus valence: Priming at SOA 300 but not at SOA 1000. *Cognition and Emotion*, 17(1), 83–99.

Hertwig, R., Herzog, S. M., Schooler, L. J., & Reimer, T. (2008). Fluency heuristic: A model of how the mind exploits a by-product of information retrieval. *Journal of Experimental Psychology: Learning, Memory, and Cognition*, 34(5), 1191.

Hsee, C. K., & Zhang, J. (2010). General evaluability theory. *Perspectives on Psychological Science*, 5(4), 343–355.

Moscovici, S., & Personnaz, B. (1980). Studies in social influence: V. Minority influence and conversion behavior in a perceptual task. *Journal of Experimental Social Psychology*, 16(3), 270–282.

Kaufmann, C., Weber, M., & Haisley, E. (2013). The role of experience sampling and graphical displays on one's investment risk appetite. *Management Science*, 59(2), 323–340.

Kelley, H. H. (1971). Moral evaluation. *American Psychologist*, 26(3), 293–300.

Koriat, A. (2008). Easy comes, easy goes? The link between learning and remembering and its exploitation in metacognition. *Memory & Cognition*, 36(2), 416–428.

Lawrence, D. H., & Festinger, S. (1962). *Deterrents and reinforcement: The psychology of insufficient reward*. Stanford University Press.

Maass, A., & Clark, R. D. (1984). Hidden impact of minorities: Fifteen years of minority influence research. *Psychological Bulletin*, 95(3), 428–450.

Mellers, B. A., Schwartz, A., Ho, K., & Ritov, I. (1997). Decision affect theory: Emotional reactions to the outcomes of risky options. *Psychological Science*, 8(6), 423–429.

Metcalfe, J., & Jacobs, W. J. (2021). Stress and imagining future selves: Resolve in the hot/cool framework. *Behavioral and Brain Sciences*, 44, 33–35.

Metcalfe, J., & Mischel, W. (1999). A hot/cool-system analysis of delay of gratification: Dynamics of willpower. *Psychological Review*, 106(1), 3–19.

Monahan, J. L., Murphy, S. T., & Zajonc, R. B. (2000). Subliminal mere exposure: Specific, general, and diffuse effects. *Psychological Science*, 11(6), 462–466.

Nosić, A., & Weber, M. (2010). How riskily do I invest? The role of risk attitudes, risk perceptions, and overconfidence. *Decision Analysis*, 7(3), 282–301.

Pleskac, T. J., & Hertwig, R. (2014). Ecologically rational choice and the structure of the environment. *Journal of Experimental Psychology: General*, 143(5), 2000–2019.

Pocheptsova, A., Labroo, A. A., & Dhar, R. (2010). Making products feel special: When metacognitive difficulty enhances evaluation. *Journal of Marketing Research*, 47(6), 1059–1069.

Reber, R., Winkielman, P., & Schwarz, N. (1998). Effects of perceptual fluency on affective judgments. *Psychological Science*, 9(1), 45–48.

Reeder, G. D., & Brewer, M. B. (1979). A schematic model of dispositional attribution in interpersonal perception. *Psychological Review*, 86(1), 61–79.

Slamecka, N. J., & Graf, P. (1978). The generation effect: Delineation of a phenomenon. *Journal of Experimental Psychology: Human Learning and Memory*, 4(6), 592.

Sparrow, B., & Wegner, D. M. (2006). Unpriming: The deactivation of thoughts through expression. *Journal of Personality and Social Psychology*, 91(6), 1009–1019.

Thaler, R. and Sunstein, C. (2008) *Nudge: Improving decisions about health, wealth, and happiness.* New Haven CT: Yale University Press.
Thorndike, E. L. (1927). The law of effect. *The American Journal of Psychology, 39*(1/4), 212–222.
Viki, G. T., & Abrams, D. (2002). But she was unfaithful: Benevolent sexism and reactions to rape victims who violate traditional gender role expectations. *Sex Roles, 47*(5), 289–293.
Whittlesea, B. W., & LeBoe, J. P. (2003). Two fluency heuristics (and how to tell them apart). *Journal of Memory and Language, 49*(1), 62–79.
Woolley, K., & Fishbach, A. (2022). Motivating personal growth by seeking discomfort. *Psychological Science, 33*(4), 510–523.
Zajonc, R. B. (1980). Feeling and thinking: Preferences need no inferences. *American Psychologist, 35*(2), 151–175.

PART II
Managing Individual Insecurity

6
THE ARC OF DISSONANCE
From Drive to Uncertainty

Joel Cooper and Logan Pearce

PRINCETON UNIVERSITY, NEW JERSEY, USA

Abstract

The arousal state of cognitive dissonance is based on our need to see the world as secure and sensible. When inconsistency is experienced, the world becomes less predictable, our actions become less comprehensible and our emotions respond with the unpleasant state of dissonance. Seeking certainty through cognitive consonance helps to reduce our sense of insecurity. We develop ways to reduce dissonance in order to restore confidence that we can understand and control our environment. Sometimes, the search for dissonance reduction can lead to suboptimal beliefs, such as the thought that the U.S. election in 2020 was stolen from Donald Trump or that COVID-19 vaccines are conspiracies of the government to invade people's minds and bodies. At other times and in other situations, dissonance reduction strategies can leverage more positive social attitudes, such as creating support for COVID-19 vaccines, as we have recently reported in our research. In this chapter, We will examine the broad arc that links insecurity to dissonance and discuss empirical research in support of that link.

Cognitive dissonance is a ubiquitous human phenomenon that has received arguably as much empirical attention as any process in social psychology. Literally, thousands of laboratory experiments have investigated the characteristics of dissonance, expanding its reach and sharpening its focus. The application of dissonance to areas of concern outside the laboratory has accelerated, such that clinical treatment programs for difficult issues such as eating disorders (Stice et al., 2021), physical exercise (Cooper & Feldman, 2020) and immunizations for COVID-19 (Pearce & Cooper, 2020) have been based on the tenets of cognitive dissonance. At its core, cognitive dissonance is straightforward.

In Leon Festinger's seminal statement of the theory, "The holding of two or more inconsistent cognitions arouses the state of cognitive dissonance, which is experienced as uncomfortable tension. This tension has drive-like properties and must be reduced" (Festinger, 1957, p. 3).

The ramifications of cognitive dissonance arousal are well known and far-reaching. We experience dissonance whenever we make choices, especially difficult ones. After making a decision, we often view the chosen option as more attractive and/or the rejected option(s) as less attractive (Brehm, 1956). We change our opinions when those opinions appear to be contradicted by our behavior (Festinger & Carlsmith, 1959). We spuriously elevate groups as more attractive if we have had to endure suffering in order to join them (Aronson & Mills, 1959; see also Hogg & Gaffney; Hirschberger, this volume). Sometimes, dissonance causes us to double down on our own behavioral commitment to our beliefs, particularly when we are confronted by the disjunction between our attitudes and our past behavior (Priolo et al., 2019; Stone et al., 1997). In addition to dissonance being experienced as a function of how we have behaved, research has shown that we can also experience dissonance vicariously on *behalf of* significant in-group members (Norton et al., 2003). Moreover, dissonance seems to affect children (Aronson & Carlsmith, 1963; Egan et al., 2007) as well as adults, suggesting that dissonance is a phenomenon that is automatically present in humans or develops quickly with cognitive maturity.

Toward Understanding Why

Much of the voluminous research on dissonance processes has involved understanding the limits and extensions of dissonance theory. Classic questions were put under the proverbial microscope: Is choice necessary for dissonance to occur? Does behavior need to result in unwanted consequences? Does behavior need to implicate the self? Often, research findings implied that dissonance might be conceptualized in different ways, or that the theory needed to be limited, expanded or modified (see Harmon-Jones, 2019). Although those have been fascinating issues in the evolution of the dissonance approach, this chapter returns to a fundamental and core question about the motivation to restore consistency among cognitions. That question is: Why?

As straightforward as the concept of dissonance seems to be, its underlying cause has been elusive. Why should inconsistent cognitions lead to a state of arousal? Why does it evoke change? In this chapter, we will consider various conceptions of the psychological foundations of cognitive dissonance. To understand the function that dissonance reduction plays for human beings in the social environment, we need to consider why dissonance occurs. Without understanding the psychological function served by dissonance, the search for the meaning of dissonance reduction leads to misconceptions and dead ends. We will begin this chapter by considering what we can glean from Festinger's

(1957) original conception and then move toward subsequent and contemporary statements of dissonance. In the end, we will suggest that dissonance arousal serves as a signal of the precariousness of our being. The tension state is a signal that the certainty with which we face the world is limited and that our understanding of reality is insecure and precarious (see also von Hippel & Merakovsky; Pyszczynski & Sundby, this volume). The resolution of dissonance is at the service of reassuring ourselves that we understand our own reality. Dissonance resolution thus restores a sense of security in what would otherwise be a chaotic environment.

In the Beginning

What did the originator of cognitive dissonance theory think of as the antecedent cause of cognitive dissonance? The fact is, Festinger was circumspect about the reason for people preferring consistency among their cognitions. In Festinger's view, dissonance functions like a drive. "Cognitive dissonance," Festinger explained, "can be seen as an antecedent condition which leads to activity oriented toward dissonance reduction just as hunger leads to activity oriented toward hunger reduction" (p. 3). Epistemically, dissonance is a tension state that arises from inconsistency. It is its own cause. As Festinger said, "If someone cared to write a certain kind of book about the hunger drive in human beings, it would turn out to be similar in nature to the present volume" (p. 4).

For Festinger, asking about the origins of dissonance was no more psychologically relevant than asking why people seek to reduce hunger or thirst. Inconsistency produces a tension state, that tension state is unpleasant and, like hunger, motivates people to find an appropriate way to reduce it. Certainly, there is a physiological explanation for the experience of hunger. Stomach and intestinal enzymes signal hunger, and the failure to respond to the drive has further consequences for disease and morbidity. Psychologically, however, hunger motivates us to engage in whatever activities satisfy that motivation. Festinger viewed dissonance in the same way. It exists as a feature of our being human. The task for psychologists was to study the fascinating and complex ways that the dissonance could be reduced.

Before the Beginning

Although Festinger was coy about committing to a reason for dissonance in his seminal narrative, his earlier writings suggest some possibilities. In many ways, dissonance theory is an outgrowth of Festinger's work with his mentor, Kurt Lewin, on the forces that affect social groups. In a publication that set an agenda for group dynamics work, Festinger (1950) authored a conceptual paper on informal communications in social groups. Festinger held that communication among group members occurs spontaneously in order to achieve uniformity about some issue or behavior (see also Crano & Hohman, this volume). The pressure toward uniformity was based on two crucial elements necessary for group functioning: One was group locomotion, which permitted groups to move toward a particular goal, and the other was establishing a social reality. A person's opinion is "correct," according to Festinger, to the extent that it is anchored in a group of people with similar beliefs, opinions and attitudes. The psychological consequence of disagreement or discordance within a group is the occurrence of a 'force' to restore consistency. Only then can a group achieve its goals and only then can group members be assured of a common social reality.

The next milestone in the development of dissonance theory was Festinger's paper on social comparison processes (Festinger, 1954). In many ways, the theory of social comparison was the direct forerunner of cognitive dissonance theory. Examining the motivational basis of social comparison may help elucidate Festinger's underlying opinion of the basis of cognitive dissonance. While the pressures toward uniformity in his earlier paper were based on the needs of the group to which an individual belonged, the focus of social comparison shifted to the individual. Surely, the individual was conceived as a member of a group, but the relevance of being in a particular group with group boundaries took on less importance. Moreover, as he shifted his emphasis from the group to the individual, the concept of people engaging in social activities to accomplish the reduction of a drive entered Festinger's thinking. To initiate the fundamental premise of social comparison theory, Festinger states, "There exists, in the human organism, a drive to evaluate his opinions and his abilities" (Festinger, 1954, p. 117). He explains that to the extent that objective means are not available for the evaluation, then social comparison must be used: When objective, non-social means are not available, Festinger proposed that people evaluate their opinions and abilities by comparison, respectively, with the opinions and abilities of others (p. 118), particularly those who are similar on relevant dimensions (see also Hirschberger; Hogg & Gaffeny, this volume).

Why make this comparison? What drives people to compare themselves with similar other people? In the end, people need to have a functional assessment of social reality. They need to gauge how well they assess their social world. The existence of discrepancies among people in the opinions they hold about the world reduces their confidence in their judgment of reality. At the individual level, this produces pressure for them to engage in persuasion processes designed to convince others to see reality as they do (Crano & Hohman, this volume). It also puts pressure on them to seek out similar others who have the same opinions as they do. And, in a less desirable circumstance, it may persuade people to change their opinion to be like others who are similar to them. In a phrase, it is important to have a view of the world that is correct. Social comparison processes facilitate the confidence that people are correct because there is no discord between their views and the views of similar others.

Festinger's transition from social comparison to cognitive dissonance is delineated convincingly in his unpublished draft paper, "Social communication and cognition," printed as an Appendix in Harmon-Jones' (2019) volume. In this paper, Festinger focused for the first time on the concept of cognition. He described cognitions as including opinions, attitudes beliefs and thoughts ... but not behavior. Behavior was an independent factor that could be consonant or dissonant with a person's cognitions. The paper is a wonderful transition between his early work and cognitive dissonance theory. Although he introduced the concept of dissonant cognition, his theoretical focus was still very much on communication. He discussed the lengths to which people will go in order to create consonance, which they do via social communication. People communicate with each other as a way of persuading them to adopt cognitions that are consistent with their own. They also communicate with others to persuade people to support their behaviors, particularly those that are discrepant from one or more cognitions.

The excursion into Festinger's insights prior to *A theory of cognitive dissonance* suggests the evolution of his thinking about the fundamental needs satisfied by the drive for consistency. Although not specifically addressed in his theory, it is fair to conclude that Festinger saw people's need to establish social reality to be the prime motivation for their communication in groups, for their social comparisons and for their attempts to manage insecurity and maintain consistency among their cognitions. "The holding of ... inaccurate appraisals ... can be punishing, or even fatal...," he wrote (Festinger, 1954, p. 117).

After Festinger: The Search for the Meaning of Dissonance

Social psychology was stimulated by Festinger's (1957) approach and by some novel predictions made by proponents of dissonance in the decade following the publication of the original work. We believe that the drive concept was Festinger's way of positioning dissonance theory in the context of what was important in experimental psychology at the time. Learning theory was the predominant force in much of psychology. Research controversies dealt with the ways in which drives were resolved in humans and especially infrahuman species. Clark Hull and his collaborators led the study of drives to understand both behavior and cognition (Hull, 1943). Simultaneously, intellectual opponents of drive theory, such as B.F. Skinner (1957) and his followers, pressed forward with research on the importance of rewards and punishments in how human and infrahuman species behave and learn.

Festinger's consideration of dissonance as a drive placed social psychology in a position where it could at least use the same conceptual language as the learning researchers. At the same time, some of the truly provocative hypotheses based on the dissonance drive confronted and contradicted the predictions that had been the bedrock of reinforcement theories. Festinger and Carlsmith's

(1959) classic study on induced compliance showed that behaviors that were highly reinforced did not lead to greater belief (or positive sentiment) toward an issue if that issue was counterattitudinal. Large reinforcements dampened the impact of behavior on attitudes rather than the more conventional logic based on learning principles. Similarly, Aronson and Mills (1959) showed that the negative expenditure of unpleasant effort led to more attraction to the goal for which people suffered, apparently at odds with straightforward predictions based on reinforcement.

Drive as Metaphor or Reality

As we saw previously in this chapter, the theory of cognitive dissonance was not the first time that Festinger invoked the drive concept. It propelled social comparison as well, but never served a central purpose in the work that followed from that theory. Dissonance was different. For the first time, the drive concept was used to assess the *magnitude* of the need to restore consistency between cognitions. Just as people can have more or less hunger, so too can they have more or less dissonance. As in the now-classic $1 and $20 experiment of Festinger and Carlsmith (1959), reward was one way to alter the magnitude of the drive and, therefore, predict differences in the magnitude of efforts designed to reduce it.

The concept of a drive was a useful metaphor. It implicated motivation – motivation that could not be reduced until the inconsistency was reduced. It implicated magnitude. Very hungry people seek to reduce hunger more than only slightly hungry people. And, like a drive, it was propelled by physiological arousal and by the sensation of unpleasantness. Given these assumptions, many fascinating predictions followed. But did dissonance truly have the characteristics of a drive, or was it just a convenient metaphor carried over from the learning literature? Pallack and Pittman (1972) and Waterman and Katkin (1967) were among the first investigators to take seriously the idea that dissonance might function like physiological drive states. They borrowed predictions from the learning literature, reasoning that drives have established patterns of facilitation and interference with learning. Their results provided partial support for the notion that dissonance was indeed a drive.

Festinger had proposed that the drive of dissonance was, like hunger, an unpleasant tension state (see also Kruglanski & Ellenberg, this volume). He put no particular credence in determining if that was actually true, but that the assumption was useful in making predictions about attitude and behavior change. Subsequent research determined that much of his speculation was accurate. Dissonance is experienced as psychologically unpleasant (Elliot & Devine, 1994). People who hold dissonant cognitions actually show increases in skin conductance, the marker of physiologically activated stress (Croyle & Cooper, 1983; Losch & Cacioppo, 1990). Moreover, the amount of physiological arousal following inconsistent cognitions determines the magnitude of attitude change that occurs following attitude-inconsistent behavior. Raising physiological arousal with an amphetamine increases attitude change, reducing arousal with a sedative reduces attitude change (Cooper et al., 1978).

Toward the Meaning of Dissonance

The notion that dissonance is genuinely experienced as a drive does not satisfy the search for its cause. What function does having such a drive serve? The notion that holding dissonant cognitions results in a true drive gives substance to Festinger's idea that reducing dissonance is not merely a preference but rather an imperative. We need to dig deeper to assess what psychological need is addressed by dissonance.

Several important modifications to Festinger's formulation of dissonance emerged in order to focus more clearly on what accounted for the dissonance drive. Aronson (1969) suggested that people experience dissonance when a person's sense of self is compromised by a dissonant act. He believed that inconsistent cognitions *per se* are not sufficient to arouse dissonance. At the root of dissonance was holding a cognition that threatened a person's sense of self-esteem, producing a sense of insecurity. Normally, people like to think of themselves as moral, competent and helpful. Only 'shnooks' lie to others, make bad choices or suffer needlessly to obtain a mediocre goal. Dissonance arises when people behave in ways that would make them feel 'shnooky' and compromise their sense of self-worth. On the other hand, if they actually believe the position that they took in an essay writing task or decide that they suffered to attain a most worthwhile goal or made good choices among choice alternatives, then their confidence in themselves as good and worthy people can be maintained. The dissonance drive is at the service of restoring a consistent and optimal view of the self.

Similar to Aronson's view, Steele (1988) believed that supporting a sense of high self-worth was at the heart of dissonance. Steele suggested that people take action whenever their self-worth is threatened. They take action to affirm their own goodness and worthiness. In the self-affirmation view, dissonant behavior is not unique as a way of casting doubt on the integrity and worthiness of the self. Anything that casts doubt on self-worthiness motivates efforts to restore the self. The self, Steele argued, is an integrated self-system with many components. If the security of the self is compromised by convincing someone to believe a position that is not true, then it can be restored by changing one's attitude such that the actor now believes that their statement is true. But that is only one way of restoring the self-system. People have numerous arrows in their quiver: If they can conjure images of how good they are at sports or art or piano or in the classroom, or just affirm some important value, then that can restore the integrity and security of the self. Direct reduction of dissonance by changing the attitude on the issue that caused the dissonance is not necessary. The means of restoring the self are manifold and fluid.

In our New Look Model of dissonance, Fazio and Cooper tackled what we thought lay at the heart of the dissonance drive (Cooper & Fazio, 1984; Cooper, 1998). We suggested that dissonance is based on being responsible for bringing about unwanted, aversive consequences. In our view, inconsistent cognitions were not necessary for dissonance to occur. Rather, what is necessary is being responsible for bringing about a state of affairs that one would rather not have brought about. We also speculated on how such circumstances could ever result in a *drive*. We suggested, admittedly without evidence, that dissonance may not be innate, but rather a drive that is learned through experience. Because of early childhood events, we may have learned that bringing about an unwanted event is punished either physically or psychologically. Generalizing this learning throughout development leads us to avoid such circumstances if possible and to experience a negative psychological state if dissonance avoidance is not possible. Although we only had vignettes rather than data to support this notion, we thought it was important to flesh out a reason that dissonance acts like a drive.

Stone and Cooper (2003) tried to reconcile some of the conflicting points of view regarding the foundation of dissonance. We argued that people become upset (i.e., experience insecurity and dissonance) when an action they have taken violates what is expected, based on a particular standard of judgment. When a self-standard is salient, any action or decision that violates a person's expectations based on their view of themselves creates dissonance. When situational standards are made salient, people become upset when their actions and decisions violate what the culture expects of them in a given situation. We reasoned that what self-esteem, self-affirmation and New Look models have in common is the realization that an important standard by which people judge themselves has been violated by their own behavior.

Taking a different view of cognitive dissonance, Harmon-Jones et al. (2009) suggested that what is unsettling and arousing about inconsistency is that it upsets people's ability to take a secure, confident and consistent orientation toward the environment. Similar to Jones and Gerard's (1967) Unequivocal Behavior Orientation, this approach takes as its foundation the notion that people need to take an unequivocal stance toward acting in the environment. It is maladaptive to be in conflict, to be unsure of how to act and what to believe. Inconsistent cognitions interfere with our action tendencies and thus create a negative emotion, motivating us to rid ourselves of the inconsistency. From the action orientation point of view, it is not inconsistency per se that causes us to be upset, but rather the effect that inconsistency has on our need to adopt an unequivocal stance toward action in the social and physical environment.

Completing the Arc: Uncertainty and the Drive to Reduce Dissonance

Uncertainty is alarming because it threatens our understanding of social reality (see also von Hippel & Merakovsky, this volume). To act in a dissonant manner is to create uncertainty about the world and about ourselves.

Add uncertainty and you can exacerbate dissonance. Add certainty and you can reduce people's needs to resolve dissonance.

How fluid is the relationship of uncertainty and insecurity to the experience of insecurity? If dissonance is a drive, can it be reduced by any goal-satisfying behavior or does it need to be related to the events that created the arousal of the drive? When people or infrahuman animals are hungry, they seek food. We may engage in other behaviors that partially satisfy our drives. For example, we can sleep. Sometimes, sleep directly satisfies one of our needs, i.e., to rest our mental and physical being. It may also serve our hunger drive but only minimally. Perhaps it allows us to ignore our hunger for some time. More satisfactory would be water. Again, water would satisfy our thirst drive but would only partially deal with the hunger that caused the unpleasant tension state. When we are hungry, we seek food. It directly satisfies the need that people are driven to satisfy. Our view is that the fluidity of dissonance reduction exists, but is only minimally satisfying, and used as a partial bridge to the desired outcome. Coming to grips with preferred modes of dissonance reduction can help us draw a tentative conclusion about the origin and resolution of the dissonance drive.

The concept of the fluidity of dissonance is based on the notion that dissonance is a subset of ways in which the integrity of the self can be undermined. Steele's self-affirmation theory was foundational in adopting the fluidity view. In self-affirmation, people's goal is to maintain the integrity of their self-system. The self-system is an integrated, holistic concept that includes but is not restricted to a person's sense of self as being consistent. People are motivated to view the integrated system as worthy. Therefore, if people's sense of self is compromised by, for example, making an imperfect choice among alternatives or making statements that contradict a belief, that threat can be overcome in other ways. Steele (1988, p. 267) referred to this process as "fluid compensation," which is a concept that has been adopted as an explanation by several other prominent approaches that have addressed cognitive dissonance. Examples of this approach were presented earlier in the chapter. People whose self-systems were compromised by making an imperfect choice reduced their need to distort their perception of the choice alternatives if they were encouraged to remember other ways that their sense of self was good and worthy (Steele & Liu, 1983). Similarly, McGregor et al. (2001) developed a model of compensatory conviction. In their studies, they encouraged participants to engage in what they referred to as an integrity-repair exercise by remembering times that they acted consistently. They found that focusing on past incidents of consistency reduced the need to alter attitudes in the wake of current inconsistent statements.

Another comprehensive approach that adopts the fluid compensation approach is Heine et al.'s (2006) meaning maintenance model. In this approach, people seek meaningful relationships among their cognitive representations of the world. Disruptions lead to a search to restore meaning. This can be done in a variety of equally effective ways that do not have to be related to the original disruption. Similarly, terror management theory argues that cognitive inconsistency is one of many ways that disturb people's view about the stability and coherence of their worldview. The resulting existential anxiety caused by attitude-discrepant behavior can be relieved by efforts to restore consistency (Friedman & Arndt, 2005; see also Pyszczynsky & Sundby, this volume). Because of the fluid nature of the compensation process, the induction of existential terror can cause people to increase their efforts at restoring consistency while the alternative resolution of the terror decreases that effort.

Forstman and Sagioglou (2020) suggested that dissonance can be relieved by exposure to religious symbols. Religion, they argued, is perceived as a gateway to certainty. Belief in the existence of God makes people more certain of their place in the world, thereby reducing the total amount of uncertainty and ultimately the need to engage in direct strategies to reduce cognitive dissonance (see also Pyszczynsky & Sundby; Fiske, this volume). In an interesting set of studies, Forstman & Sagioglou (2020) showed that participants who chose items (candy bars) in a typical free-choice study (e.g., Brehm, 1956) were less inclined to distort their post-choice evaluations if they were exposed, even subliminally, to the Christian symbol of the cross than those who were not exposed to the religious symbol. They argued that the certainty provided by the religious link obviated the need to reduce dissonance by choice re-evaluation.

Re-inventing the Meaning of Dissonance: Certainty about Social Reality

In our view, the basis for the drive we call cognitive dissonance is the need to maintain certainty about social reality. When inconsistent cognitions shatter our certainty about social reality, psychological processes are activated to protect against the ensuing uncertainty. Festinger was focused on the consequences of the drive to restore consistency, but *A theory of cognitive dissonance* was silent about the underlying reasons. However, we believe that the extant data are consistent with the notion that dissonance is specifically caused by disruptions to our certainty about social reality. Reducing dissonance is not just one way of resolving other motivational needs. It is not just one way of protecting ourselves from meaninglessness or uplifting the integrity of the self. Reducing dissonance *may* respond to those needs as well, but satisfying the dissonance need is more focused on the need to restore consistency and maintain our certainty about the meaning of reality.

Viewing dissonance resolution as a subset of broader motivations does not fully comport with the extant data and also distracts from a fuller understanding of the meaning of dissonance. Steele's influential self-affirmation theory is illustrative. There can be little doubt that people seek an integrated self that they feel is worthy and robust. The question is whether the search for self-integrity through self-affirmation is as satisfying a way to resolve the dissonance drive as is a change of cognitions to directly restore consistency. There are substantial data to show that it is not. For example, Aronson et al. (1995) had people write a counterattitudinal essay. When given an opportunity to change their attitudes toward the issue or to receive information extolling and affirming an aspect of their personality, the participants overwhelmingly chose to reduce their dissonance directly. They preferred to make their attitudes consistent with their behavior rather than affirming their self-systems with positive and glowing information about a personality trait (see also Blanton et al., 1997; Galinsky et al., 2000; Stone et al., 1997).

Like the drive of hunger, dissonance seeks a direct resolution. Related ways to bolster aspects of mortality or one's self-esteem may be helpful ways to reduce the unpleasant tension, but they are not equally fungible. Reductions in uncertainty that are specifically related to the dissonant cognitions are sought, appreciated and serve to resolve the tension (e.g., McKimmie et al., 2003; also Hogg & Gaffney; Krueger & Gruening, this volume). Other reductions in uncertainty serve to reduce tension, but much like the way that water reduces hunger in the absence of food, it is a suboptimal solution.

And so we return to the beginning. A theoretical arc connects the current state of understanding of cognitive dissonance with Festinger's insights written years prior to writing dissonance theory. Festinger eloquently described the need for a shared social reality as a fundamental human need. *Uncertainty* about that social reality serves as the most plausible basis for the dissonance drive. Resolving uncertainty is best accomplished by altering or adding cognitions in order to restore consistency and confidence in the social world (https://library.princeton.edu/).

References

Aronson, E. (1968). Dissonance theory: Progress and problems. In R. P. Abelson et al. (Eds.) *Cognitive consistency: A sourcebook* (pp. 5–27). Chicago: Rand McNally.

Aronson, E. & Carlsmith, J. M. (1963). The effect of severity of threat on devaluation of forbidden behavior. *Journal of Abnormal and Social Psychology, 66,* 584–588.

Aronson, E. & Mills, J. M. (1959). The effect of severity of initiation on liking for a group. *Journal of Abnormal & Social Psychology, 59,* 177–181.

Aronson, J., Blanton, H. & Coper, J. (1995). From dissonance to disidentification: Selectivity in the self-affirmation process. *Journal of Personality and Social Psychology*, *68*, 986–996.

Blanton, H., Cooper, J. Skirnuk, I. and Aronson, J (1997). When bad things happen to good feedback: Exacerbating the need for self justification with self affirmations. *Personality and Social Psychology Bulletin*, *23*, 684–692.

Brehm, J. W. (1956). Postdecision changes in the desirability of alternatives. *Journal of Abnormal and Social Psychology*, *52*, 384–389.

Cooper, J. (1998). Unlearning cognitive dissonance: Toward an understanding of the development of cognitive dissonance. *Journal of Experimental Social Psychology*, *34*, 562–575.

Cooper, J. & Fazio, R. H. (1984). A New Look at dissonance theory. In L. Berkowitz (Ed.) *Advances in experimental social psychology, Vol 17*, Orlando Fl: Academic, 229–264.

Cooper, J. & Feldman, L. A. (2020). Helping the couch potato: A cognitive dissonance approach to increasing exercise in the elderly. *Journal of Applied Social Psychology*, *50*, 33–40.

Cooper, J., Zanna, M. P. & Taves, P. (1978). Arousal as a necessary condition for attitude change following induced compliance. *Journal of Personality and Social Psychology*, *36*, 1101–1106.

Egan, L. C., Bloom, P. & Santos, L. R. (2010). Choice-induced preferences in the absence of choice: Evidence from a blind two choice paradigm with oun children and capuchin monkeys. *Journal of Experimental Social Psychology*, *46*, 204–207.

Ellliot, A. & Devine, P. G. (1994). On the motivational nature of cognitive dissonance: Dissonance as psychological discomfort. *Journal of Personality and Social Psychology*, *67*, 382–394.

Festinger, L. (1950). Informal social communication. *Psychological Review*, *57*, 271–282.

Festinger, L. (1954). A theory of social comparison processes. *Human Relations*, *7*, 117–140.

Festinger, L. (1957). *A theory of cognitive dissonance*. Evanston, Ill: Row Peterson.

Festinger, L. & Carlsmith (1959). Cognitive consequences of forced compliance. *Journal of Abnormal & Social Psychology*, *58*, 203–210.

Friedman, R. S. & Arndt, J. (2005). Reexploring the connection between terror management theory and dissonance theory. *Personality & Social Psychology Bulletin*, *31*, 1217–1225.

Fortsman, M. & Sagioglou, C. (2020). Religious concept activation attenuates cognitive dissonance reduction in free-choice and induced compliance paradigms. *Journal of Social Psychology*, *160*, 75–91.

Galinsky, A., Stone, J. & Cooper, J. (2000). The reinstatement of dissonance and psychological discomfort following failed affirmations. *European Journal of Social Psychology*, *30*, 123–147.

Harmon-Jones, E. (Ed) (2019). *Cognitive dissonance: Reexamining a pivotal theory in psychology*, 2nd edition. Washington, DC: APA.

Harmon-Jones, E., Amodio, D. M. & Harmon-Jones, C. (2009). Action based model of dissonance: A review, integration and expansion of conceptions of cognitive conflict. In M. P. Zanna (Ed.) *Advances in experimental social psychology*, *41* (pp. 119–166).

Heine, S. J., Proulx, T. & Vohs, K. D. (2006). The meaning maintenance model: On the coherence of social motivations. *Personality and Social Psychology Review*, *2*, 85–110.

Hull, C. (1943). *Principles of behavior: An introduction to behavior theory.* New York: Appleton-Century.

Jones, E. E. & Gerard, H. B. (1967). *Foundations of social psychology.* New York: Wiley.

McGregor, I., Zanna, M. P., Holmes, M. & Spencer, S. (2001). Compensatory conviction in the face of personal uncertainty: Going to extremes and being oneself. *Journal of Personality and Social Psychology, 80,* 472–488.

McKimmie, B., Terry, D. & Hogg, M. A. (2003). I'm a hypocrite but so is everyone else: Group support and the reduction of cognitive dissonance. *Group Dynamics, 7,* 214–224.

Norton, M. I., Monin, B., Cooper, H. J. & Hogg, M. A. (2003). Vicarious dissonance: Attitude change from the inconsistency of others. *Journal of Personality and Social Psychology, 85,* 47–62.

Pallack, M. S. & Pittman, T. S. (1972). General motivational effects of dissonance arousal. *Journal of Personality and Social Psychology, 21,* 349–358.

Pearce, L. & Cooper, J. (2020). Fostering COVID-19 safe behaviors using cognitive dissonance. *Basic and Applied Social Psychology, 43,* 267–282.

Priolo, D., Pelt, A., Bauzel, R., rubens, L., Voisin, D. & Fointiat, V. (2019). *Personality and Social Psychology Bulletin, 45,* 1681–1701.

Skinner, B. F. (1957). *Schedules of reinforcement.* New York: Appleton-Century.

Steele, C. M. & Liu, T. J. (1983). Dissonance processes as self-affirmation. *Journal of Personality and Social Psychology, 45,* 5o–19.

Steele, C. M. (1988). The psychology of self-affirmation: Sustaining the integrity of the self. In L. Berkowitz (Ed.) *Advances in experimental social psychology, 21,* San Diego: Academic Press, 261–302.

Stice, E., Rohde, P., Gau, J. M., Butryn, M. L., Shaw, H., Cloud, K. & D'Adamo, L. (2021). Enhancing efficacy of a dissonance-based obesity and eating disorder prevention program: Experimental therapeutics. *Journal of Consulting and Clinical Psychology, 89,* 793–804.

Stone, J., Aronson, E. Crain, A. L., Winslow, M. P. & Fried, C. B. (1994). Inducing hypocrisy as a means of encouraging young adults to use condoms. *Personality & Social Psychology Bulletin, 20,* 116–128.

Stone, J., Wiegand, A., Cooper, J. & Aronson, E. (1997). When exemplification fails: Hypocrisy and the motive for self-integrity. *Journal of Personality and Social Psychology, 72,* 54–65.

Waterman, C. & Katkin, E. S. (1967). Energizing (dynamogernic) effect of cognitive dissonance on task performance. *Journal of Personality and Social Psychology, 6,* 126–131.

7
PERSUASION AS A SOP TO INSECURITY

William D. Crano

CLAREMONT GRADUATE UNIVERSITY, USA

Zachary Hohman

TEXAS TECH UNIVERSITY, TEXAS, USA

Abstract

The role of insecurity has been a consistent feature of psychological theorizing from the field's origins (Wundt, 1900) and is reflected today in our most influential models. Theories of cognitive consistency presupposed the need to be right, to fit in, or reduce inconsistencies among beliefs or behaviors. In theories of inter- and intra-group behavior, being a part of a group plays a role in our self- and social identities, which affect our relations with others. The group provides cover and a rationale for beliefs and behaviors that may appear ridiculous, immoral, or inhuman to those outside the group. An underlying motive of contemporary social psychological theorizing implicitly or explicitly assumes the need to resolve or forestall insecurity, and by doing so, establish a reality shared and reinforced by others who think and act as we do. In this chapter, we outline some of the paradigms (e.g., consistency, persuasion, and group process models) that take the fundamental human need for certainty as nearly axiomatic. We end with a discussion of the utility of the insecurity orientation in the authors' research on ambivalence, attitudes, and persuasion.

Dictionaries are consistent in their definition of *insecurity* as describing a state of self-uncertainty or self-doubt or as involving feelings of personal inadequacy or low self-confidence. While there exist many sources of insecurity – e.g., food, housing, identity, and financial – all involve an element of perceiving the self in a less-than-optimal light. As such, the study of insecurity arguably should be a consistent focus of social psychologists, for surely the study of the individual in context is a central feature of the expanding realm of proper exploration in the field. Yet, although at its most basic definitional level insecurity belongs firmly in the realm of social psychology, in practice, it is more

DOI: 10.4324/9781003317623-9

about the focus of clinical, rather than social psychology. In part, this division makes sense, because insecurity, if sufficiently severe, can render an individual incapable of negotiating even the most basic issues of everyday life, and this clearly is a proper focus of clinical psychology. However, by its very definition, insecurity can permeate almost every aspect of social life, and thus, clearly falls in the realm of social psychology. For this reason alone, this monograph on the psychology of insecurity, as part of this Symposium, is both timely and perhaps overdue.

The recognition of insecurity in social psychology is not forced, because insecurity is acknowledged directly or indirectly as a major motivational factor in many of our most central theories. The acknowledgment of centrality is evident when considering some of social psychology's long-standing preoccupations and theoretical models. For example, Festinger's (1954) social comparison theory, one of his many key theoretical contributions to the field (Suls et al., 2020), proposes that we are driven to know how to act or to evaluate the quality of our actions in settings lacking clear and objective (or widely acknowledged) evidentiary standards. In this theory, we gain such knowledge through a process of comparing ourselves with like others. This is the motive engine of Festinger's (1957; see also Cooper & Pearce, this volume) social comparison – we want to know how we stack up with others who share similar preconceptions, skills, abilities, attitudes, etc., especially on issues that involve subjective judgments that might affect other's evaluations of ourselves. Later research that indicated we sometimes care more for comparisons with unlike others does not diminish the utility or importance of the theory (e.g., Goethals, 1986; Goethals & Darley, 1977; Gorenflo & Crano, 1989). We want to be right or contextually competent, or to see how far from right or competent we are. We attend to these issues because their implications are critical in dealing with the social context in which we find ourselves. They also are linked to the anxiety that may arise from insecurity about others' evaluations and of our place in our social groups and the larger society that we negotiate in our daily lives.

Role of Insecurity in Theories of Cognitive Consistency

The heyday of intense focus on cognitive consistency models has passed, its demise demarcated arguably by the publication of the classic *Theories of cognitive consistency: A sourcebook* by Abelson et al. (1968). After some years of intense activity, the field appeared ready to move on to other concerns. The sourcebook seems to have closed a chapter on the intense focus on consistency as a motive force in human behavior (but see Gawronski & Brannon, 2019), but whatever one's historic judgment, the study of cognitive consistency unarguably left its mark on much of what followed in social psychology.

An important question some researchers neglected in our apparent quest for cognitive consistency was its prime cause, which we identify here as perceived or anticipated insecurity. Many begged the question, why would a person be concerned with the consistency of beliefs or the consistency of beliefs with behaviors? This makes sense to some degree, but it disregards the carrying costs, cognitive and otherwise, of persistent cognitive inconsistency. Obviously, most scientifically informed individuals would be expected to feel some degree of discomfort when holding mutually contradictory positions. However, survey research on studies of "non-attitudes" (e.g., see Converse, 1964; 1970) suggests respondents from randomly sampled populations often have little hesitancy in expressing contradictory attitudes (Bassili & Fletcher, 1991; Feick, 1989; Sniderman et al., 2001). How this squares with the central assumption of cognitive consistency models is unclear and unlikely to be addressed in our short chapter.

Let us return instead to Festinger (1957) and his relatively newer *Theory of Cognitive Dissonance*, which eclipsed social comparison theory in citation counts and TV scripts. Dissonance theory was based on the idea that inconsistencies among beliefs or between beliefs and behaviors had negative drive-like properties of the same force as those of hunger or thirst (see Cooper & Pearce, this volume). The "drive-like" emphasis focused critics and supporters on the causes of the apparent need for consistency between cognitions or behaviors, whose absence caused upset or disequilibrium, and which resulted in changing cognitions or behaviors to restore cognitive balance, equilibrium, or more colloquially, peace of mind (Cooper et al., 1978). However, we might ask why inconsistencies involving a lack of fit among cognitions or conflicts among cognitions and actions should take on such importance. The answer may be found in Festinger's earlier theory of social comparison, which was focused on the need to be right rather than the need to be cognitively consistent. Perhaps Festinger had anticipated the zeitgeist of the coming cognitive revolution in social psychology and adjusted his approach to become more in tune with the times by changing the presumptive fundamental motive from a self-defensive desire to look good and be on the right side of issues (social comparison) to a concern with the interface of need and cognition? Motives and speculations aside, the unpleasantness of dissonant cognitions does affect cognitive equilibrium, which by definition and experimental evidence is unpleasant and which also may lead to greater or lesser degrees of felt insecurity (see van den Bos, this volume).

Gawronski (2012, p. 652) argued that the dissonance craze of the 1960s was in some ways a lost opportunity because it focused attention narrowly on "dissonance related changes in attitudes and alternative accounts that attribute such changes to mechanisms of ego-defense," rather than on cognitive consistency, a "core motive" in human behavior. He argued that cognitive inconsistency was an "epistemic cue for errors in one's system of beliefs." We agree with Gawronski's perceptive analysis and believe the theorized search for errors, which is evident as a driving force in much of the social psychological literature, reflects an even more fundamental desire to avoid the insecurities cognitive conflicts entail (see Forgas, this volume). Whether such conflicts arise because of clashes between beliefs, potential norm violations, or fears that an established cost-benefit analysis has turned sour, the details of the specific conflict are largely irrelevant. What is evident is that insecurity lies at the core of them all. As we shall argue, the fundamental role of insecurity may be found at the core of other areas of intense concern in social psychology.

Insecurity as a Basis for Attitude Change and Social Influence

Social Influence. An influential research series in the early days of social influence, anticipating and co-terminus with Festinger's classics, was created by Asch (1952, 1956, 1961), whose work on independence and conformity still carries weight nearly a century after its inception (e.g., see Koriat et al., 2020; Moscovici & Personnaz, 1980; Pettigrew, 2018). Asch's prototypical study placed a naïve participant in a situation in which his or her visual perceptions were in direct conflict with the reported perceptions of a group of fellow participants. The correct judgment was trivially apparent, yet in study after study, surprisingly to many in the field, Asch (1956) found that an unexpected proportion of his naïve subjects denied the obvious reality of their own senses and retreated to the more comfortable role of agreeing with their clearly incorrect peers. Finding security in agreement with our fellows may be more important than being right if social insecurities lie at the heart of such judgments. The trick in much research of this type is to know with whom naïve research participants identify as their fellows (see also Hogg & Gaffney, this volume).

This explanation for seemingly inexplicable behavior helps us understand the still-startling results of Stanley Milgram (1963, 1975), whose naïve participants proved willing to shock a total stranger into stunned silence or worse at the behest of another stranger, the experimenter. As noted, it pays to know whom one's fellows are. In the case of Milgram's research, it was the researcher who hovered over the naïve shocker, not the anonymous individual in another room who was on the receiving end of the shocks.

In a study designed to determine the factors that influenced the importance attached to others' views versus one's own directly experienced perceptions, Crano (1970) exposed participants to a simple judgment task in which each judgment across 30 trials was made by both the naïve subject and a confederate, whose judgments of the number of items projected on a screen for one second – much too quickly to count – were 10% higher than the true number. Preliminary subjects who responded without socially supplied information tended, on average, to underestimate the actual number by 10%. Thus, on average, confederates' judgments in the experimental task were approximately 20% greater than the judgments of the naïve participants with whom they were paired.

Results indicated experimental participants tended to weigh direct perceptual inputs *and* those supplied by a fellow respondent (the confederate) – but not equally across contexts. Participants weighed the confederate's judgments significantly more heavily than their own perceptions when paired with another who was thought to have great skill in the judgment task, and thus, whose expertise likely was considerably greater than their own. However, participants put significantly greater weight on their own direct perceptions when making judgments in contexts in which the social influence source, the confederate, was understood to be deficient in the task. In both circumstances, the confederate's judgments were given some weight in the participant's judgment – they were never ignored completely, but the weight placed on the other's judgments varied as a function of differences in the assumed competence of the influence source. In other research, it was found that when dealing with judgments of extreme ambiguity, subjects tended to rely heavily on others, even those whose presumptive skills in the task were no greater than their own (Crano & Hannula-Bral, 1994; Fiedler et al., 2019). Consistent with Campbell's (1990) view, we believe these "conforming" behaviors were the product of the desire to be accurate, motivated by the insecurity of appearing incompetent, not the simple expedient of agreeing with others to avoid censure.

Majority/Minority influence. In interpersonal contexts, we are considerably more inclined to attend to in-group information sources, as might be inferred from Festinger's (1954) social comparison theory. Research on majority and minority influence strongly supports this observation (Crano & Alvaro, 1998a, 1998b; 2014; Martin & Hewstone, 2008). Studies have established that in-group message sources are considerably more effective than out-group sources (Crano, 2010; Maass & Clark, 1984; Maass et al., 1982). Furthermore, immediate influence effects in response to majority-status sources are usually greater than changes induced by out-group minorities (Mackie, 1987).

With in-group minorities, the situation is considerably more complex, but the pattern is clearly discernable when viewed from the vantage point of insecurity avoidance. The majority almost always enjoys a persuasive advantage (Mackie, 1987), as would be expected, to assuage the insecurity of potential ostracism (for disagreeing), but when the majority is no longer surveilling, the individual can adopt behaviors that may be inconsistent with majority demands. Compliance based on powerful others' surveillance is common, but it does not necessarily imply internalization of the majority's dictates (Crano, 2012). The majority's advantage is strong, but often not persistent; the immediate majority advantage would be expected to assuage the insecurity of potential ostracism, but when the majority is no longer surveilling, the individual can adopt behaviors that may be inconsistent with majority demands but that help reduce cognitive inconsistencies, and hence insecurities (see also Fiedler & McCaughey, this volume).

Minorities usually are not immediately effective when attempting to influence members of the majority group. However, many studies have shown immediate *indirect* minority influence effects, which involve the in-group minority's influence on attitudes that are related to the focal attitude but that may never have been mentioned in the minority's persuasive appeal. Such changes can occur in the absence of any apparent change of focal attitude, i.e., the attitude that is the focus of the minority's appeal (Alvaro & Crano, 1996, 1997; Martin & Hewstone, 2008; Pérez & Mugny, 1987). Such influence effects are found almost exclusively when the minority is in-group. Conforming to out-group minorities can occur, but only in highly restricted circumstances that usually involve issues of fact rather than opinion (Crano & Hannula-Bral, 1994). More complete theoretical treatments are beyond the scope of this chapter, but the pattern of findings relating minority advantage in contexts involving objective judgments and the majority's advantage in judgments involving subjective judgments are discussed elsewhere (e.g., see Butera et al., 2017; Crano & Alvaro, 1998a, b; 2014; Crano & Seyranian, 2007, 2009; Martin et al., 2008).

Persuasion Models

Social norms theory. The social norms model is built on the assumption that norm misperceptions of relevant others motivate behavior. This model has proved a popular foundation for studies of health-relevant behavior, especially the behavior of youth (e.g., Ajzen & Schmidt, 2020; Dempsey et al., 2018; Handren et al., 2016). The relation of this general approach to considerations of insecurity proposes that if behavior is controlled to some degree by the assumed norms of one's peers or significant others, the basis of such concerns must be identified. The most obvious and plausible hypothesis is that behaving in concert with one's peers lends comfort and security, whereas acting non-normatively signals the likelihood of conflict and its attendant insecurities. This reasoning leads to the proposition that appraisals may be made largely in the service of insecurity avoidance or reduction.

The apparent dependence on norm appraisals does not suggest that such appraisals are infallibly accurate. Indeed, it is common in prevention research to try to modify incorrect norm-based assumptions and thus move respondents toward healthier behavior. For example, in a study of inhalant misuse, Crano et al. (2008) measured middle-school children's use of inhalant substances and their estimates of their friends' use. Relatively few of the respondents had used inhalants themselves.[1] However, estimates of friends' usage by those who themselves had used these substances were significantly greater than the estimates of those who had not, and this result was found in both 6th- and 7th-grade school children (approximate ages, 12- and 13-year-olds – see Figure 7.1). A reasonable interpretation of this result is that the participants had helped

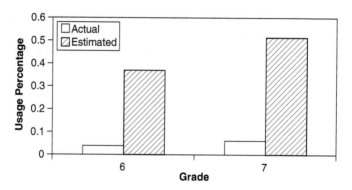

FIGURE 7.1 Proportion of inhalant users in 6th and 7th grade, actual (self-reported) and estimated (friends). Differences in both grades between actual and estimated usage were statistically significant (both $p < .0001$).

moderate the insecurities involved in inhalant misuse by greatly overestimating the proportion of their peers engaged in the same behavior (Crano et al., 2008).

Theory of Reasoned Action. Fishbein and Azjen's theory of reasoned action (TRA) (e.g., Fishbein & Ajzen, 1975, 2010) acknowledged the importance of attitudes toward an action and the perceived (i.e., subjective) norms of significant others (family, close friends, etc.) regarding the considered action as predictive of the intention to perform the action and ultimately the actual performance of the action. The TRA has been applied widely and successfully in a host of applications (Rise et al., 2010; Schulze & Wittmann, 2003; Steinmetz et al., 2016), and most meta-analytic studies have supported the predictive validity of the model and its later revision, the theory of planned behavior (TPB: e.g., Albarracín et al., 2001; Sheeran & Taylor, 1999; Tyson et al., 2014). Variations have been found in the predative efficiency of subjective norms in the models, but we believe they may be a function of variations in the significance of the "significant others," which both attitude models postulate as indispensable.

Echoes of this requirement are found in our earlier discussion of the varying capacity of in-group and out-group minorities to effect social influence. In the context of the TRA and TPB, this requirement is converted to the requirement that others' normative expectations be transmitted by *significant* others if they are to influence behavior. Why others' significance weighs heavily in the process is that it signals approval or disapproval of individuals and groups to which the individual actor owes allegiance, and which provide social validation. Social validation, in turn, is crucial to self-esteem – losing it is threatening to the self (e.g., Hales & Williams, 2021; Hartgerink et al., 2015; Rudert et al., 2019; see also Hogg & Gaffney, this volume). This has been recognized from the very beginnings of social psychology (e.g., Wundt & Schaub, 1916). Indeed, William James (1890, p. 430) considered social isolation "a mode of torture too cruel and unnatural for civilized countries to adopt" (also see Hales et al., 2019; Wesselmann et al., 2021).

Ambivalence

People often simultaneously hold negative *and* positive evaluations of an attitude object. For example, one might enjoy the sensation of using illicit or risky substances and at the same time realize that doing so is a health danger. Consequences of attitude ambivalence are numerous: ambivalent attitudes are highly susceptible to persuasion (Bassili, 1996), less predictive of behavior (Armitage & Conner, 2000; Sparks et al., 2004), less accessible in memory (Bargh et al., 1992; Krosnick, 1989), and associated with a variety of negative outcomes (Brown & Farber, 1951; Crano et al., 2019; Kaplan, 1972; Newby-Clark et al., 2002; Priester, 2002; Rydell et al., 2008).

A particularly negative outcome of attitude ambivalence is the psychological discomfort associated with simultaneously holding positive and negative evaluations (Newby-Clark et al., 2002). For example, if people believe they will have more friends if they do something dangerous but also fear the negative health consequences of the dangerous action, they will experience aversive discomfort (Hass et al., 1992), of which the individual may or may not be consciously aware (Petty & Briñol, 2009). A result of the tension is that the person will feel insecure about how they should respond to the attitude object. Thus, ambivalence can result in insecurity. Insecurity associated with ambivalence is experienced as undesirable, and accommodations are made to reduce it (Priester, 2002).

To reduce ambivalence, research suggests that people pay close attention to information that facilitates resolution (Briñol et al., 2006; Clark et al., 2008; Hohman et al., 2016; Maio et al., 1996). For example, Maio and associates (1996) showed that ambivalent participants elaborated a persuasive message more intensely and were more persuadable than non-ambivalent participants. Other research demonstrates that people with highly ambivalent attitudes use consensus information to resolve their ambivalence (Hodson et al., 2001). It seems evident that people seek information in the social context to reduce the insecurity aroused by ambivalence.

A valued source for resolving ambivalence is one's social group (see Hogg, 2006; Hirschberger, this volume). Groups help define how members should think and act by supplying (social) norms for appropriate behavior that define the expected attitudes and actions of group members (Hogg & Reid, 2006). Because attitudes are fundamental features of group life, individuals look to their groups for correct attitudes when highly ambivalent. Given that social norms help define group-related attitudes, they provide a viable means of reducing ambivalence. This idea is supported by research that demonstrates, in the absence of concrete information, people look to similar others for what to think and how to behave (Festinger, 1954; Hogg & Turner, 1987; Suls & Miller, 1977; Suls & Wheeler, 2012; Turner et al., 1987).

Taken together, people who experience insecurity from ambivalence should be motivated to reduce it by adhering to the social norms of important groups.

Aligning attitudes with the group will transform the ambivalent attitude positively or negatively, depending on group norms, thereby reducing ambivalence. Social norms also should moderate the relation between ambivalence and behavioral intentions: norms should be stronger predictors of intentions for people holding highly ambivalent attitudes. Across four studies, we now have found consistent and strong evidence that social norms reduce the insecurity associated with attitude ambivalence, as will be shown in the pages that follow.

Research on the Ambivalence-Insecurity Association

Adolescent marijuana ambivalence and friend norms. Hohman et al. (2014) conducted a secondary analysis of data from a longitudinal study to determine if ambivalence about marijuana use rendered adolescents more persuadable by their friends' norms about marijuana use. Analysis revealed that friends' norms significantly predicted intentions to use marijuana (cross-sectionally, $d = .19$) and use of marijuana in the previous year (longitudinally, $d = .22$) when adolescents were highly ambivalent; however, for those of low ambivalence, norms did not significantly predict intentions or behaviors. These results suggest that adolescents use social norms to reduce ambivalence, but the study did not manipulate ambivalence or social norms, limiting our ability to make causal determinations. Our next study addressed these limitations.

College tobacco use and social norms. Hohman et al. (2016) conducted a study assessing the moderating effects of attitude ambivalence on the relationship between social norms, attitudes, and behavioral intentions to use tobacco. We predicted that people would use social norms to reduce attitude ambivalence, and that reduced ambivalence would lead to changes in attitudes and behavioral intentions. Participants were exposed to persuasive communications designed to influence attitude ambivalence and social norms regarding tobacco use. The persuasive communication contained negative and positive information about tobacco use (ambivalent condition), or only negative information about tobacco use (univalent condition). As well, the communication contained either social normative information (against the use of tobacco use) or no social normative information.

Analysis indicated that providing a social norm antagonistic to tobacco use significantly reduced ambivalence among participants reading the ambivalent message. There was a significant decrease in tobacco attitudes (from pre- to post-manipulations) for participants reading the ambivalent message who were provided the anti-tobacco use norm ($d = .40$). Ambivalent participants also expressed significantly lower intentions to use tobacco when provided anti-tobacco social norms ($d = .42$). There were no changes in the non-ambivalent condition from pre- to post-manipulations on tobacco attitudes nor difference on intentions to use tobacco. These results point to the causal role of social norms in mediating the effects of attitude ambivalence on subsequent behavior.

Ambivalence, physiological arousal, and endocrine arousal. Hohman and associates conducted a study to assess the role of psychophysiological responses in ambivalence (Hohman et al., 2017). Specifically, we examined the physiological markers associated with the insecurity feature of ambivalence. We predicted that participants high in ambivalence would have an increased heart rate, salivary cortisol, and salivary alpha amylase compared to those low in ambivalence. Participants viewed a commercially created 30-second public service announcement (PSA) about prescription drug misuse. PSAs were pilot tested to identify those that led to high or low ambivalence. While watching, the PSA participants' heart rates were measured, and before and after the PSA, participants provided a saliva sample to measure cortisol and alpha amylase. Results demonstrated that heart rate ($d = .50$) spiked while watching the high ambivalence (versus the low ambivalence PSA). As well, we found significant increases in cortisol ($d = .98$) and alpha amylase ($d = .70$) for the ambivalent PSA and no change for the low ambivalence PSA. Taken together, these results support the idea that feeling ambivalent is negative and experienced at a physiological level.

Ambivalence and personalized norm feedback intervention. In a recent study, Hohman et al. (2022) examined if inducing attitude ambivalence about alcohol use before the personalized norm feedback (PNF) intervention would reduce intentions to drink more than the intervention by itself. PNF (Neighbors et al., 2004) is an intervention developed to reduce binge drinking on college campuses. PNF reduces binge drinking by correcting students' misperceptions that other students drink more than they do. The PNF interventions work by using social norms to induce people to change their behaviors. We predicted that inducing ambivalence before the PNF intervention would make it more effective because ambivalence should motivate people to pay more attention to and use the social norms provided in PNF to reduce their ambivalence. To test this prediction, we manipulated participants' alcohol ambivalence (by writing about things that made them ambivalent about their drinking or not), provided them with the typical PNF intervention (Miller et al., 2013), and then measured their intentions to drink in the next month. Results revealed that participants in the ambivalent condition intended to drink fewer drinks than the univalent condition ($d = .25$). Overall, this study supports the notion that social norms help alleviate the insecurity associated with ambivalence.

Applications of Insecurity-Attitude Association

Our emphasis on the flight from insecurity that seems to lie at the base of much of what drives behavior is focused on a fundamental concern of social psychology, the study of attitudes and persuasion. Why should we care when we learn that many (anonymous) peers hold beliefs at odds with our own?

Why should we be willing to change an established behavior merely because a popular "influencer" whom we have never met suggests we should? In seeking the underlying motivation of attitude or behavior change in response to such sources, the answer we have proposed is the same: The avoidance of insecurity arising from a contradiction between belief and reality, or between a belief or action that is inconsistent with presumed normatively appropriate behavior suggested by an influential source, is a central motivator of change. The motivation to assuage insecurity may affect not only our thoughts and attitudes but also our behaviors, even those that may prove dangerous or self-destructive. We are motivated to change if we are persuaded our attitudes or behaviors are not in accord with many others who hold alternate ideas, however attributed, especially if the attitude involves a vested outcome (Crano, 1995, 1997; Donaldson et al., 2020). The change is in part a response involving the attainment of a desired outcome, but it is motivated by a search for certainty, a sop to our insecurities. We are aware that this statement clearly implies a motivated search for factors that might attenuate uncertainty. Such a search also could be the result of an unnoticed superficial anchor stimulus, which also could be operative in uncertainty reduction behavior. We prefer the more motivated alternative but acknowledge data favoring both alternatives could be brought to bear (e.g., see Hütter & Fiedler, 2019).

Research emphasizing the importance of normative effects has been shown in a recent series of secondary data analyses of adolescents' risky behavior involving dangerous substance use. In a comprehensive secondary analysis of data from the Monitoring the Future archive (see Miech et al., 2021), Handren et al. (2016) found parental involvement with their children (total N = 14,785) played a major role in their children's underage alcohol use. Involvement affected children's alcohol use directly and indirectly through their children's perceived peer norms. Parents' involvement with their children was *directly* associated with their offspring's self-esteem, self-derogation, perception of peer norms regarding alcohol use, and most importantly, their children's alcohol use. Adolescents' perceptions of peer norms, in turn, were directly linked to alcohol disapproval, perceived risk of alcohol, and alcohol use, which was directly related, as might be expected, to heavy episodic drinking behavior (i.e., bingeing). All these relations were clearly statistically significant at $p < .001$.

Following this research, Donaldson et al. (2016) conducted a path analysis involving four waves of panel data collected over 12–14 years, which showed parental behaviors played a major role in their children's binge drinking behavior and alcohol-related difficulties with the law many years later. Data (N = 7857) on parents and children were drawn from the National Longitudinal Study of Adolescent Health (AddHealth). Adolescent binge drinking was defined in this archive with two variables: How often they got drunk or very high on alcohol in the past year, and on how many consecutive days they consumed five or more drinks, a common measure of binge drinking (e.g., Guilamo-Ramos et al., 2005; Shin et al., 2009). These variables were assessed on Likert-type scales (0 = no binging to 7 frequent binging), and their average was used as an indicator of severity of binge drinking behavior.

The parental predictor variables of interest in the analysis were parental monitoring, warmth, parental alcohol use, and parents' estimates of their children's behavior with respect to alcohol use in the past year (before the first year's measurement). The analysis revealed that all four parental behaviors were significantly associated with their children's binge drinking in adolescence and later as well, when the children had reached young adulthood. This is an important result because binge drinking in young adulthood predicted the likelihood of the (now) young adults having an arrest record. All these relations were statistically significant at $p < .001$. Note that in addition to causing the problems described here, binge drinking is an important health problem on other grounds. Excessive use of alcohol is responsible for more than 40,000 deaths in the US each year (Centers for Disease Control, 2012).

Expanding our lens to the problem of adolescent nicotine use, a dangerous problem, Coleman et al. (2021) performed a longitudinal analysis focused on the impact of peer and family e-cigarette (e-cig) use on adolescent participants' own use, their subsequent use of tobacco cigarettes, and their subsequent cannabis use. The central predictors in the study were family and peer e-cig use in year 1, adolescents' e-cig and tobacco cigarette use in year 2, and cannabis use in year 3. Data from this three-year longitudinal panel study (N = 859) were analyzed in a longitudinal path analysis, which revealed the powerful effects of peer and familial e-cig use on the participants' own e-cig and tobacco cigarette adoption in the following year, which in turn predicted their initiation of cannabis in the third year. All relations reported were statistically significant.

The upshot of all three of these large-scale studies is that parents and peers can strongly influence the immediate and long-term behaviors of youth. Parents' continuing influence on their children's behavior is obvious. Based on recent research (Donaldson et al., 2021), we have speculated that this influence may be moderated by the attachment style that characterizes the parent-child interaction of our participants, especially as it relates to the constructs of parental monitoring and parental warmth. The various combinations of parental warmth and monitoring had powerful effects on young adolescents' susceptibility to risky behavior, especially behavior relating to the use of psychoactive substances. Parental behavior was postulated as related to variations in children's attachment with their parents (e.g., Ainsworth, 1989, 2010; Ainsworth & Bowlby, 1991), which we have postulated as related to child and adolescent social insecurity, as was indicated in our earlier findings (e.g., Donaldson et al., 2016). This program of parent-child substance use research has indicated that parents whose reasonable disciplinary rules are agreed upon with their children and enforced (monitoring) in a warm and encouraging fashion (i.e., fostering a sense of warmth and openness to their children's self-disclosures) bode well for children's social development and, not incidentally, their resistance to normative inducements to engage in risky behavior. All combinations of monitoring and warmth, other than close monitoring and warm parent-child relations, do not encourage the best outcomes (see also Mikulincer & Shaver, this volume). These results comport with findings reported over the years by Kerr, Stattin, and their associates (Kerr & Stattin, 2000, 2003; Stattin & Kerr, 2000; Stattin et al., 2011), and suggest to us that insecurities arising from the earliest parental behavior vis a vis their children have powerful effects on a host of positive *and* negative outcomes in their offspring. These findings admit to no other recommendation than consistent parental monitoring accompanied by high warmth, which fosters children's self-disclosures to parents without fear of angry rebuke. Parent-child relations that follow this mode have been shown in a host of our and others' studies to result in secure, confident children and, later, adults. All other combinations involving monitoring and warmth do not foster security, and as such, are at best a second alternative, or worse. On the other hand, if we take the view that some insecurity can be energizing and perhaps sought out (see Fiedler & McCaughey, this volume), it becomes conceivable that some variation in parental relations with their children may also produce exemplary outcomes – and there are many such examples readily available. Of course, these outcomes perhaps came at a cost to both parent and child, and it is not certain what a more consistently cordial parent-child relation might have produced.

Conclusion

In this chapter, we have made the argument that insecurity has been an underlying thread that runs throughout the history of social psychology and is reflected today in our most influential theories. Theories of cognitive consistency presupposed the need to be right, to fit in, or to reduce inconsistencies among beliefs or behaviors. In theories of inter- and intra-group behavior, being a part of a group plays a major role in our self- and social identities, which affect who we think we are, and how we relate to others. The group provides cover and a rationale for beliefs and behaviors that may appear ridiculous, immoral, or inhuman to those outside the group. An underlying motive of contemporary social psychology derives from a need to resolve or forestall insecurity and, by doing so, establish a reality shared and reinforced by others who think and act as we do. We are excited to see where the field takes the study of insecurity in the future.

Note

1 Lifetime inhalant use of young respondents (8th-grade students, approximately 14 years of age) reported in 2021 in the Monitoring the Future study was 11.3%, a decline from the prior years (Johnston et al., 2022).

References

Abelson, R. P., Aronson, E., McGuire, W. J., Newcomb, T. M., Rosenberg, M. J., & Tannenbaum, P. H. (Eds.). (1968). *Theories of cognitive consistency: A sourcebook.* Rand-McNally.

Ainsworth, M. S. (1989). Attachments beyond infancy. *American Psychologist, 44*(4), 709–716. https://doi.org/10.1037/0003-066X.44.4.709

Ainsworth, M. D. S. (2010). Security and attachment. In R. Volpe (Ed.), *The secure child: Timeless lessons in parenting and childhood education.* (pp. 43–53). IAP Information Age Publishing. https://search.ebscohost.com/login.aspx?direct=true&AuthType=sso&db=psyh&AN=2010-01404-003&site=ehost-live&scope=site&custid=s8438901

Ainsworth, M. S., & Bowlby, J. (1991). An ethological approach to personality development. *American Psychologist, 46*(4), 333–341. https://doi.org/10.1037/0003-066X.46.4.333

Ajzen, I. (2012). The theory of planned behavior. In P. A. M. Van Lange, A. W. Kruglanski, & E. T. Higgins (Eds.), *Handbook of theories of social psychology., Vol. 1* (pp. 438–459). Sage Publications Ltd. https://doi.org/10.4135/9781446249215.n22

Ajzen, I., & Schmidt, P. (2020). Changing behavior using the theory of planned behavior. In M. S. Hagger, L. D. Cameron, K. Hamilton, N. Hankonen, & T. Lintunen (Eds.), *The handbook of behavior change.* (pp. 17–31). Cambridge University Press. https://search.ebscohost.com/login.aspx?direct=true&AuthType=sso&db=psyh&AN=2020-64356-002&site=ehost-live&scope=site&custid=s8438901

Albarracín, D., Johnson, B. T., Fishbein, M., & Muellerleile, P. A. (2001). Theories of reasoned action and planned behavior as models of condom use: A meta-analysis. *Psychological Bulletin, 127*(1), 142–161. https://doi.org/10.1037/0033-2909.127.1.142

Alvaro, E. M., & Crano, W. D. (1996). Cognitive responses to minority- or majority-based communications: Factors that underlie minority influence. *British Journal of Social Psychology, 35*(1), 105–121. https://doi.org/10.1111/j.2044-8309.1996.tb01086.x (Minority influences)

Alvaro, E. M., & Crano, W. D. (1997). Indirect minority influence: Evidence for leniency in source evaluation and counterargumentation. *Journal of Personality and Social Psychology, 72*, 949–964. https://doi.org/10.1037/0022-3514.72.5.949

Armitage, C. J., & Conner, M. (2000). Attitudinal ambivalence: A test of three key hypotheses. *Personality and Social Psychology Bulletin, 26*(11), 1421–1432. https://doi.org/10.1177/0146167200263009

Asch, S. E. (1952). *Social psychology.* Prentice-Hall, Inc. https://doi.org/10.1037/10025-000

Asch, S. E. (1956). Studies of independence and conformity: I. A minority of one against a unanimous majority. *Psychological Monographs: General and Applied, 70*(9), 1–70. https://doi.org/10.1037/h0093718

Asch, S. E. (1961). Issues in the study of social influences on judgment. In I. A. Berg & B. M. Bass (Eds.), *Conformity and deviation* (pp. 143–158). Harper and Brothers. https://doi.org/10.1037/11122-005

Bargh, J. A., Chaiken, S., Govender, R., & Pratto, F. (1992). The generality of the automatic attitude activation effect. *Journal of Personality and Social Psychology, 62*(6), 893–912. https://doi.org/10.1037/0022-3514.62.6.893

Bassili, J. N. (1996). Meta-judgmental versus operative indexes of psychological attributes: The case of measures of attitude strength. *Journal of Personality and Social Psychology, 71*(4), 637–653. https://doi.org/10.1037/0022-3514.71.4.637

Bassili, J. N., & Fletcher, J. F. (1991). Response-time measurement in survey research: A method for CATI and a new look at nonattitudes. *Public Opinion Quarterly, 55*(3), 331–346. https://doi.org/10.1086/269265

Bem, D. J. (1972). Self-perception theory development of self-perception theory was supported primarily by a grant from the National Science Foundation (GS 1452) awarded to the author during his tenure at Carnegie-Mellon University. In L. Berkowitz (Ed.), *Advances in Experimental Social Psychology* (Vol. 6, pp. 1–62). Academic Press. https://doi.org/https://doi.org/10.1016/S0065-2601(08)60024-6

Briñol, P., Petty, R. E., & Wheeler, S. C. (2006). Discrepancies between explicit and implicit self-concepts: Consequences for information processing. *Journal of Personality and Social Psychology, 91*(1), 154–170. https://doi.org/10.1037/0022-3514.91.1.154

Brown, J. S., & Farber, I. E. (1951). Emotions conceptualized as intervening variables—with suggestions toward a theory of frustration. *Psychological Bulletin, 48*(6), 465–495. https://doi.org/10.1037/h0058839

Butera, F., Falomir-Pichastor, J. M., Mugny, G., & Quiamzade, A. (2017). Minority influence. In S. G. Harkins, K. Williams, & J. M. Berger (Eds.), *The Oxford handbook of social influence* (pp. 317–338). Oxford University Press.

Buunk, A. P., & Gibbons, F. X. (2007). Social comparison: The end of a theory and the emergence of a field. *Organizational Behavior and Human Decision Processes, 102*(1), 3–21. https://doi.org/https://doi.org/10.1016/j.obhdp.2006.09.007

Campbell, D. T. (1963). Social attitudes and other acquired behavioral dispositions. In S. Koch (Ed.), *Psychology: A study of a science. Study II. Empirical substructure and relations with other sciences. Volume 6. Investigations of man as socius: Their place in psychology and the social sciences.* (pp. 94–172). McGraw-Hill. https://doi.org/10.1037/10590-003

Campbell, D. T. (1990). Asch's moral epistemology for socially shared knowledge. In I. Rock (Ed.), *The legacy of Solomon Asch: Essays in cognition and social psychology.* (pp. 39–55). Lawrence Erlbaum Associates, Inc. http://search.ebscohost.com/login.aspx?direct=true&AuthType=sso&db=psyh&AN=1990-97618-003&site=ehost-live&scope=site&custid=s8438901

Clark, J. K., Wegener, D. T., & Fabrigar, L. R. (2008). Attitudinal ambivalence and message-based persuasion: Motivated processing of proattitudinal information and avoidance of counterattitudinal information. *Personality and Social Psychology Bulletin, 34*(4), 565–577. https://doi.org/10.1177/0146167207312527

Coleman, M., Donaldson, C. D., Crano, W. D., Pike, J. R., & Stacy, A. W. (2021). Associations between family and peer e-cigarette use with adolescent tobacco and marijuana usage: A longitudinal PATH analytic approach. *Nicotine & Tobacco Research, 23*(5), 849–855. https://search.ebscohost.com/login.aspx?direct=true&AuthType=sso&db=psyh&AN=2022-14076-011&site=ehost-live&scope=site&custid=s8438901

Converse, P. E. (1964). The nature of belief systems in mass publics. In D. Apter (Ed.), *Ideology and discontent* (pp. 206–261). Free Press.

Converse, P. E. (1970). Attitudes and nonattitudes: Continuation of a dialog. In E. R. Tufte (Ed.), *The quantitative analysis of social problems* (pp. 168–189). Addison-Wesley.

Cooper, J., Zanna, M. P., & Taves, P. A. (1978). Arousal as a necessary condition for attitude change following induced compliance. *Journal of Personality and Social Psychology, 36*(10), 1101–1106. https://doi.org/10.1037/0022-3514.36.10.1101

Crano, W. D. (1970). Effects of sex, response order, and expertise in conformity: A dispositional approach. *Sociometry, 33*, 239–252. https://doi.org/10.2307/2786155

Crano, W. D. (1995). Attitude strength and vested interest. In R. E. Petty & J. A. Krosnick (Eds.), *Attitude strength: Antecedents and consequences* (pp. 131–157). Lawrence Erlbaum Associates, Inc. http://search.ebscohost.com/login.aspx?direct=true&AuthType=sso&db=psyh&AN=1995-98997-006&site=ehost-live&scope=site&custid=s8438901

Crano, W. D. (1997). Vested interest, symbolic politics, and attitude-behavior consistency. *Journal of Personality and Social Psychology, 72*(3), 485–491. https://doi.org/10.1037/0022-3514.72.3.485

Crano, W. D. (2010). Majority and minority influence and information processing: A theoretical and methodological analysis. In R. Martin & M. Hewstone (Eds.), *Minority influence and innovation: Antecedents, processes, and consequences* (pp. 53–77). Psychology Press.

Crano, W. D. (2012). *The rules of influence.* St. Martin's Press.

Crano, W. D., & Alvaro, E. M. (2014). Social factors that affect the processing of minority-sourced persuasive communications. In J. P. Forgas, Vincze, O., & Laszlo, J. (Eds.), *Social cognition and communication* (pp. 297–312). New York: Psychology Press.

Crano, W. D., & Alvaro, E. M. (1998a). The context/comparison model of social influence: Mechanisms, structure, and linkages that underlie indirect attitude change. In W. Stroebe & M. Hewstone (Eds.), *European review of social psychology,* Vol. 8. (pp. 175–202). John Wiley & Sons Inc. http://search.ebscohost.com/login.aspx?direct=true&db=psyh&AN=1998-06032-006&site=ehost-live

Crano, W. D., & Alvaro, E. M. (1998b). Indirect minority influence: The leniency contract revisited. *Group Processes & Intergroup Relations, 1*(2), 99–115. https://doi.org/10.1177/1368430298012001

Crano, W. D., Donaldson, C. D., Siegel, J. T., Alvaro, E. M., & O'Brien, E. K. (2019). Selective invalidation of ambivalent pro-marijuana attitude components. *Addictive Behaviors, 97,* 77–83. https://doi.org/10.1016/j.addbeh.2019.05.020

Crano, W. D., Gilbert, C., Alvaro, E. M., & Siegel, J. T. (2008). Enhancing prediction of inhalant abuse risk in samples of early adolescents: A secondary analysis. *Addictive Behaviors, 33*(7), 895–905. https://doi.org/10.1016/j.addbeh.2008.02.006

Crano, W. D., & Hannula-Bral, K. A. (1994). Context/categorization model of social influence: Minority and majority influence in the formation of a novel response norm. *Journal of Experimental Social Psychology, 30*(3), 247–276. https://doi.org/10.1006/jesp.1994.1012

Crano, W. D., & Prislin, R. (2006). Attitudes and persuasion. *Annual Review of Psychology, 57,* 345–374. https://doi.org/10.1146/annurev.psych.57.102904.190034

Crano, W. D., & Prislin, R. (2008). *Attitudes and attitude change.* Psychology Press. http://search.ebscohost.com/login.aspx?direct=true&db=psyh&AN=2008-09973-000&site=ehost-live

Crano, W. D., & Seyranian, V. (2007). Majority and minority influence. *Social and Personality Psychology Compass, 1,* 572–589. https://doi.org/10.1111/j.1751-9004.2007.00028.x

Crano, W. D., & Seyranian, V. (2009). How minorities prevail: The contxt-comparison-leniency contract model. *Journal of Social Issues, 65,* 335–363. Special Issue: Michele Marie Grossman Alexander.doi.org.ccl.idm.oclc.org/10.1111/j.1540-4560.2009.01603.x

Dempsey, R. C., McAlaney, J., & Bewick, B. M. (2018). A critical appraisal of the social norms approach as an interventional strategy for health-related behavior and attitude change. *Frontiers in Psychology, 9.* https://doi.org/10.3389/fpsyg.2018.02180

Donaldson, C. D., Alvaro, E. M., Ruybal, A. L., Coleman, M., Siegel, J. T., & Crano, W. D. (2021). A rebuttal-based social norms-tailored cannabis intervention for at-risk adolescents. *Prevention Science.* https://doi-org.ccl.idm.oclc.org/10.1007/s11121-021-01224-9

Donaldson, C. D., Handren, L. M., & Crano, W. D. (2016). The enduring impact of parents' monitoring, warmth, expectancies, and alcohol use on their children's future binge drinking and arrests: A longitudinal analysis. *Prevention Science, 17,* 606–614.

Donaldson, C. D., Siegel, J. T., & Crano, W. D. (2020). Preventing college student nonmedical prescription stimulant use: Development of vested interest theory-based persuasive messages. *Addictive Behaviors, 108*. https://doi.org/10.1016/j.addbeh.2020.106440

Feick, L. F. (1989). Latent class analysis of survey questions that include don't know responses. *Public Opinion Quarterly, 53*(4), 525–547. https://doi.org/10.1086/269170

Festinger, L. (1954). A theory of social comparison processes. *Human Relations, 7*, 117–140. https://doi.org/10.1177/001872675400700202

Festinger, L. (1957). *A theory of cognitive dissonance.* Stanford University Press. http://search.ebscohost.com/login.aspx?direct=true&db=psyh&AN=1993-97948-000&site=ehost-live

Festinger, L., Riecken, H. W., & Schachter, S. (1956). *When prophecy fails.* University of Minnesota Press. https://doi.org/10.1037/10030-000

Festinger, L., Riecken, H. W., & Schachter, S. (1964). *When prophecy fails: A social and psychological study of a modern group that predicted the destruction of the world.* Harper Torchbooks. https://search.ebscohost.com/login.aspx?direct=true&AuthType=sso&db=psyh&AN=1965-01410-000&site=ehost-live&scope=site&custid=s8438901

Fiedler, K., Hütter, M., Schott, M., & Kutzner, F. (2019). Metacognitive myopia and the overutilization of misleading advice. *Journal of Behavioral Decision Making, 32*(3), 317–333. https://doi.org/10.1002/bdm.2109

Fishbein, M., & Ajzen, I. (1975). *Belief, attitude, intention and behavior.* Addison-Wesley.

Fishbein, M., & Ajzen, I. (2010). *Predicting and changing behavior: The reasoned action approach.* Psychology Press. http://search.ebscohost.com/login.aspx?direct=true&AuthType=sso&db=psyh&AN=2009-17267-000&site=ehost-live&scope=site&custid=s8438901

Gawronski, B. (2012). Back to the future of dissonance theory: Cognitive consistency as a core motive. *Social Cognition, 30*(6), 652–668. https://doi.org/10.1521/soco.2012.30.6.652 (Threat-compensation in social psychology: Is there a core motivation?)

Gawronski, B., & Brannon, S. M. (2019). What is cognitive consistency, and why does it matter? In E. Harmon-Jones (Ed.), *Cognitive dissonance: Reexamining a pivotal theory in psychology.*, 2nd ed. (pp. 91–116). American Psychological Association. https://doi.org/10.1037/0000135-005

Goethals, G. R. (1986). Social comparison theory: Psychology from the lost and found. *Personality and Social Psychology Bulletin, 12*(3), 261–278. https://doi.org/10.1177/0146167286123001

Goethals, G. R., & Darley, J. M. (1977). Social comparison theory: An attributional approach. In J. M. Suls & R. L. Miller (Eds.), *Social comparison processes: Theoretical and empirical perspectives* (pp. 1977). Hemisphere.

Gorenflo, D. W., & Crano, W. D. (1989). Judgmental subjectivity/objectivity and locus of choice in social comparison. *Journal of Personality and Social Psychology, 57*(4), 605–614. https://doi.org/10.1037/0022-3514.57.4.605

Guilamo-Ramos, V., Jaccard, J., Turrisi, R., & Johansson, M. (2005). Parental and school correlates of binge drinking among middle school students. *American Journal of Public Health, 95*, 894–899.

Hales, A. H., & Williams, K. D. (2021). Social ostracism: Theoretical foundations and basic principles. In P. A. M. Van Lange, E. T. Higgins, & A. W. Kruglanski (Eds.), *Social psychology: Handbook of basic principles.*, 3rd ed. (pp. 337–349). The Guilford Press. https://search.ebscohost.

com/login.aspx?direct=true&AuthType=sso&db=psyh&AN=2020-23032-017&site=e-host-live&scope=site&custid=s8438901

Handren, L. M., Donaldson, C. D., & Crano, W. D. (2016). Adolescent alcohol use: Protective and predictive parent, peer, and self-related factors. *Prevention Science, 17*, 862–871.

Hartgerink, C. H. J., van Beest, I., Wicherts, J. M., & Williams, K. D. (2015). The ordinal effects of ostracism: A meta-analysis of 120 cyberball studies. *PLOS One, 10*(5). https://doi.org/10.1371/journal.pone.0127002

Hass, R. G., Katz, I., Rizzo, N., Bailey, J., & Moore, L. (1992). When racial ambivalence evokes negative affect, using a disguised measure of mood. *Personality and Social Psychology Bulletin, 18*(6), 786–797. https://doi.org/10.1177/0146167292186015

Hodson, G., Maio, G. R., & Esses, V. M. (2001). The role of attitudinal ambivalence in susceptibility to consensus information. *Basic and Applied Social Psychology, 23*(3), 197–205. https://doi.org/10.1207/153248301750433678

Hogg, M. A. (2006). Social identity theory. In P. J. Burke (Ed.), *Contemporary social psychological theories* (pp. 111–136). Stanford University Press. https://search.ebscohost.com/login.aspx?direct=true&AuthType=sso&db=psyh&AN=2006-07094-006&site=ehost-live&scope=site&custid=s8438901

Hogg, M. A., & Reid, S. A. (2006). Social identity, self-categorization, and the communication of group norms. *Communication Theory, 16*(1), 7–30. https://doi.org/10.1111/j.1468-2885.2006.00003.x

Hogg, M. A., & Turner, J. C. (1987). Social identity and conformity: A theory of referent information influence. In W. Doise & S. Moscovici (Eds.), *Current issues in European social psychology, Vol. 2* (pp. 139–182). Cambridge University Press Editions de la Maison des Sciences de l'Homme. https://search.ebscohost.com/login.aspx?direct=true&AuthType=sso&db=psyh&AN=1987-98772-003&site=ehost-live&scope=site&custid=s8438901

Hogg, M. A. (2012). Uncertainty-identity theory. In P. A. M. Van Lange, A. W. Kruglanski, & E. T. Higgins (Eds.), *Handbook of theories of social psychology., Vol. 2* (pp. 62–80). Sage Publications Ltd. https://doi.org/10.4135/9781446249222.n29

Hohman, Z. P., Crano, W. D., & Niedbala, E. M. (2016). Attitude ambivalence, social norms, and behavioral intentions: Developing effective antitobacco persuasive communications. *Psychology of Addictive Behaviors, 30*(2), 209–219. https://doi.org/10.1037/adb0000126

Hohman, Z. P., Peabody, J., & Neighbors, C. (2022). Personalized norm feedback and ambivalence. *Addictive Behaviors Reports, 16*, 1–5. https://doi.org/10.1016/j.abrep.2022.100461

Hohman, Z. P., Crano, W. D., Siegel, J. T., & Alvaro, E. M. (2014). Attitude ambivalence, friend norms, and adolescent drug use. *Prevention Science, 15*(1), 65–74. https://doi.org/10.1007/s11121-013-0368-8

Hohman, Z. P., Keene, J. R., Harris, B. N., Niedbala, E. M., & Berke, C. K. (2017). A biopsychological model of anti-drug PSA processing: Developing effective persuasive messages. *Prevention Science, 18*(8), 1006–1016. https://doi.org/10.1007/s11121-017-0836-7

Hovland, C. I., Janis, I. L., & Kelley, H. H. (1953). *Communication and persuasion: Psychological studies of opinion change*. Yale University Press. http://search.ebscohost.com/login.aspx?direct=true&db=psyh&AN=1964-00891-000&site=ehost-live

Hütter, M., & Fiedler, K. (2019). Advice taking under uncertainty: The impact of genuine advice versus arbitrary anchors on judgment. *Journal of Experimental Social Psychology, 85*. https://doi.org/10.1016/j.jesp.2019.103829

James, W. (1890). *The principles of psychology, Vol I*. Henry Holt and Co. https://doi.org/10.1037/10538-000

Johnston, L. D., Miech, R. A., O'Malley, P. M., Bachman, J. G., Schulenberg, J. E., & Patrick, M. E. (2022). *Monitoring the Future national surveyresults on drug use 1975–2021: Overview, key findings on adolescent drug use*. Institute for Social Research, University of Michigan.

Kaplan, K. J. (1972). On the ambivalence-indifference problem in attitude theory and measurement: A suggested modification of the semantic differential technique. *Psychological Bulletin, 77*(5), 361–372. https://doi.org/10.1037/h0032590

Koriat, A., Undorf, M., & Schwarz, N. (2020). The effects of group conformity on the prototypical majority effect for confidence and response latency. *Social Cognition, 38*(5), 447–469. https://doi.org/10.1521/soco.2020.38.5.447

Kerr, M., & Stattin, H. (2000). What parents know, how they know it, and several forms of adolescent adjustment: Further support for a reinterpretation of monitoring. *Developmental Psychology, 36*(3), 366–380. https://doi.org/10.1037/0012-1649.36.3.366

Kerr, M., & Stattin, H. (2003). Straw men, untested assumptions, and bi-directional models: A response to Capaldi and Brody. In A. C. Crouter & A. Booth (Eds.), *Children's influence on family dynamics: The neglected side of family relationships*. (pp. 181–187). Lawrence Erlbaum Associates Publishers. https://search.ebscohost.com/login.aspx?direct=true&AuthType=sso&db=psyh&AN=2003-02704-013&site=ehost-live&scope=site&custid=s8438901

Krosnick, J. A. (1989). Attitude importance and attitude accessibility. *Personality and Social Psychology Bulletin, 15*(3), 297–308. https://doi.org/10.1177/0146167289153002

Kuhn, T. S. (1996). *The structure of scientific revolutions., 3rd ed*. University of Chicago Press. https://doi.org/10.7208/chicago/9780226458106.001.0001

Maass, A., & Clark, R. D. (1984). Hidden impact of minorities: Fifteen years of minority influence research. *Psychological Bulletin, 95*(3), 428–450. https://doi.org/10.1037/0033-2909.95.3.428

Maass, A., Clark, R. D., & Haberkorn, G. (1982). The effects of differential ascribed category membership and norms on minority influence. *European Journal of Social Psychology, 12*, 89–104. https://doi.org/10.1002/ejsp.2420120107

Mackie, D. M. (1987). Systematic and nonsystematic processing of majority and minority persuasive communications. *Journal of Personality and Social Psychology, 53*(1), 41–52. https://doi.org/10.1037/0022-3514.53.1.41

Maio, G. R., Bell, D. W., & Esses, V. M. (1996). Ambivalence in persuasion: The processing of messages about immigrant groups. *Journal of Experimental Social Psychology, 32*(6), 513–536. https://doi.org/10.1006/jesp.1996.0023

Maloney, E. K., Lapinski, M. K., & Witte, K. (2011). Fear appeals and persuasion: A review and update of the extended parallel process model. *Social and Personality Psychology Compass, 5*(4), 206–219. https://doi.org/10.1111/j.1751-9004.2011.00341.x

Martin, R., & Hewstone, M. (2008). Majority versus minority influence, message processing and attitude change: The source-context-elaboration model. In M. P. Zanna (Ed.), *Advances in experimental social psychology* (Vol. 49, pp. 238–326). Academic.

Martin, R., Hewstone, M., & Martin, P. Y. (2008). Majority versus minority influence: The role of message processing in determining resistance to counter-persuasion. *European Journal of Social Psychology, 38*(1), 16–34. https://doi.org/10.1002/ejsp.426

Miech, R. A., Johnston, L. D., O'Malley, P. M., Bachman, J. G., Schulenberg, J. E., & Patrick, M. E. (2021). *Monitoring the Future national survey results on drug use, 1975–2020: Volume I, Secondary school students*. Ann Arbor: Institute for Social Research, The University of Michigan. Available at http://monitoringthefuture.org/pubs.html#monographs

Milgram, S. (1963). Behavioral Study of obedience. *The Journal of Abnormal and Social Psychology, 67*(4), 371–378. https://doi.org/10.1037/h0040525

Milgram, S. (1975). *Obedience to authority: An experimental view*. Harper & Row.

Miller, M. B., Leffingwell, T., Claborn, K., Meier, E., Walters, S., & Neighbors, C. (2013). Personalized feedback interventions for college alcohol misuse: An update of Walters & Neighbors (2005). *Psychology of Addictive Behaviors, 27*(4), 909–920. https://doi.org/10.1037/a0031174

Moscovici, S., & Personnaz, B. (1980). Studies in social influence: V. Minority influence and conversion behavior in a perceptual task. *Journal of Experimental Social Psychology, 16*, 270–282. https://doi.org/10.1016/0022-1031(80)90070-0

Mugny, G. (1984). Compliance, conversion and the Asch paradigm. *European Journal of Social Psychology, 14*(4), 353–368. https://doi.org/10.1002/ejsp.2420140402

Neighbors, C., Larimer, M. E., & Lewis, M. A. (2004). Targeting misperceptions of descriptive drinking norms: Efficacy of a computer-delivered personalized normative feedback intervention. *Journal of Consulting and Clinical Psychology, 72*(3), 434–447. https://doi.org/10.1037/0022-006X.72.3.434

Nesci, D. A. (2018). *Revisiting Jonestown: An interdisciplinary study of cults*. Lexington Books/Rowman & Littlefield. https://search.ebscohost.com/login.aspx?direct=true&AuthType=sso&db=psyh&AN=2018-03509-000&site=ehost-live&scope=site&custid=s8438901

Newby-Clark, I. R., McGregor, I., & Zanna, M. P. (2002). Thinking and caring about cognitive inconsistency: When and for whom does attitudinal ambivalence feel uncomfortable? *Journal of Personality and Social Psychology, 82*(2), 157–166. https://doi.org/10.1037/0022-3514.82.2.157

Pérez, J. A., & Mugny, G. (1987). Paradoxical effects of categorization in minority influence: When being an outgroup is an advantage. *European Journal of Social Psychology, 17*, 157–169. https://doi.org/10.1002/ejsp.2420170204

Pettigrew, T. F. (2018). The emergence of contextual social psychology. *Personality and Social Psychology Bulletin, 44*(7), 963–971. https://doi.org/10.1177/0146167218756033

Petty, R. E., & Briñol, P. (2009). Implicit ambivalence: A meta-cognitive approach. In R. E. Petty, R. H. Fazio, & P. Briñol (Eds.), *Attitudes: Insights from the new implicit measures* (pp. 119–163). Psychology Press.

Priester, J. R. (2002). Sex, drugs, and attitudinal ambivalence: How feelings of evaluative tension influence alcohol use and safe sex behaviors. In W. D. Crano & M. Burgoon (Eds.), *Mass media and drug prevention: Classic and contemporary theories and research*. (pp. 145–162). Lawrence Erlbaum Associates Publishers. https://search.ebscohost.com/login.aspx?direct=true&AuthType=sso&db=psyh&AN=2001-05966-006&site=ehost-live&scope=site&custid=s8438901

Prislin, R., & Crano, W. D. (2022). Sociability: A foundational construct in social psychology. In J. P. Forgas, W. D. Crano, & K. Fiedler (Eds.), *The psychology of sociability: Understanding human attachment*. Routledge.

Rise, J., Sheeran, P., & Hukkelberg, S. (2010). The role of self-identity in the theory of planned behavior: A meta-analysis. *Journal of Applied Social Psychology, 40*(5), 1085–1105. https://doi.org/10.1111/j.1559-1816.2010.00611.x

Rudert, S. C., Greifeneder, R., & Williams, K. D. (2019). *Current directions in ostracism, social exclusion, and rejection research.* Routledge/Taylor & Francis Group. https://doi.org/10.4324/9781351255912

Rydell, R. J., McConnell, A. R., & Mackie, D. M. (2008). Consequences of discrepant explicit and implicit attitudes: Cognitive dissonance and increased information processing. *Journal of Experimental Social Psychology, 44*(6), 1526–1532. https://doi.org/10.1016/j.jesp.2008.07.006

Schulze, R., & Wittmann, W. W. (2003). A meta-analysis of the theory of reasoned action and the theory of planned behavior: The principle of compatibility and multidimensionality of beliefs as moderators. In R. Schulze, H. Holling, & D. Böhning (Eds.), *Meta-analysis: New developments and applications in medical and social sciences* (pp. 219–250). Hogrefe & Huber Publishers.

Sheeran, P., & Taylor, S. (1999). Predicting intentions to use condoms: A meta-analysis and comparison of the theories of reasoned action and planned behavior. *Journal of Applied Social Psychology, 29*(8), 1624–1675. https://doi.org/10.1111/j.1559-1816.1999.tb02045.x

Shin, S., Edwards, E., Heeren, T., & Amodeo, M. (2009). Relationship between multiple forms of maltreatment by a parent or guardian and adolescent alcohol use. *American Journal on Addictions, 18,* 226–234.

Sniderman, P. M., Tetlock, P. E., & Elms, L. (2001). Public opinion and democratic politics: The problem of nonattitudes and the social construction of political judgment. In J. H. Kuklinski (Ed.), *Citizens and politics: Perspectives from political psychology.* (pp. 254–288). Cambridge University Press. https://doi.org/10.1017/CBO9780511896941.013

Sparks, P., Harris, P. R., & Lockwood, N. (2004). Predictors and predictive effects of ambivalence. *British Journal of Social Psychology, 43*(3), 371–383. https://doi.org/10.1348/0144666042037980

Stattin, H., & Kerr, M. (2000). Parental monitoring: A reinterpretation. *Child Development, 71*(4), 1072–1085. https://doi.org/10.1111/1467-8624.00210

Stattin, H., Persson, S., Burk, W. J., & Kerr, M. (2011). Adolescents' perceptions of the democratic functioning in their families. *European Psychologist, 16*(1), 32–42. https://doi.org/10.1027/1016-9040/a000039 (Special Section: Family Dynamics)

Steinmetz, H., Knappstein, M., Ajzen, I., Schmidt, P., & Kabst, R. (2016). How effective are behavior change interventions based on the theory of planned behavior? A three-level meta-analysis. *Zeitschrift für Psychologie, 224*(3), 216–233. https://doi.org/10.1027/2151-2604/a000255 (Hotspots in Psychology 2016)

Suls, J., Collins, R. L., & Wheeler, L. (2020). *Social comparison, judgment, and behavior.* Oxford University Press. https://doi.org/10.1093/oso/9780190629113.001.0001

Suls, J., & Wheeler, L. (2012). Social comparison theory. In P. A. M. Van Lange, A. W. Kruglanski, & E. T. Higgins (Eds.), *Handbook of theories of social psychology (Vol 1)* (pp. 460–482). Sage Publications Ltd. http://search.ebscohost.com/login.aspx?direct=true&db=psyh&AN=2011-21800-022&site=ehost-live&scope=site

Suls, J. M., & Miller, R. L. (1977). *Social comparison processes: Theoretical and empirical perspectives.* Hemisphere. https://search.ebscohost.com/login.aspx?direct=true&

AuthType=sso&db=psyh&AN=1979-03513-000&site=ehost-live&scope=site&custid=s8438901

Turner, J. C., Hogg, M. A., Oakes, P. J., Reicher, S. D., & Wetherell, M. S. (1987). *Rediscovering the social group: A self-categorization theory.* Blackwell.

Tyson, M., Covey, J., & Rosenthal, H. E. S. (2014). Theory of planned behavior interventions for reducing heterosexual risk behaviors: A meta-analysis. *Health Psychology, 33*(12), 1454–1467. https://doi.org/10.1037/hea0000047, 10.1037/hea0000047.supp (Supplemental)

Ulman, R. B., & Abse, D. W. (1983). The group psychology of mass madness: Jonestown. *Political Psychology, 4*(4), 637–661. https://doi.org/10.2307/3791059

Wesselmann, E. D., Williams, K. D., Ren, D., & Hales, A. H. (2021). Ostracism and solitude. In R. J. Coplan, J. C. Bowker, & L. J. Nelson (Eds.), *The handbook of solitude: Psychological perspectives on social isolation, social withdrawal, and being alone.*, 2nd ed. (pp. 209–223). Wiley Blackwell. https://doi.org/10.1002/9781119576457.ch15

Wright, J. D., & Esses, V. M. (2019). It's security, stupid! Voters' perceptions of immigrants as a security risk predicted support for Donald Trump in the 2016 US presidential election. *Journal of Applied Social Psychology, 49*(1), 36–49. https://doi.org/10.1111/jasp.12563

Wundt, W. M. (1900). *Volkerpsychologie. Eine Untersuchung der Entwickelungsges etze von Sprache, Mythus und Sitte.* Verlag von Wilhelm Engelmann.

Wundt, W. M. (1907). *Outlines of Psychology (3nd Revision, translated by C. H. Judd).* W. Engelmann.

Wundt, W. M., & Schaub, E. L. (1916). *Elements of folk psychology: Outlines of a psychological history of the development of mankind.* George Allen & Unwin. https://doi.org/10.1037/13042-000

8
SELF-HANDICAPPING IN THE FACE OF UNCERTAINTY

The Paradox That Most Certainly Is

Edward R. Hirt, Samantha L. Heiman, and Julie R. Eyink
INDIANA UNIVERSITY-BLOOMINGTON, USA

Sean McCrea
UNIVERSITY OF WYOMING, USA

Abstract

Thirty years ago, Higgins, Snyder, & Berglas published an influential edited book in which they described self-handicapping as "the paradox that isn't." They argued that while self-handicapping may appear to be a self-defeating behavior destined to bring about the very task failure that the handicapper fears, the attributional benefits accrued from this strategy far outweigh the potential performance costs. However, if self-handicapping is so beneficial, then why doesn't everyone habitually use this self-protective strategy? In this chapter, we explore the role of uncertainty in increasing and decreasing the likelihood of self-handicapping behaviors at multiple stages, including (a) the role of concern about potential task failure, (b) the consideration of the believability of potential handicaps, and (c) projections about the reactions of potential audience members. Taking into account the plethora of ways that uncertainty can both promote and undermine the decision to self-handicap provides a fuller picture of the true paradox that exists when individuals opt to employ this strategy as opposed to other potential self-protective strategies within the pantheon of the "self-zoo".

Imagine two men, Pablo and Ryan, who are finalists for the same job. The night before his interview, Pablo reviews information about the company and

DOI: 10.4324/9781003317623-10

practices his responses to questions. He goes to sleep early so that he is well rested for the next day. On the other hand, the night before his interview, Ryan goes out partying with friends late into the night. Ryan is groggy and hung over the next day, and it costs him dearly as he blows his interview and doesn't get the job.

Why would Ryan party and stay out late before his interview? It seems almost as if he was purposely sabotaging his chances to succeed. As social psychologists, we know that scenarios such as this one, paradoxical as they might seem, occur with some regularity. The label we apply to instances of self-sabotaging behavior when people experience insecurity or uncertainty is **self-handicapping** (Jones & Berglas, 1978). Originally coined by Berglas and Jones (1978), self-handicapping refers to a class of self-protective behaviors whereby an individual "attempts to reduce a threat to esteem by actively seeking or creating inhibitory factors that interfere with performance and thus provide a persuasive causal explanation for failure" (Arkin & Baumgardner, 1985, p. 170). By prospectively acquiring or creating some obstacle or handicap for success, the individual has a potential excuse for poor performance. If the person performs poorly, they can blame the failure on the handicap rather than on a lack of ability (e.g., "I was too sick to perform well"). Conversely, if the person performs well, they can attribute their success to their exceptional ability since they were able to overcome the debilitating handicap. Thus, self-handicapping affords the individual the opportunity to reduce insecurity by augmenting attributions of ability in the event of task success as well as to discount attributions to lack of ability following task failure (cf. Kelley, 1973). This attributional win-win is offset by the increased likelihood of task failure caused by the handicap. Therefore, self-handicapping represents a tradeoff of poorer performance for the short-term maintenance of a desired self-image (Baumeister & Scher, 1988; Rhodewalt, 2008).

The literature shows individuals use a wide range of self-handicaps, including ingesting alcohol (e.g. Tucker et al., 1981; Higgins & Harris, 1988) or other performance-inhibiting drugs (e.g. Berglas & Jones, 1978; Kolditz & Arkin, 1982), withdrawing effort (e.g. Harris & Snyder, 1986; Tice, 1991; Hirt et al., 1991), listening to distracting music (e.g. Shepperd & Arkin, 1989; Rhodewalt & Davison 1986), and reporting high levels of stress (Hirt et al., 1991), test anxiety (Smith et al., 1982; Greenberg et al., 1984), illness (Smith et al., 1983), bad mood (Baumgardner et al., 1985), debilitating testing conditions such as excessive noise or poor lighting (Shepperd & Arkin, 1989), and even prejudice (Eyink et al., in prep). Indeed, it is impressive how many different behaviors have been identified under the umbrella of self-handicapping in the face of insecurity. But not all self-handicaps are created equal! We can separate handicaps into two fundamentally different classes: (1) **behavioral self-handicaps**, which refer to active forms of self-sabotaging behavior, such as alcohol use or effort withdrawal; and (2)

claimed self-handicaps, which refer to reports of debilitating circumstances, such as illness or stress (Arkin & Baumgardner, 1985; Leary & Shepperd, 1986; Hirt et al., 1991).

Given the attributional tradeoffs inherent in both behavioral and claimed self-handicapping, it makes sense why Higgins et al. (1990) described handicapping as "the paradox that isn't." Specifically, they argued that while self-handicapping may appear to be self-defeating, the attributional benefits accrued from this strategy far outweigh the potential performance costs. Indeed, the self-protective benefits of this strategy create a situation where you are covered whether you succeed or fail! Thus, a natural question remains: Why then doesn't everyone self-handicap when feeling insecure? In this chapter, we attempt to answer that question. First, we explore the roots from which this strategy emanates, focusing primarily on the role of uncertainty in motivating its employment. While there is a strong consensus that uncertainty about performance and concern about potential task failure serve as a powerful motivator of this strategy, we show that there are many ways in which uncertainty can also deter its use. Specifically, we'll discuss how uncertainty regarding the believability of potential handicaps and the reactions of potential audience members can discourage the use of self-handicapping in response to a threatening upcoming performance. Taking into account the plethora of ways that uncertainty can both promote and undermine the decision to self-handicap provides a fuller picture of the true paradox that exists when individuals opt to employ this self-protective strategy (see also Cooper & Pearce, this volume).

The Roots of Self-handicapping

Despite the many avenues by which people can self-handicap, we have yet to explain why some individuals opt to handicap in performance situations while others do not. Early research focused on identifying key individual differences that elicit this behavior. A logical candidate was self-esteem, and many assumed that low-self-esteem individuals were the ones who engaged in self-handicapping. However, Harris and Snyder (1986) identified that it is the **uncertainty** of one's self-esteem and the resulting insecurity from that uncertainty that motivates self-handicapping actions (Harris & Snyder, 1986). Indeed, individuals uncertain in their self-esteem are more likely to engage in self-handicapping (Newman & Wadas, 1997), as well as in other motivated strategies for managing insecurity (see also Crano & Hohman; Hirschberger; Hogg & Gaffney, this volume). Relatedly, Jones and Rhodewalt developed the Self-Handicapping Scale (SHS), a face-valid, 25-item self-report measure of individuals' chronic tendencies to make excuses and use self-handicaps. Many studies have utilized this scale and found that it reliably identifies those individuals most likely to self-handicap (see Rhodewalt, 1990, for a review).

But as is often the case within social psychology, researchers have examined both personal and situational antecedents of self-handicapping behavior. One of the most robust situational factors shown to encourage self-handicapping is **non-contingent success feedback** (Berglas & Jones, 1978). In these paradigms, participants complete unsolvable problems and afterwards receive bogus success feedback. They then perform a second, similar task under the expectation that they should be able to replicate their earlier success. This situation evokes a great amount of uncertainty on the part of participants, who have no idea how they performed so well on the initial task and feel ill-equipped to reproduce that same level of success on the subsequent task. This leads the self-handicapper to entertain the genuine possibility of failure and consider what that would mean for their self-concept. Research consistently demonstrates that non-contingent success promotes self-handicapping behavior (e.g., Jones & Berglas, 1978; Greenberg et al., 1984; Tucker et al., 1981). This feedback by itself is not sufficient to produce self-handicapping – the task must also be important and personally relevant (e.g., Shepperd & Arkin, 1989; Tice, 1991; Rhodewalt et al., 1984; Smith et al., 1983). On tasks that are not deemed highly self-relevant, individuals experiencing great uncertainty do not go to the trouble of self-handicapping.

But clearly, uncertainty is at the heart of these aforementioned personal and situational variables. The clearest evidence demarcating the role of uncertainty in promoting self-handicapping behavior comes from studies demonstrating the mediating role of **evaluative concern** on self-handicapping. Hirt et al. (2000) created a three-item measure ($\alpha = .70$; "I am uncertain about how well I will do," "I am confident that I will perform well (reverse scored)," and "I expect to do poorly on this test") and found it to mediate self-handicapping behaviors. Later studies have identified other antecedent variables that increase both evaluative concern and subsequent self-handicapping behavior. For instance, Hirt et al. (2000) found that public self-focus increased self-handicapping by elevating evaluative concern among high self-handicapping men. Subsequently, Hendrix and Hirt (2009) illustrated that a prevention focus likewise motivated greater self-handicapping by inducing significantly higher levels of evaluative concern. Conversely, interventions to minimize uncertainty (i.e., forming an if-then plan to ignore worries and tell yourself, "I can do it!" reduce subsequent self-handicapping (Thürmer et al., 2013).

Based on the literature, it seems clear that many variables previously identified as correlating with self-handicapping may also do so through evaluative concern. For instance, Rhodewalt (1994) reported a link between handicapping and theories of intelligence (cf. Dweck & Leggett, 1988). He argued that fixed theorists, who believe failure is indicative of a lack of innate intelligence, would experience greater self-threat when confronted with an important and challenging performance and thus be more prone to self-handicap. This argument clearly hinges on evaluative concern – the fixed theorist will experience more anxiety about meeting performance standards and will therefore be more likely to handicap. Relatedly, self-handicapping is associated with the pursuit of performance goals over mastery goals (Elliot et al., 2006). Individuals with a performance goal focus on achieving particular outcomes (e.g., earning an A), while those with mastery goals concentrate on learning over time. Performance goals increase self-handicapping because the threat of potential failure to reach the desired outcome generates considerable anxiety and evaluative concern, leading to self-protective behavior. Work by Blascovich and Mendes (2001) has illustrated the many negative consequences of framing performances as threats rather than challenges.

Members of marginalized groups (e.g., low perceived SES) may resort to self-handicapping due to evaluative concern stemming from a low sense of belonging (Spencer et al., 2016; Wondra & McCrea, 2022; see also Hogg & Gaffney, this volume). Finally, there is a relationship between locus of control and self-handicapping (Arazzini, Stewart, & De George-Walker, 2014). Specifically, self-handicappers express a more external locus relative to non-handicappers (see also Krueger and Gruening, this volume). While this relationship has not received a great deal of attention among researchers, we argue that it again illustrates the powerful role of evaluative concern in motivating self-handicapping. Individuals who believe they lack personal control over performance outcomes are likely to be more threatened and insecure about impending performance situations (e.g., experience high evaluative concern) and may resort to self-handicapping in order to salvage their self-esteem in the face of great uncertainty about their chances for success.

Why Self-handicapping May Be Preferred Over Other Forms of Self-protection

While this analysis provides us with a profile of the specific antecedents that promote self-handicapping in the face of insecurity, an important yet unanswered question is why individuals might choose to self-handicap over other forms of self-protection. The literature has outlined a host of different self-protection strategies, ranging from self-affirmation (Steele, 1988) and dissonance-reduction processes (Zanna & Cooper, 1974; Cooper & Pearce, this volume) to downward social comparison (Wills, 1991) and other self-evaluation maintenance processes (Tesser, 1988; see also Crano & Hohman; Fiedler & McCaughey, this volume). In fact, Tesser et al. (1996) labeled this constellation of self-protective mechanisms the "self-zoo" and hypothesized that these various processes would substitute for each other since they all served the same fundamental self-protective goal. Tesser et al. (1996) found evidence for this substitutability in a series of studies: for example, when participants

Self-Handicapping in the Face of Uncertainty **135**

engaged in downward social comparison, they were less likely to self-affirm later. Similarly, Steele and Liu (1983) illustrated that self-affirmation reduced later rationalization and other dissonance-reduction processes.

In response to this impressive body of work, McCrea and Hirt (2001) examined whether these other self-protective mechanisms would effectively substitute for self-handicapping. They posited that self-handicapping might be unique among the other mechanisms in the "self-zoo" in that it allows the handicapper to maintain beliefs about competence in the threatened domain. Self-handicapping affords the attribution that one has more ability than one's performance shows and thus preserves competence beliefs. Alternative mechanisms, like self-affirmation and downward social comparison, may effectively protect global feelings of self-worth, but they do not safeguard competence beliefs in the threatened ability domain. To test this hypothesis, McCrea and Hirt (2001) had introductory psychology students report their amount of effort in the class (e.g., class attendance, keeping up with the course reading, taking good notes) prior to an important midterm exam. To no surprise, high-trait self-handicappers engaged in more self-handicapping (i.e., reported less effort) than low self-handicappers and performed significantly worse on their next test. Despite their poor performance, high-self-handicappers maintained their self-esteem, consistent with the work of Rhodewalt and colleagues (1991). Importantly, McCrea and Hirt (2001) also asked students to rate their ability in several different domains, including academics, athletics, social, artistic/ musical, and psychology. While no differences were found in other domains, high-self-handicapping males reported significantly higher psychological ability than any other group. In fact, these individuals rated themselves in the top 10% of psychological ability, despite a mean test performance in the D range! Moreover, analyses revealed that these specific beliefs about psychological ability mediated the preservation of self-esteem following poor performance. Accordingly, McCrea and Hirt (2002) concluded that self-handicapping provides the opportunity to maintain beliefs about the specific threatened ability despite objectively poor performance.

Thus, there seems to be something unique about self-handicapping that other self-protective mechanisms do not afford. Individuals who are especially insecure about competence in a particular domain might choose self-handicapping over other self-protective strategies because self-handicapping uniquely serves this ability-protection goal. Consequently, we would expect that other self-protective mechanisms cannot effectively substitute for and eliminate the need to self-handicap. Indeed, work by McCrea and Hirt (2011) has shown specifically that a prior opportunity to engage in self-affirmation does not reduce later self-handicapping. While we have not yet studied whether other self-protective mechanisms such as dissonance reduction and downward social comparison processes can substitute for self-handicapping, we would confidently expect that they would not. Nonetheless, these expectations await empirical validation.

Summary

At this point, we have a solid handle on the antecedents likely to promote self-handicapping behavior. The fundamental precursors are feelings of uncertainty and insecurity derived from a threatening upcoming performance in an important domain. In this situation, self-handicappers start to question their ability to succeed and feel a lack of outcome control. Their thoughts focus on the possibility of failure, and they worry about what that failure would indicate about their ability in that domain. With no other option available that would satisfy this nexus of concerns, their attention is directed toward salvaging the situation by controlling the attributions made about their performance and to creating a viable handicap to excuse potential poor performance.

The Tradeoff That Is: It's All in the Execution

To this point, we have discussed antecedents that lead to self-handicapping. The assumption of this analysis has been that if these conditions are satisfied, the individual is going to opt to self-handicap to alleviate their uncertainty and evaluative concern. We also have noted self-handicapping's inherent downside: the increased likelihood of task failure. But the attributional and self-esteem benefits of the strategy, coupled with the ability to protect competence beliefs in the threatened domain, seem to make it a worthy tradeoff. So were Higgins et al. (1990) correct? Is self-handicapping "the tradeoff that isn't?"

We believe there is far more to the story than has been considered so far. First, we have found evidence that self-handicapping requires a considerable amount of thought and cognitive resources to successfully pull off (Eyink et al., 2017). Once an individual begins to contemplate the notion of self-handicapping, there are several additional steps that must be completed before executing a successful handicap. Most notably, the handicapper must identify something that can serve as a viable excuse for poor performance in the given context. Individuals prone to self-handicap are more likely to think about ways to fail an upcoming performance, and these thoughts determine subsequent behavior (McCrea & Flamm, 2012). But as we have already discussed, not all handicaps are created equal. Some may be more believable than others, in general or in particular situations. In selecting their excuse, the handicapper must consider what would be most effective or persuasive to their intended audience. Thus, the handicapper must be able to take an outsider's perspective to determine what the audience would or would not accept as a valid excuse for poor performance. Finally, the handicapper must be able to work out the details of the presentation and timing of the handicap. The handicapper must ensure that the audience is aware of the handicap so that the audience can use that information to affect their performance attributions. After all, a handicap can only work if the audience knows about it and considers it a viable excuse for poor performance.

While each of these stages is essential for successful execution of this strategy, surprisingly little attention has been given to aspects of the process other than the initial inclination toward self-handicapping. We find this lack of consideration unfortunate because it ignores some very real and very important concerns that the handicapper must consider. As in any interpersonal interaction, we must be able to reliably anticipate and predict how others will react to effectively navigate our relationships with others (see also Mikulincer & Shaver; Krueger & Gruening, this volume). Specifically, we argue that uncertainties arise at each of these later steps, which can undermine the handicap's successful and effective execution. Uncertainties at these latter stages can dissuade individuals from self-handicapping, as there is the potential for the strategy to backfire if one cannot (1) confidently determine which (if any) handicaps might "work" in that situation for a given audience, and (2) devise a successful plan for the execution of the strategy, one that fosters awareness of the handicap without arousing disbelief or suspicion of ulterior motives. In the remainder of this chapter, we turn our attention to the elaboration of these additional steps in the self-handicapping process, highlighting the challenges that exist for the handicapper at each stage and the potential uncertainties that might discourage the further consideration of the strategy.

Motives Underlying Self-handicapping

The astute reader will note that we've mentioned the term **audience** multiple times above. But who is the intended audience of the handicap? This highlights a key issue within the literature: identifying the primary driver of self-handicapping. Berglas and Jones (1978) posited that the main handicapping motive is protecting self-esteem from the threat of potential failure. This implies the intended audience is **the self** – that the fear of learning they lack an important ability pushes them to engage in an elaborate self-protective ruse. Several lines of work support this perspective. Strube and Roemmele (1985) found that self-handicappers prefer tests with nondiagnostic feedback in order to avoid potentially threatening diagnostic feedback. Further, self-handicappers are able to maintain a positive sense of self-esteem despite task failure (Rhodewalt et al., 1991). Finally, McCrea and Hirt's (2001) findings that self-handicapping protects competence beliefs suggest that handicappers prioritize self-beliefs in ability over all else. In sum, the evidence that self-handicapping is motivated by maintaining self-esteem seems quite solid.

Alternatively, Kolditz and Arkin (1982) argued that self-handicapping is primarily an impression management strategy. According to these researchers, the intended audience for self-handicapping is **others**. That is, self-handicappers may be aware of their own deficiencies but are primarily interested in maintaining others' favorable impressions of them. Kolditz and Arkin (1982) found that participants self-handicapped more in public than in private conditions, providing some evidence for the impression management perspective. Since then, however, studies have found mixed results: although self-handicapping clearly preserves others' beliefs that the handicapper has greater ability than their performance shows, it comes at an interpersonal cost (Luginbuhl & Palmer, 1991; Rhodewalt et al., 1995). In these studies, participants read about a target who either did or did not self-handicap before an important test. Participants attributed the handicapped target's performance less to ability (showing evidence that they made the desired attributions about the target's performance), while also perceiving the handicapped target as less concerned about performance, less motivated, and less desirable as a study partner. Thus, although the handicap "worked" to protect others' ability beliefs, the attributional benefit was offset by steep interpersonal costs.

Interestingly, work in our lab (Hirt et al., 2003; McCrea, Hirt, Hendrix, et al., 2008; McCrea, Hirt, & Milner, 2008) has denoted an important moderating factor in audience reactions to self-handicapping – gender. Specifically, women are more tempered in the attributional benefits they ascribe to the handicapper and express far more negative interpersonal consequences to self-handicapping than do men, sex differences that may well have evolutionary origins (see also von Hippel & Merakovsky, this volume). These findings parallel some consistent, but perplexing, gender differences found to affect self-handicapping behavior itself. Throughout many studies, men have been shown to handicap more often than women, though these gender differences emerge only in the use of behavioral forms of self-handicapping (Harris & Snyder, 1986; Hirt et al., 1991; McCrea, Hirt, & Milner, 2008; Rhodewalt, 1990). Conversely, men and women have been shown to use claimed self-handicaps equally (Hendrix & Hirt, 2009; Hirt et al., 1991; Koch et al., 2003). Paralleling these findings, McCrea, Hirt, Hendrix, et al. (2008) found that men tend to score higher on the behavioral subscale of the SHS, whereas women tend to score higher on the claimed subscale of the SHS (see also Jaconis et al., 2016).

These gender differences have been perplexing, as nothing in Jones and Berglas' (1978) original formulation would lead us to expect that men should be more prone to engage in this behavior. Initially, we thought that the source of these gender differences may reside in the differential effectiveness of behavioral self-handicaps for men and women. Dweck et al. (1978) demonstrated that teachers attribute girls' failure to a lack of ability but attribute boys' failure to a lack of effort. Thus, it's possible that behavioral handicaps, like effort withdrawal, may not work as effectively for women as for men, given that women's failures are blamed on ability rather than effort. To address this possibility, we (Hirt et al., 2003) had participants read a scenario about a target named Chris who had an important test the following day. We varied multiple factors about the scenario, including: (1) Chris' gender; (2) if Chris self-handicapped (went to a movie with friends) or not (spent the evening studying); and (3) Chris' grade on the test. Participants then evaluated Chris on a number of interpersonal dimensions (e.g., liking, similarity, and desire to have as a friend) and made performance attributions.

To our surprise, target gender had no effect – participants rated the male and female versions of Chris similarly. However, the gender of the participant profoundly affected reactions. Women rated the self-handicapping Chris far more negatively than men and were less likely to give Chris "the benefit of the doubt" by discounting lack of ability as the cause of failure. Women were also more likely to attribute dispositional motives such as laziness or a lack of self-control to the self-handicapping Chris than were men, and they were the only ones who thought the target might be acting out of a desire to excuse failure (i.e., self-handicap). Men endorsed more situational motivations for Chris' actions, attributing the behavior to peer pressure, anxiety, the need for a study break, or the erroneous belief that Chris was adequately prepared. Importantly, these differences in inferred motives mediated the gender differences in the evaluations of the target. These basic findings have also been observed with a different type of behavioral handicap, namely, alcohol use (Kretchmann, 2008).

Recently, we have replicated and extended these gender differences in audience reactions to self-handicappers using a reverse correlation procedure (Hirt, Hughes, Eyink, Heiman, & Gray, 2022). Participants read a scenario describing a male target named Chris who did or did not self-handicap prior to an important performance. Participants were then given several pairs of faces (which had random noise imposed on a base face) and asked which looked more like their mental image of Chris. Composite faces were then created for the self-handicapper and non-self-handicapper for both male and female participants. Subsequently, another set of participants served as raters and evaluated the composite faces on a series of dimensions (e.g., trustworthy, competent, likeable, and dominant). Results indicated that raters (male and female raters alike) evaluated the self-handicapping composite face generated by female participants less favorably on multiple dimensions (less trustworthy, less competent, and less likeable) than all of the other composites (the self-handicapping composite face generated by men, or either of the non-self-handicapping composite faces).

Thus, it seems clear that women envision a self-handicapper as a more undesirable and incompetent person than do men.

It appears from these results that women feel that expending effort is normative and that effort withdrawal is unacceptable. Certainly, there is considerable research that supports the notion that women value effort more than men. For example, female students report studying harder, procrastinating less, and adopting more effortful learning goals and strategies than male students (Ablard & Lipschultz, 1998; Cooper et al., 1991; Stricker et al., 1993; Zimmerman & Martinez-Pons, 1990). Furthermore, in a series of studies, we (McCrea, Hirt, Hendrix, et al., 2008) developed a measure called the Worker scale to assess the extent to which someone views themselves as a hard worker and personally values this characteristic. The Worker scale consistently shows gender differences (McCrea, Hirt, Hendrix, et al., 2008; McCrea, Hirt, & Milner, 2008), such that women place a higher value on effort than men. Importantly, the effort beliefs indexed by the Worker scale can explain the robust gender differences in behavioral self-handicapping – women who value effort (i.e., score high on the Worker scale) do not engage in behavioral forms of self-handicapping, even when given the opportunity to do so. Moreover, the Worker scale also mediates the gender differences in reactions to a self-handicapping target. Because women more frequently value hard work, they not only act according to that norm and refrain from behavioral self-handicapping, but they also evaluate those who violate that norm (i.e., behavioral self-handicappers) more harshly than men (cf. Hirt & McCrea, 2009).

While these gender differences are certainly interesting in their own right, they beg the question of whether self-handicappers (1) are aware that women strongly dislike handicapping, and if so, (2) modulate their behavior accordingly. Early research in our lab seemed to indicate a lack of awareness. For example, using an empathic accuracy paradigm, Devers and Hirt (2009) found male handicappers (vs. non-handicappers) were more lenient in their own evaluations of a handicapping target and were less accurate in projecting others' responses to that behavior. Similarly, McCrea et al. (2009) found that self-handicappers expect others to be more critical of a target who tries his best and fails than do non-self-handicappers. Thus, it seems that male handicappers may not perceive how negatively their actions are viewed by others, and this lack of insight may be critical to understanding why men persist despite the interpersonal costs.

However, the conclusion that men are clueless might be premature. In her dissertation, Milner (2007, Study 3) examined whether self-handicappers adjusted their handicapping behavior contingent upon their audience. Male and female participants came to the lab with either a male or female friend, supposedly to take an intelligence test. A researcher informed all participants that practice affected their test scores, and all participants then had the opportunity to practice in full view of their friends. The amount of practice served as the key index of behavioral self-handicapping. Results indicated that men practiced less, or handicapped more, in the presence of a male peer than a female peer. Female participants' practice behavior was unaffected by the gender of the peer; in all cases, female participants practiced a great deal, displaying no evidence of self-handicapping. Thus, it appears that men are at least partially aware that women do not look favorably on individuals who withdraw effort and modulate their behavior accordingly.

One might wonder whether these behavioral adjustments result in any desirable outcomes for the handicapper. To examine this (Hirt et al. 2022), we also asked the peers to rate the participant who practiced for an upcoming exam. Female peers rated both men and women who practiced more far more favorably than those who did not, providing further support that women derogate self-handicapping. However, in this case, we observed these effects in person rather than with a hypothetical written scenario (Hirt et al., 2003). Conversely, male peers rated men who practiced less more favorably than men who practiced a lot. Intriguingly, male peers preferred women who practiced more over women who practiced less. This shows a clear double standard in men's evaluation of self-handicapping – while male peers condoned self-handicapping from other males, they derogated that same behavior from females.

So while the evidence to date on these questions is somewhat mixed, our data implies that self-handicappers do attempt to anticipate the reaction of their audience and adjust their behavior accordingly (akin to what we see in relationship behaviors – see Murray & Lafranche, and in Arriaga and Kumashiro, this volume). This process takes a great deal of cognitive resources, and to no surprise, Eyink and Hirt (2017) found that individuals engage in self-handicapping significantly more when they have their full contingent of mental resources (e.g., at their peak time of the day) than when they are depleted (at off-peak times). These data reinforce the fact that self-handicapping is a **strategic** behavior for managing insecurity, and its success is contingent on how well handicappers can correctly forecast what will work for their intended audience (for further discussion of the value of human foresight, see von Hippel & Merakovsky; and Fiedler & McCaughey, this volume). If they are poorly calibrated and mispredict the response from their audience, their self-handicapping behavior could backfire on them. But if they can correctly anticipate the likely reaction of their audience, it can be an effective self-protection strategy and perhaps even have some unanticipated benefits. Anecdotally, we have seen evidence that if self-handicappers play their cards right and choose the right self-handicap, they might not only acquire the desired attributional benefits but may also gain sympathy from the audience. This suggests that if the audience sees the handicap as sincere and legitimate, then the handicapper may be viewed as a poor unfortunate soul who has incurred these debilitating conditions through no fault of their own (and thus deserves compassion and sympathy).

But how do you know what that optimal or perfect self-handicap is? Behavioral self-handicaps like effort withdrawal seem dangerous, for you could be perceived (by women at least) as just lazy or a "blow-off." Other behavioral self-handicaps like alcohol and drug use suffer from similar concerns. On the one hand, then, it might seem that claimed self-handicaps like stress, illness, or injury are preferable, since they may carry fewer interpersonal consequences and baggage. But we would argue that there are several counterarguments to this view. First, claimed handicaps are far more dubious in nature. Is the person really sick, stressed, or injured? Who knows?? Behavioral handicaps undeniably sabotage one's chances to perform well and are thus evoke far less uncertainty. Second, behavioral handicaps are often observable to the intended audience and thus do not require the handicapper to point them out. Conversely, claimed handicaps require that the handicapper must bring them up and draw attention to them, an act that can arouse considerable suspicion among skeptical audiences. Many lines of work demonstrate the challenges faced by those claiming unfair or debilitating conditions and the potential backlash they can face. For example, work by Kaiser and colleagues shows that stigmatized group members who make discrimination claims are labeled complainers and troublemakers – regardless of whether the claim is valid or not (e.g., Kaiser & Miller, 2001; Kaiser & Miller, 2003). Thus, the decision of what handicap to use is a complicated one, and there appear to be several factors operating here that could ultimately discourage an individual from employing this strategy.

Conclusions

We began this chapter by outlining a set of antecedent conditions that promote self-handicapping behavior as a strategy to manage insecurity and uncertainty. Much work has focused on identifying these conditions that encourage individuals to self-handicap in the face of self-threat. Individual difference variables (e.g., uncertain self-esteem, fixed mindset, prevention focus, and external locus of control) contribute to the likelihood of employing this strategy, as well as situational factors (e.g. non-contingent success feedback and task importance). Paramount among those factors is a looming uncertainty about one's ability to live up to expectations and perform well. Once individuals entertain the possibility of failure, the need for self-protection is engaged. We argued that among the pantheon of self-protective strategies in the "self-zoo" (Tesser et al., 1996), self-handicapping is unique, for it can protect both perceptions of competence in the threatened domain as well as overall feelings of self-worth. Thus, while some might be perplexed at a talented individual self-sabotaging prior to an important performance (like a sporting event or a job interview), others

(Higgins et al., 1990) have argued that the attributional benefits of self-handicapping far outweigh the potential costs.

While this may be true, we contend that contemplation of self-handicapping is but one step in a very complex process. Self-handicapping is a strategic behavior that requires sufficient cognitive resources for its successful execution (Eyink et al., 2017). For this strategy to work, an individual must be able to identify available handicaps (McCrea & Flamm, 2012) and select one that can viably excuse poor performance. The plausibility of that handicap is a function of both the beliefs of the intended audience (i.e., what that audience sees as persuasive) and the convincing execution of the handicap prior to, during, and after the performance situation. Uncertainties at any of these points could undermine success and discourage the handicapper from implementing this strategy. We discussed several specific areas of concern. Behavioral forms of self-handicapping may be more believable and arouse less suspicion since they are observable and undoubtedly affect performance negatively. Claimed self-handicaps are more dubious in that the audience must accept the word of the handicapper regarding the reported obstacle(s) for their performance. Thus, all other things being equal, handicappers might be drawn to use behavioral handicaps over claimed handicaps. However, behavioral handicaps (particularly effort withdrawal) incur strong negative repercussions from women, who attribute pejorative interpersonal attributes (e.g., less likeable, less trustworthy, and less competent) to those who don't try their best. It seems that self-handicappers may be aware of these gendered audience reaction differences and adjust their behavior accordingly. We suspect there are other audience beliefs that handicappers take into account in deciding how to execute self-handicapping "in the wild" and are beginning to explore these factors in our lab.

Still, we are left with the ultimate question – is self-handicapping a paradox or not? Given the many stages that must be successfully traversed in order to "pull it off," we suspect that it is often self-defeating. For some, it can be quite an effective strategy to manage insecurity. But like Marlon Brando's character in *On the Waterfront* (Kazan, 1954), one might be left with the counterfactual belief that "I could have been a contender," especially when one's excuses are not well received by the audience. The answer comes down to the question of tradeoffs. Even suspicious audience members might have to admit that a self-handicapper could have more ability than his performance showed. Thus, we (like them) can always wonder what might have been if things had worked out differently. But are those upward counterfactuals worth the costs incurred from the use of this strategy (Zuckerman et al., 1998; Urdan, 2004; Zuckerman & Tsai, 1995)? Even under the best of circumstances, the self-handicapper is seen as a victim of unfortunate events, but more commonly, people end up blaming victims for their misfortunes (Lerner & Simmons, 1966). For this reason, many of us would concur that the tradeoffs are simply not worth it. But as we have seen in this volume, insecurity and uncertainty lead people to do an awful lot of bizarre and surprising things. We hope that this work spawns further efforts toward the goal of understanding the roots of this paradoxical self-protective behavior.

References

Ablard, K. E., & Lipschultz, R. E. (1998). Self-regulated learning in high-achieving students: Relations to advanced reasoning, achievement goals, and gender. *Journal of Educational Psychology, 90*(1), 94–101. https://doi.org/10.1037/0022-0663.90.1.94

Arazzini Stewart, M., & De George-Walker, L. (2014). Self-handicapping, perfectionism, locus of control and self-efficacy: A path model. *Personality and Individual Differences, 66,* 160–164.

Arkin, R. M., & Baumgardner, A. H. (1985). Self-handicapping. In J. H. Harvey & G. Weary (Eds.), *Attribution: Basic issues and applications* (pp. 169–202). New York: Academic Press.

Baumeister, R. F., & Scher, S. J. (1988). Self-defeating behavior patterns among normal individuals: Review and analysis of common self-destructive tendencies. *Psychological Bulletin, 104*(1), 3–22. https://doi.org/10.1037/0033-2909.104.1.3

Baumgardner, A. H., Lake, E. A., & Arkin, R. M. (1985). Claiming mood as a self-handicap: The influence of spoiled and unspoiled public identities. *Personality and Social Psychology Bulletin, 11*(4), 349–357. https://doi.org/10.1177/0146167285114001

Berglas, S., & Jones, E. E. (1978). Drug choice as a self-handicapping strategy in response to noncontingent success. *Journal of Personality and Social Psychology, 36*(4), 405–417. https://doi.org/10.1037/0022-3514.36.4.405

Blascovich, J., & Mendes, W. B. (2001). Challenge and threat appraisals. In J. P. Forgas (Ed.), *Feeling and thinking: The role of affect in social cognition* (pp. 59–82). Cambridge: Cambridge University Press.

Cooper, H., Baumgardner, A. H., & Strathman, A. (1991). Do students with different characteristics take part in psychology experiments at different times of the semester? *Journal of Personality, 59*(1), 109–127. https://doi.org/10.1111/j.1467-6494.1991.tb00770.x

Courtenay, W. H., McCreary, D. R., & Merighi, J. R. (2002). Gender and ethnic differences in health beliefs and behaviors. *Journal of Health Psychology, 7,* 219–231.

Devers, E., & Hirt, E. R. (2009). *Examining the empathic accuracy of self-handicappers: Unskilled and unaware of it?* Unpublished manuscript.

Dweck, C. S., Davidson, W., Nelson, S., & Enna, B. (1978). Sex differences in learned helplessness: II. The contingencies of evaluative feedback in the classroom and III. An experimental analysis. *Developmental Psychology, 14*(3), 268–276. https://doi.org/10.1037/0012-1649.14.3.268

Dweck, C. S., & Leggett, E. L. (1988). A social-cognitive approach to motivation and personality. *Psychological Review, 95*(2), 256–273. https://doi.org/10.1037/0033-295X.95.2.256

Eagly, A. H., & Steffen, V. J. (1984). Gender stereotypes stem from the distribution of women and men into social roles. *Journal of Personality & Social Psychology, 46,* 735–754.

Elliot, A. J., Cury, F., Fryer, J. W., & Huguet, P. (2006). Achievement goals, self-handicapping, and performance attainment: A mediational analysis. *Journal of Sport & Exercise Psychology, 28*(3), 344–361.

Eyink, J., Hirt, E. R., Hendrix, K. S., & Galante, E. (2017). Circadian variations in the use of self-handicapping: Exploring the strategic use of stress as an excuse. *Journal of Experimental Social Psychology, 69*, 102–110. https://doi.org/10.1016/j.jesp.2016.07.010

Feick, D. L., & Rhodewalt, F. (1997). The double-edged sword of self-handicapping: discounting, augmentation, and the protection and enhancement of self-esteem. *Motivation and Emotion, 21*(2), 147–163. https://doi.org/10.1023/A:1024434600296

Greenberg, J., Pyszczynski, T., & Paisley, C. (1984). Effect of extrinsic incentives on use of test anxiety as an anticipatory attributional defense: Playing it cool when the stakes are high. *Journal of Personality and Social Psychology, 47*(5), 1136–1145. https://doi.org/10.1037/0022-3514.47.5.1136

Harris, R. N., & Snyder, C. R. (1986). The role of uncertain self-esteem in self-handicapping. *Journal of Personality and Social Psychology, 51*(2), 451–458. https://doi.org/10.1037/0022-3514.51.2.451

Hendrix, K. S., & Hirt, E. R. (2009). Stressed out over possible failure: The role of regulatory fit on claimed self-handicapping. *Journal of Experimental Social Psychology, 45*(1), 51–59. https://doi.org/10.1016/j.jesp.2008.08.016

Higgins, R. L., & Harris, R. N. (1988). Strategic "alcohol" use: Drinking to self-handicap. *Journal of Social and Clinical Psychology, 6*(2), 191–202. https://doi.org/10.1521/jscp.1988.6.2.191

Higgins, R. L., Snyder, C. R., & Berglas, S. (1990). *Self-handicapping: The paradox that isn't.* New York: Plenum.

Hirt, E. R., Deppe, R. K., & Gordon, L. J. (1991). Self-reported versus behavioral self-handicapping: Empirical evidence for a theoretical distinction. *Journal of Personality and Social Psychology, 61*(6), 981–991. https://doi.org/10.1037/0022-3514.61.6.981

Hirt, E. R., Eyink, J. R., Crawford, J., & Milner, B. (2022). *"I know what boys (and girls) like": Gender differences in audience reactions to self-handicapping.* Unpublished manuscript.

Hirt, E. R., Hughes, C., Eyink, J. R., Heiman, S. L., & Gray, S. (2022). *Look at my face and what do you see?: Gender differences in mental images of a self-handicapper.* Unpublished manuscript.

Hirt, E. R., & McCrea, S. M. (2002). Positioning self-handicapping within the self-zoo: Just what kind of animal are we dealing with here? In J. P. Forgas & K. D. Williams (Eds.). *The social self: Cognitive, Interpersonal, and Intergroup Perspectives.* (pp. 97–126). New York: Psychology Press.

Hirt, E. R., & McCrea, S. M. (2009). Man smart, woman smarter? Getting to the root of gender differences in self-handicapping. *Social and Personality Psychology Compass, 3*(3), 260–274. https://doi.org/10.1111/j.1751-9004.2009.00176.x

Hirt, E. R., McCrea, S. M., & Boris, H. I. (2003). "I know you self-handicapped last exam": Gender differences in reactions to self-handicapping. *Journal of Personality and Social Psychology, 84*(1), 177–193. https://doi.org/10.1037/0022-3514.84.1.177

Hirt, E. R., McCrea, S. M., & Kimble, C. E. (2000). Public self-focus and sex differences in behavioral self-handicapping: Does increasing self-threat still make it "just a man's game?". *Personality and Social Psychology Bulletin, 26*(9), 1131–1141. https://doi.org/10.1177/01461672002611009

Jaconis, M., Boyd, S. J., Hartung, C. M., McCrea, S. M., Lefler, E. K., & Canu, W. H. (2016). Sex differences in claimed and behavioral self-handicapping and ADHD symptomatology in emerging adults. *ADHD Attention Deficit and Hyperactivity Disorders, 8*(4), 205–214. DOI 10.1007/s12402-016-0200-y.

Jones, E. E., & Berglas, S. (1978). Control of attributions about the self through self-handicapping strategies: The appeal of alcohol and the role of underachievement. *Personality and Social Psychology Bulletin, 4*(2), 200–206. https://doi.org/10.1177/014616727800400205

Jones, E. E., & Rhodewalt, F. (1982). *The self-handicapping scale*. Princeton, NJ: Princeton University.

Kaiser, C. R., & Miller, C. T. (2001). Stop complaining! The social costs of making attributions to discrimination. *Personality and Social Psychology Bulletin, 27*(2), 254–263. https://doi.org/10.1177/0146167201272010

Kaiser, C. R., & Miller, C. T. (2003). Derogating the victim: The interpersonal consequences of blaming events on discrimination. *Group Processes & Intergroup Relations, 6*(3), 227–237. https://doi.org/10.1177/13684302030063001

Kazan, E. (1954). *On the Waterfront*. Columbia Pictures.

Kelley, H. H. (1973). The processes of causal attribution. *American Psychologist, 28*(2), 107–128. https://doi.org/10.1037/h0034225

Kimble, C. E., Funk, S. C., & DaPolito, K. L. (1990). The effects of self-esteem certainty on behavioral self-handicapping. *Journal of Social Behavior & Personality, 5*(3), 137–149.

Koch, K. A., Hirt, E. R., & McCrea, S. M. (2003). *Public self-focus and claimed self-handicapping*. Unpublished Manuscript.

Kolditz, T. A., & Arkin, R. M. (1982). An impression management interpretation of the self-handicapping strategy. *Journal of Personality and Social Psychology, 43*(3), 492–502. https://doi.org/10.1037/0022-3514.43.3.492

Kretchmann, J. E. (2008). *Individual differences in observer's reactions to alcohol use as a behavioral self-handicapping strategy*. Unpublished Manuscript, University of Konstanz, Germany.

Leary, M. R., & Shepperd, J. A. (1986). Behavioral self-handicaps versus self-reported handicaps: A conceptual note. *Journal of Personality and Social Psychology, 51*(6), 1265–1268. https://doi.org/10.1037/0022-3514.51.6.1265

Lerner, M. J., & Simmons, C. H. (1966). Observer's reaction to the "innocent victim": Compassion or rejection? *Journal of Personality and Social Psychology, 4*(2), 203–210. https://doi.org/10.1037/h0023562

Luginbuhl, J., & Palmer, R. (1991). Impression management aspects of self-handicapping: Positive and negative effects. *Personality and Social Psychology Bulletin, 17*(6), 655–662. https://doi.org/10.1177/0146167291176008

McCrea, S. M., & Flamm, A. (2012). Dysfunctional anticipatory thoughts and the self-handicapping strategy. *European Journal of Social Psychology, 42*, 72–81. DOI: 10.1002/ejsp.845.

McCrea, S. M., & Hirt, E. R. (2001). The role of ability judgments in self-handicapping. *Personality and Social Psychology Bulletin, 27*(10), 1378–1389. https://doi.org/10.1177/01461672012710013

McCrea, S. M., & Hirt, E. R. (2011). Limitations on the substitutability of self-protective processes: Self-handicapping is not reduced by related-domain self-affirmations. *Social Psychology, 42*(1), 9–18. https://doi.org/10.1027/1864-9335/a000038

McCrea, S. M., Hirt, E. R., Hendrix, K. L., Milner, B. J., & Steele, N. L. (2008). The worker scale: Developing a measure to explain gender differences in behavioral self-handicapping. *Journal of Research in Personality, 42*(4), 949–970. https://doi.org/10.1016/j.jrp.2007.12.005

McCrea, S. M., Hirt, E. R., & Milner, B. J. (2008). She works hard for the money: Valuing effort underlies gender differences in behavioral self-handicapping. *Journal of Experimental Social Psychology, 44*(2), 292–311. https://doi.org/10.1016/j.jesp.2007.05.006

McCrea, S. M., Hirt, E. R., & Myers, A. L. (2008). *On the anticipatory nature of self-handicapping behavior.* Unpublished Manuscript.

McCrea, S. M., Myers, A. M., & Hirt, E. R. (2009). Self-handicapping as an anticipatory self-protection strategy. In E. P. Lamont (Ed.). *Social psychology: New research* (pp. 31–53). Hauppuage, NY: Nova Science.

Milner, B. J. (2007). *Individual differences in peer relationships: The role of self-handicapping.* Unpublished doctoral dissertation, Indiana University.

Newman, L. S., & Wadas, R. F. (1997). When stakes are higher: Self-esteem instability and self-handicapping. *Journal of Social Behavior & Personality, 12*(1), 217–232.

Pro Football Hall of Fame. (2021). *The 1998 NFL Draft.* Retrieved July 13, 2022, from https://www.profootballhof.com/football-history/nfl-draft-history/1990/98/

Pro Football Reference. (n.d.a). *Peyton Manning.* Retrieved July 13, 2022, from https://www.pro-football-reference.com/players/M/MannPe00.htm

Pro Football Reference. (n.d.b). *Ryan Leaf.* Retrieved July 13, 2022, from https://www.pro-football-reference.com/players/L/LeafRy00.htm

Rhodewalt, F. (1990). Self-handicappers: Individual differences in the preference for anticipatory, self-protective acts. In R. L. Higgins, C. R. Snyder & S. Berglas (Eds.), *Self-handicapping: The paradox that isn't. The Plenum series in social/clinical psychology* (pp. 69–106). New York, NY: Plenum Press.

Rhodewalt, F. (1994). Conceptions of ability, achievement goals, and individual differences in self-handicapping behavior: On the application of implicit theories. *Journal of Personality, 62*(1), 67–85. https://doi.org/10.1111/j.1467-6494.1994.tb00795.x

Rhodewalt, F. (2008). Self-handicapping: On the self-perpetuating nature of defensive behavior. *Social and Personality Psychology Compass, 2*(3), 1255–1268. https://doi.org/10.1111/j.1751-9004.2008.00117.x

Rhodewalt, F., & Davison, J. (1986). Self-handicapping and subsequent performance: Role of outcome valence and attributional certainty. *Basic and Applied Social Psychology, 7*(4), 307–322. https://doi.org/10.1207/s15324834basp0704_5

Rhodewalt, F., Morf, C., Hazlett, S., & Fairfield, M. (1991). Self-handicapping: The role of discounting and augmentation in the preservation of self-esteem. *Journal of Personality and Social Psychology, 61*(1), 122–131. https://doi.org/10.1037/0022-3514.61.1.122

Rhodewalt, F., Sanbonmatsu, D. M., Tschanz, B., Feick, D. L., & Waller, A. (1995). Self-handicapping and interpersonal trade-offs: The effects of claimed self-handicaps on observers' performance evaluations and feedback. *Personality and Social Psychology Bulletin, 21*(10), 1042–1050. https://doi.org/10.1177/01461672952110005

Rhodewalt, F., Saltzman, A. T., & Wittmer, J. (1984). Self-handicapping among competitive athletes: The role of practice in self-esteem protection. *Basic and Applied Social Psychology, 5*(3), 197–209. https://doi.org/10.1207/s15324834basp0503_3

Sheppard, J. A., & Arkin, R. M. (1989). Determinants of self-handicapping: Task importance and the effects of preexisting handicaps on self-generated handicaps.

Personality and Social Psychology Bulletin, 15(1), 101–112. https://doi.org/10.1177/0146167289151010

Snyder, C. R., Smith, T. W., Augelli, R. W., & Ingram, R. E. (1985). On the self-serving function of social anxiety: Shyness as a self-handicapping strategy. *Journal of Personality and Social Psychology, 48*(4), 970–980. https://doi.org/10.1037/0022-3514.48.4.970

Smith, T. W., Snyder, C. R., & Perkins, S. C. (1983). The self-serving function of hypochondriacal complaints: Physical symptoms as self-handicapping strategies. *Journal of Personality and Social Psychology, 44*(4), 787–797. https://doi.org/10.1037/0022-3514.44.4.787

Snyder, C. R., Ford, C. E., & Hunt, H. A. (1985). *Excuse-making: A look at sex differences.* Paper presented at the American Psychological Association, Los Angeles, CA.

Stricker, L. J., Rock, D. A., & Burton, N. W. (1993). Sex differences in predictions of college grades from scholastic aptitude test scores. *Journal of Educational Psychology, 85*(4), 710–718. https://doi.org/10.1037/0022-0663.85.4.710

Steele, C. M. (1988). The psychology of self-affirmation: Sustaining the integrity of the self. In L. Berkowitz (Ed.), *Advances in experimental social psychology, Vol. 21. Social psychological studies of the self: Perspectives and programs* (pp. 261–302). Academic Press.

Steele, C. M., & Liu, T. J. (1983). Dissonance processes as self-affirmation. *Journal of Personality and Social Psychology, 45*(1), 5–19. https://doi.org/10.1037/0022-3514.45.1.5

Strube, M. J., & Roemmele, L. A. (1985). Self-enhancement, self-assessment, and self-evaluative task choice. *Journal of Personality and Social Psychology, 49*(4), 981–993. https://doi.org/10.1037/0022-3514.49.4.981

Swim, J. K., & Sanna, L. J. (1996). He's skilled, she's lucky: A meta-analysis of observers' attributions for women's and men's successes and failures. *Personality and Social Psychology Bulletin, 22*(5), 507–519. https://doi.org/10.1177/0146167296225008

Tesser, A. (1988). Toward a self-evaluation maintenance model of social behavior. In L. Berkowitz (Ed.), *Advances in experimental social psychology, Vol. 21. Social psychological studies of the self: Perspectives and programs* (pp. 181–227). Academic Press.

Tesser, A., Martin, L. L., & Cornell, D. P. (1996). On the substitutability of self-protective mechanisms. In P. M. Gollwitzer & J. A. Bargh (Eds.), *The psychology of action: Linking cognition and motivation to behavior* (pp. 48–68). The Guilford Press.

Thürmer, J. L., McCrea, S. M., & Gollwitzer, P. M. (2013). Regulating self-defensiveness: If-then plans prevent claiming and creating performance handicaps. *Motivation and Emotion, 37*, 712–725. DOI: 10.1007/s11031-013-9352-7.

Tice, D. M. (1991). Esteem protection or enhancement? Self-handicapping motives and attributions differ by trait self-esteem. *Journal of Personality and Social Psychology, 60*(5), 711–725. https://doi.org/10.1037/0022-3514.60.5.711

Tucker, J. A., Vuchinich, R. E., & Sobell, M. B. (1981). Alcohol consumption as a self-handicapping strategy. *Journal of Abnormal Psychology, 90*(3), 220–230. https://doi.org/10.1037/0021-843X.90.3.220

Urdan, T. (2004). Predictors of academic self-handicapping and achievement: Examining achievement goals, classroom goal structures, and culture. *Journal of Educational Psychology, 96*(2), 251–264. https://doi.org/10.1037/0022-0663.96.2.251

Wills, T. A. (1991). Similarity and self-esteem in downward comparison. In J. Suls & T. A. Wills (Eds.), *Social comparison: Contemporary theory and research* (pp. 51–78). Lawrence Erlbaum Associates, Inc.

Wondra, T. K., & McCrea, S. M. (2022). Collective self-doubt: Does subjective SES predict behavioral self-handicapping tendency in college students? *Social Psychology of Education, 25*(1), 129–167.

Zanna, M. P., & Cooper, J. (1974). Dissonance and the pill: An attribution approach to studying the arousal properties of dissonance. *Journal of Personality and Social Psychology, 29*(5), 703–709. https://doi.org/10.1037/h0036651

Zimmerman, B. J., & Martinez-Pons, M. (1990). Student differences in self-regulated learning: Relating grade, sex, and giftedness to self-efficacy and strategy use. *Journal of Educational Psychology, 82*(1), 51–59. https://doi.org/10.1037/0022-0663.82.1.51

Zuckerman, M., Kieffer, S. C., & Knee, C. R. (1998). Consequences of self-handicapping: Effects on coping, academic performance, and adjustment. *Journal of Personality and Social Psychology, 74*(6), 1619–1628. https://doi.org/10.1037/0022-3514.74.6.1619

Zuckerman, M., & Tsai, F.-F. (2005). Costs of self-handicapping. *Journal of Personality, 73*(2), 411–442. https://doi.org/10.1111/j.1467-6494.2005.00314.

9
STRATEGY, TRUST, AND FREEDOM IN AN UNCERTAIN WORLD

Joachim I. Krueger

BROWN UNIVERSITY, RHODE ISLAND, USA

David J. Grüning

HEIDELBERG UNIVERSITY, GERMANY
GESIS – LEIBNIZ INSTITUTE FOR THE SOCIAL SCIENCES, GERMANY

Abstract

Insecurities may arise in situations of strategic interaction, interpersonal trust, and intrapersonal experiences of free will. The uncertainties associated with these contexts are comprehensible and manageable if ineliminable. Whether uncertainty turns into insecurity is a question of temperament and attitude. We focus on the uncertainties a self-aware social creature faces. We suggest, among other points, the following: First, strategic reasoning is characterized by, depending on the context, a tolerance or an aversion to surprise. Second, acts of trust as opposed to acts of distrust increase trust and thus decrease insecurity if there is any trust to begin with. Third, belief in the freedom of the will is negatively related to uncertainty aversion and to social perceptions of competence and morality. Fourth, radical freedom, creativity, personal growth, and strategic advantage may require randomness and a tolerance of insecurity.

Uncertainty becomes insecurity if we allow it to upset us. –Hoca Camide

Humans live their lives suspended between states of knowledge and ignorance. Knowing – or hoping – that knowledge is power and that ignorance is no bliss, most humans seek knowledge, and they do so in part because knowledge reduces states of insecurity (see also von Hippel, this volume). Importantly, there are also instances where it is wise to choose ignorance (Krueger et al., 2020; Hertwig & Engel, 2020), a surprising qualification to the general epistemic drive to seek knowledge, to which we shall return.

The quest for knowledge takes various forms. In the simplest case, there are risks, that is, known probabilities with which certain outcomes will occur, and there are methods to manage these risks (Gigerenzer, 2015). In more complex

DOI: 10.4324/9781003317623-11

cases, there is uncertainty, that is, states in which the probabilities of the outcomes are not known (Gigerenzer, 2020; Knight, 1921). States of uncertainty are most likely and most poignant when they depend on the unpredictable behavior of other people (Gigerenzer, 2022). Such unpredictability is present especially when these others have their own strategic interests and when they are able and willing to deploy the arts of deception. These limitations to knowing what will happen can make the soundest mind feel insecure. Yet, although feelings or insecurity are generally unpleasant, they may be psychologically useful (see also Fiedler & McCaughey and Kruglanski & Ellenberg, this volume); again, a point to which we shall return.

In this chapter, we explore four ways in which uncertainty and its concomitant states of insecurity are tractable challenges. In the first section, we approach this challenge from a game-theoretic perspective. A simple guessing game serves as the context for the exploration of biased and optimal strategies when a player is faced with an opponent's unknown choice. We show that a theory of mind and social projection in particular can clear some of the fog of uncertainty (see for other examples, Krueger & Grüning, 2021; Krueger et al., 2022; Kruglanski & Ellenberg; van Prooijen, this volume). In the second section, we turn to the trust game, where uncertainty is amplified due to an interplay between strategies of coordination and discoordination. We show that acts of trust can reduce uncertainty by allowing the trustor to estimate the prevalence of trustworthiness more accurately (see also van den Bos, this volume). In the third section, we explore the association between folk beliefs about free will and individual differences in uncertainty intolerance. Furthermore, we ask how free-will believers compared with skeptics are socially evaluated (Krueger et al., 2020). Finally, we explore how human freedom may grow from a strategic randomization of options and actions, that is, from a stance of self-imposed deliberate uncertainty. Whether such strategies reduce or increase a person's sense of insecurity likely depends on the person's character. Freedom, we suspect, is not for everyone.

Strategy and Uncertainty

Individuals need to think strategically when their interests conflict with the interests of other, also potentially thoughtful, individuals (Krueger, 2007a). If the unpredictability of the natural or artificial world did not offer enough risk and uncertainty, strategic competition between reasoning agents adds a host of motives that make success difficult. The complexity of social interaction is liable to induce feelings of stress and insecurity (see also, Mikulincer & Shaver, this volume). This disutility of affect, together with potentially pointless attempts to reason through the entire strategic event horizon, suggests that there is wisdom in letting go (see also, Kruglanski & Ellenberg, this volume). The Prussian strategist Helmuth von Moltke the Elder noted that calculated planning can go up to the first contact with the enemy. From then on, one must assume that the opponent is equally able to anticipate strikes and counterstrikes. Ultimately, flexibility, luck, and the weather are decisive.

What is the hallmark of strategic reasoning? Fiske and Taylor (1984) noted that 'other people think back,' and that this thinking back is what separates social cognition from other types of cognition. What is more, people know that other people think back and that these others know that they know *et cetera et ad nauseam*. But perhaps there need not be any nausea. If it is true that an infinite recursion of the 'I-know-you-know-I-know' variety might lead to insecurity and despair, it is also true that few mortals can dive deeply into the rabbit hole of reflected thought. Insecurity can be managed by limiting recursive thought or by deliberately choosing to act randomly.

To explore these issues, we used a thought experiment described by Dixit and Nalebuff (2008) as our starting point. In that experiment, a participant must try to guess a number randomly drawn from the range of 1–100. If the first guess is correct, there is a premium of $100. Otherwise, there is only a note saying whether the target number is larger or smaller than the guess. If the second guess is correct, the premium falls to $80. If the guess is wrong again, the game continues until a guess is correct or the money has run out, whichever comes first. The optimal strategy is to always guess the middle number of the available range, which minimizes the residual range. The social version of this game is more complex. Here, a human chooser has selected the number to be guessed, which raises the question of whether the split-range strategy is still optimal. The guesser must worry that the chooser anticipates this strategy and therefore avoids the numbers this method generates. Can the guesser account for the chooser's ability to anticipate their own guesses? And what is a guesser who is mindful of this complication to do?

Dixit and Nalebuff (2008) argued that the engineering version of the game imposes a tractable risk, whereas the social version creates uncertainty as the guesser and the chooser seek to anticipate each other's moves. As there are no logical grounds for either player to assume that they are able to think one step further than the opponent, it appears that the game's mentalizing element collapses, with the result that the chooser resorts to selecting the target number randomly and the guesser, who anticipates this, returns to the split-range method, which is optimal under these circumstances. All told, the guesser in this strategic game may not be worse off in terms of the game's expected value but will be more distressed than a guesser in the engineering version of the game. The same holds true for the chooser. If they had a choice about which version of the game to play, both players might prefer the engineering version in order to avoid the uncertainties and insecurities that come with the social game. As we will see, however, most respondents ignore this advice, thereby revealing a fundamental social preference. Perhaps it is more fun to play against a person than against a robot, the insecurities notwithstanding. If victory is sweeter and defeat is bitterer, social games expose a titillating kind of insecurity-seeking.

Now consider a simplified one-shot version of this game. The chooser picks a number from 1 to 6, and the guesser tries to match it. If the numbers match, the guesser receives 60 Forint; otherwise, the chooser receives 12 Forint to equalize the game's expected value. When both players are invited to actively think of a number and consider what the other player may think (they think, etc.) the game is strategic. If, however, a player casts a die to obtain a random number, the game is merely aleatory. Any mentalization about the other player's approach to the game evaporates.

We showed research participants four versions of this game (Grüning & Krueger, 2021), which were constructed to constrain players to either actively think about a number or to randomly select one. Participants then indicated how much they would pay for the right to play the game as a guesser. The critical result, with data pooled over two experiments, was a significant interaction effect. As shown in Figure 9.1, if the chooser thought of a number, participants were willing to pay more if they could also think of a number, $d = .36$. When the chooser was said to randomize the number selection, the difference was trivially small, $d = .13$.

This pattern is reasonable if people, regardless of their role, are biased in their selection of numbers. If a chooser is more likely to think of a mid-range number than of a high or low number, guessers can improve their chances of matching that number simply by choosing the number that comes to mind. They can act strategically without trying to replicate the chooser's reasoning. Using the heuristic of social projection, they can capitalize on the value their own ideas have in predicting the ideas of others (Krueger, 2007b). Choosers who are

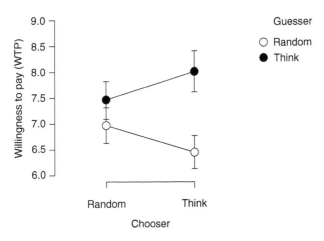

FIGURE 9.1 Guesser's willingness to pay when thinking or randomizing vs. thinking or randomizing chooser.

meta-cognitively rational and not overconfident know this (Moore, 2020); they randomize (e.g., by throwing a die) even when they have the prerogative to think about what number they wish to bet on. They rationally refuse to think.

Randomization is a minimax strategy guarding against exploitation, but it does not improve the chooser's prospects of winning. Our data suggest that in the role of choosers, participants also prefer to think of a number when the other player thinks of a number (Grüning & Krueger, in preparation). Because only guessers are rewarded when numbers match, the think-think scenario can be advantageous for them, and they seem to be confident in their ability to utilize this potential. In contrast, a preference for the think-think scenario undercuts the choosers' material interests. Respondents do not seem to realize this. Believing in the superiority of their own strategic insights, they end up hurting themselves.

Figure 9.2 shows that thoughtful number generation is biased toward the mid-range (Grüning & Krueger, 2021). When participants were instructed to predict the *least popular* number generated by the sample of participants given the same instructions, they preferred high or low numbers (dashed lines). Whereas the first result demonstrates the familiar population stereotype in number generation, the second result reveals myopic meta-cognition. Respondents appeared to think just one step ahead without realizing that others do too. That is, they understood the population stereotype favoring mid-range numbers, and they acted in contradistinction to it. They failed to see, however, that other participants did the same. This is an interesting result because it shows that social projection is myopic. Participants projected only one step ahead and failed to projectively realize that others would do the same. They thought they could outsmart others. Had they understood the recursive nature of the number generation task, they would have randomized.

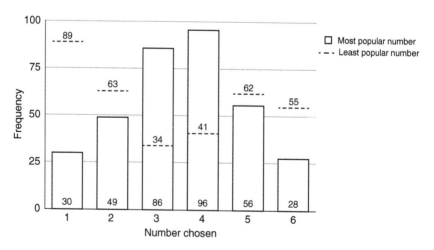

FIGURE 9.2 Frequencies of chosen numbers when instructed to select the most vs. least popular number from 1 to 6.

These findings shed light on uncertainty aversion (vs. tolerance) and the psychology of insecurity. Whereas game theory uses randomness to find Nash equilibria when there are no dominant strategies, the theory treats all players the same, at least if the payoffs are symmetrical (von Neumann & Morgenstern, 1947). Game theory teaches players how not to get exploited; it does not teach them how to beat an opponent who is being coached by the same game theorist. After John von Neumann proved that there is no winning strategy in the game of rock, paper, scissors (Fisher, 2008), cunning gamers understood that in order to win, they had to step outside of the game (Schelling, 1960). History teaches that those who are the first to change the rules stand to win. Epaminondas of Thebes routed the Spartans at Leuctra when he flouted convention by deepening the left wing of his phalanx. Likewise, Field Marshal Moltke invented new stratagems and won. His nephew, Moltke the Younger, did not, and his armies were thrown back at the River Marne.

Surprise is self-eliminating. Once a new trick has been played, it is no longer new. The element of surprise has to continually renew itself, a point that is as trivial as it is profound. The search for an effective surprise does not change the self-limiting logic of recursive mentalization. Any would-be innovator must worry that opponents have even newer tricks of their own. The possibility of being surprised in a strategic environment creates insecurity, which constantly renews itself.

The chooser's challenge in the number game is discoordination. Although the guesser seeks to find a matching number, the chooser wants to foil the effort. Discoordination entails uncertainty and perhaps insecurity. Less obviously, games of coordination do the same. Suppose both the chooser and the guesser get rewarded if their numbers match. These two players are, as it were, members of the same team, but there is no reason to think that they could do better by engaging in so-called 'team reasoning' (Gold & Colman, 2020) than by social projection alone, that is, by picking a number that comes to mind and assuming that this number also comes to the other player's mind. With team reasoning, players would think of a number and, assuming that the other player is also a team reasoner, think that the other would also think of that number. Then, they would, in an exercise of pointless recursion, redouble their commitment to that number. If projection is a sufficient condition of team reasoning, it is unclear how team reasoning proper can improve over the advantages already bestowed by social projection (Krueger, 2014).

We may ask if a coordination version of the number-guessing game is more attractive than the conventional discoordination version. We suspect that a failure to coordinate hurts more than a loss in a competitive discoordination game. In a coordination game, there is no point in blaming an opponent who wants to win. The guesser may only fault the other for not trying hard enough to achieve a common goal. Disappointment at the low performance of a presumed ally is akin to a loss of trust; it is more painful than the upset felt at the expectedly high performance of a known foe.

Trust and Uncertainty

The number-guessing game, like many behavioral dilemmas, is symmetrical in the sense that both parties act without knowing the other's strategy. Both have to simulate it, and then simulate the other's simulation of it, and so forth. Certain sequential games break this symmetry. Here, only the first mover faces the strategic challenge of figuring out what the other will do. The second mover simply needs to respond to the first and play his or her preferences. If the game allows foreplay, the second mover may attempt to signal certain preferences to elicit the most favorable opening from the first mover (Karakowsky et al., 2020).

This sequentiality complicates the uncertainty faced by the first mover. Consider the trust game (Berg et al., 1995). The trustor has a small endowment, say 10 Forint, and a choice between keeping it and transferring it to the trustee. If the money is transferred, it triples in value (other multipliers have also been used). In turn – and the trustor knows this – the trustee then has a choice between keeping the money and splitting it with the trustor, in which case both players end up better off (Evans & Krueger, 2009). Being the second mover, the trustee is free to act on his or her social preferences. If there is foreplay and an opportunity to signal trustworthiness or make promises, the trustee might do so if the trustee thinks the trustor will not consider such talk to be cheap. One might say that the trustee must trust the trustor to trust, but this is not proper trust unless the trustee invests resources that might be lost if the trustor fails to trust.

The trustor faces a more taxing psychological challenge because it is not known whether the trustee is poised to play a coordination game or a discoordination game, that is, whether the trustee will reward or punish the trustor's trust (Krueger et al., 2008). Not knowing what kind of game they are in, trustors experience greater uncertainty than players in the numbers game. As the trust game sets the stage for conflicting emotions, it is likely that trustors experience greater insecurity than players in a number-guessing game (see also Kruglanski & Ellenberg, this volume). Game theorists refer to behavioral choice options as 'strategies.' Trusting is a strategy, and so is distrusting. Recall that in the numbers game, a guesser can reduce uncertainty with the split-range strategy. As we shall see, trustors may also try to reduce uncertainty, which could be their reason for trusting in the first place. The trustee has no uncertainty once the trustor has moved.

The trustor faces the challenge of reducing uncertainty by predicting the trustee's trustworthiness via repeated updating and cue utilization (Grüning & Salmen, 2021). Of course, trustors can eliminate uncertainty entirely by keeping the money, that is, by claiming game-theoretic rationality or by appealing to biases such as loss aversion, the certainty effect, or the endowment effect. A more enlightened trustor, however, may realize that there is an opportunity to learn from trusting and thereby reduce uncertainty. At the limit, a trustor who always trusts in a first encounter can eventually generate a highly accurate estimate of the prevalence of trustworthiness in a population. In contrast, someone who never trusts remains locked in a worldview of suspicion. As a consequence, a paranoiac is not only lonely and highly insecure but also ill-informed about others.

Trustors who accept the social risk of trusting will sometimes be disappointed. By statistical necessity, very high estimates of trustworthiness likely overestimate the actual prevalence of trustworthiness, whereas very low estimates of trustworthiness underestimate this prevalence. Since low expectations of trustworthiness will often end in decisions not to trust, these expectations cannot be updated with experiential data; the estimates of trustworthiness remain too low. Conversely, high expectations are updated by experience and can thereby achieve a high level of calibration (Moore, 2020). When the expectations of optimistic and pessimistic trustors are pooled, the population of trustors underestimates the level of trustworthiness in the population of trustees (Denrell & Le Mens, 2012; Prager et al., 2018). We are, in other words, more distrustful than we need to be (Evans & Krueger, 2016).

How much can optimistic trustors learn from accepting the risk of defection and betrayal? We modeled the trustors' information gain with Bayes's Theorem, assuming that an optimistic trustor begins with a prior probability of the trustee being trustworthy. Let this perceived prior probability of trait trustworthiness, $p(T)$, be greater than .5, for otherwise few trustors would trust. The trustor updates $p(T)$ depending on whether the trustee rewards trust, r, or betrays it, $-r$. Let us also assume that the trustor is interested in the trait of trustworthiness and knows that trustworthy trustees (for whom $p(T)$ is high) on occasion fail to reward trust and that untrustworthy trustees (for whom $p(T)$ is low) sometimes reward it. The trait of trustworthiness translates into corresponding behavior with a bit of unreliability, or with trembling hands as it were. For simplicity, we assume that $p(-r|T) = 1 - p(r|T)$.

Before the trustor trusts, the degree of certainty is the probability of facing a trustworthy other, $p(T)$. If trust is rewarded, the posterior probability is $p(T|r)$; after betrayal, it is $p(T|-r)$. We assume that the former occurs with the prior probability of trust, $p(T)$, and the latter with its complement, $1 - p(T)$. The change from prior to posterior *certainty* is written as

$$\underbrace{p(T) \times \underbrace{p(T|r)}_{\text{Posterior certainty, reciprocation}} + (1 - p(T)) \times \underbrace{p(T|-r)}_{\text{Posterior certainty, no reciprocation}}}_{\text{Probabilities of occurrence}} - \underbrace{p(T)}_{\text{Prior certainty}}$$

where the subtracted term is the prior degree of certainty.

Figure 9.3 shows the net increments in perceived trustworthiness, that is, the difference expressed in the formula, for $.6 \leq p(T) \leq .95$. Differences in the trustee's reliability (i.e., 1 − tremor) are shown in the 5 concave functions.

Besides the finding that the consequences of trusting behavior increase the trustor's certainty regarding trustworthiness, there are two nonlinearities. First, the magnitude of the certainty gain is largest if the prior expectation of trustworthiness falls in the mid-range between maximum uncertainty ($p(T) = .5$) and maximum certainty ($p(T) = 1$). Second, this effect is stronger for high levels of trait-behavior reliability ($p(r|T) = .8$) than for very high or low reliability. These are higher-order uncertainties, which the trustor on the street may not be cognizant of, but they are mathematically compelling and socially describable.

We now see that there is a hitherto neglected reason for why people trust: they stand to learn information about their social world, information that would remain hidden if trustors refused to take social risks (see also Hirschberger; Hogg & Gaffney; van den Bos, this volume). To trust is to forage the social world for productive patches of exchange (Stephens et al., 2007). Under the conditions we modeled, trustors will gain, despite occasional setbacks, confidence that others are trustworthy, and their estimates of the prevalence of trustworthiness will gain precision. Critically, the first of these gains is not obtained when trustors set out with inflated estimates of trustworthiness. Here, repeated experiences with a mix of trustworthy and untrustworthy trustees will result in updated beliefs that are lower than the prior beliefs. The logic of regression requires this (Fiedler & Krueger, 2012). This is why parents teach their children not to trust others indiscriminately.

Trust is multi-determined. Material and informational interests affect it. A recent study sheds light on how trustors strategize (Evans et al., 2021). Trust

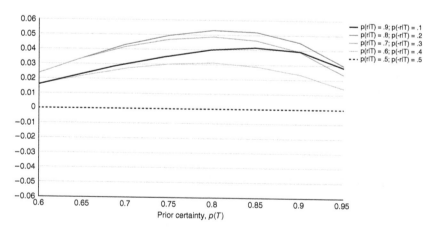

FIGURE 9.3 Association between prior certainty, $p(T)$, and posterior gain of certainty.

increased with perceived trustor-trustee similarity, and three psychological variables mediated this effect. Similarity heightened the expectation of reciprocity, increased liking of the trustee, and increased a sense of moral obligation to trust. Only the first mediator, expectation, picks up on the experience of uncertainty and the hope for its reduction. This is why expectations should be most responsive to experience. Interpersonal liking should follow such that rewarders are liked more than betrayers irrespective of self-other similarity (Clement & Krueger, 1998; see also Mikulincer & Shaver, and Murray & Lamarche, this volume). Finally, normative beliefs regarding one's obligation to trust should be most resistant to learning from experience, which is why trustors with strong normative beliefs are most vulnerable to exploitation. In conclusion, there is a tension between the informational benefits of trusting and the dangers of principled or consistent trust. Trust remains an irreducible social dilemma; it can be described, modeled, and managed, but it cannot be eliminated (Krueger & Evans, 2013).

Free-Will Belief, Determinism, and Uncertainty

Players in the numbers game and in the trust game might experience their own choice and agency as freely willed. 'I could,' they might say, 'have acted differently.' A cheeky response to this assertion is, 'Why didn't you?' Such rhetoric would give no comfort to either the free-will believer or the skeptic. If the chooser had chosen differently, the same claim and the same response could be made. It's much like arguing whether the number 3 is random. Any individual act or event in time eliminates its alternatives, and this is so both under the doctrine of free will and under the doctrine of determinism. It is so even under the doctrine of randomness (MacIntyre, 1957). When the die falls and shows the number 6, it may be said that things could have been different, but they weren't. Yet, the human ability to engage in counterfactual ideation is critical for many reasoning tasks (Dawes, 1988) and for causal inference in particular (Morgan, 2014; Roese, 1995). When counterfactuals are marshaled to defend the doctrine of free will, we submit, the ability to reason about that which did not happen is being abused to support an absurd claim.

Determined critics of freedom of the will see the belief in this freedom as equivalent to visual illusions (e.g., Pinker, 2021). It may seem that we *will* our actions and that we could have acted differently, but no science supports this view (Seth, 2021). As Emil Cioran (2012/1973, p. 92) put the dilemma: "I *feel* I am free but I *know* I am not" (italics in the original). Still, the visual illusion analogy has its limitations. Whereas illusions, such as Doktor Poggendorf's, fool most of us, many people reject the libertarian doctrine of free will, as documented by survey research into individual differences in the degree to which people embrace free-will beliefs (Grüning & Krueger, 2023; Nadelhoffer et al., 2014; see also Sarkissian et al., 2010; Genschow & Lange, 2022; Westfal et al., 2021).

We must therefore ask if the belief in the freedom of free will offers psychological benefits to some people (see also Cooper & Pearce, this volume).

One indication of free-will belief's functionality is its correlation with certain religious values. Some Abrahamic religions, such as Judaism (Schimmel, 1977) and Catholicism, insist on the freedom of the will, which allows them to view transgressions as blameworthy and punishable (see also Fiske, this volume, for similar patterns in other cultures). Ironically, however, the presumed necessity of free will for attributions of responsibility and the delivery of just deserts may itself be an illusion. Calvinists, for example, get along without the notion of free will and without having stopped to punish transgressors (Krueger, 2022). Arguably, some of these ideas are matters of theological dogma, and ordinary people show greater flexibility in what they profess (Murray et al., 2021).

An hypothesis regarding the psychological functionality of free-will belief is that it is linked to the experience of – and aversion to – uncertainty. To explore this possibility, we asked 62 participants to respond to some items from the Free-Will Inventory (e.g., "I always have the ability to do otherwise," Nadelhoffer et al., 2014) and the Intolerance of Uncertainty Scale (e.g., "Unforeseen events upset me greatly"; Buhr & Dugas, 2002). The questions were framed to refer to the participants' personal perspectives (see Appendix B for all items). We found a small negative correlation, $r = -.24$, suggesting that stronger free-will beliefs are associated with a higher tolerance of uncertainty. In other words, a belief in the freedom of the will may protect the believer from facing some of the uncertainties of human existence or the uncertainties that come from interactions with other humans who prefer to make themselves unpredictable.

As part of the functionality of social beliefs is grounded in the reputations they support, we asked a sample of observers to judge target individuals in terms of their competence and their morality given the targets' beliefs about free will (weak vs. strong) and their degree of uncertainty aversion (low vs. high; see also Hogg & Gaffney, this volume). We presented four target person profiles to each of 180 participants, who then rated each target on their competence (competent, intelligent, rational) and morality (benevolent, moral, principled).

Figure 9.4 shows that a target expressing a strong belief in free will was rated both as more competent, $d = .35$, and as more moral, $d = .67$, than a target expressing a weak belief, suggesting a strong positive halo effect. 'People that believe in their freedom to choose,' respondents seem to be inferring, 'have to be capable of doing so in the first place, and they are, then, also aware of the responsibility for their actions.'

Figure 9.5 shows the results for uncertainty aversion (a facet of uncertainty intolerance). This belief was only associated with judgments of competence, $d = .35$, but not with judgments of morality. Perhaps, people, in a reverse inference, believe that an uncertain mind is an incapable mind. Those who are not able or skilled have to deal with more uncertainties in life. For the moral

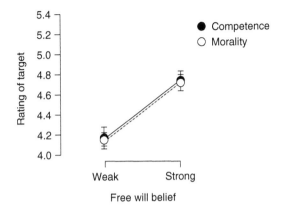

FIGURE 9.4 Participants' rating of the competence and morality of a target person that displays a high vs. low degree of belief in free will.

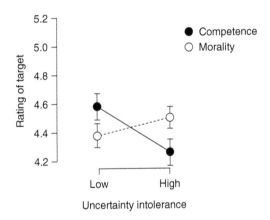

FIGURE 9.5 Participants' rating of the competence and morality of a target person that displays a high vs. low uncertainty intolerance.

virtue of a person, their uncertainty aversion is irrelevant, a null result that is hard to interpret.

A believer in the FWB appears as more competent *and* more moral than a skeptic, whereas an uncertain person signals incompetence but lacks definite cues about moral virtues like their benevolence toward others.

Randomness and Resistance

It is not for us to resolve the free-will debate (see Pleasants, 2019 for a balanced philosophical review). As definitive proof in its favor is lacking, we might either settle for a faith-based belief in free will or we might look elsewhere to ground human freedom. The former strategy does not heal our insecurities; instead, it might deepen them. While Hirt, Eyink, & Heilman (this volume) discuss self-handicapping as a means to reduce personal responsibility in the face of uncertainty, we now consider a radical alternative: freedom through randomness. The time has come to meet the Dice Man.

George Cockcroft, under the pen name Luke Rhinehart (1971), related the adventures of his hero, a New York City psychiatrist turned gambler. Disillusioned by the ego psychology and the human relations movement of his day, which urged clients and encounter groupies to discover their one true self, Rhinehart took a different tack. Instead of one true self, he reasoned, there is an assemblage of egos, desires, tendencies, and potentialities vying to be expressed (Seth, 2021). The most dominant ego – tautologically – wins, and when it wins repeatedly, it creates the illusion of a unitary self (Krueger et al., 2017). Meanwhile, the other egos see their wishes ignored or suppressed. The result is civilized discontent and the neurotic feeling that something is amiss. Ödön von Horváth (1929/1978, p. 67) captured this sentiment when lamenting that "*Ich bin nämlich eigentlich ganz anders, aber ich komme nur so selten dazu.*" [I am actually quite different, but I just rarely get around to it].

Rhinehart's solution was just as ingenious. He opened the door to randomness to give his alter egos a chance. Beginning with a single die and one toss, he laid out several options for action and their probabilities. For example, he would take the trash out if the die yielded a 1 or a 2, and otherwise watch football. Things got complicated – and more fun – fast. Rhinehart introduced variable odds involving several dice, and he included risqué behavioral options that had some motivational appeal, such as having a go with pretty Arlene upstairs. His only rules were that there had to be some otherwise suppressible wishes (though they would be assigned long odds) and that, once the dice had spoken, he had to act as told. The Dice Man thus began to explore himself with a precommitment to obey chance. He expanded his within-person variance of behavior, and he broke the causal link between intention – or a single will – and behavior.

The Dice Man solved the problem of the freedom of the will by breaking the box. He proved that he could have acted otherwise, that is, if the dice had yielded different numbers, but this 'otherwise' was a matter of chance and not a matter of a libertarian will changing itself. This way, the Dice Man found a creative response to the question of freedom. As competing egos were freed, so was 'the person.' But no unitary person remained. Uncertainty, it appears, can be engineered, and the experience can be both liberating and frightening. Uncertainty and insecurity, we recall, are not the same. Whether individuals are comfortable playing the Dice Man's game is for them to find out. Many people will recoil from the experiment, leaving us to conclude that uncertainty and insecurity are empirically related even if they are conceptually distinct.

As noted before, randomization is the game theorist's defense against exploitation. In the Dice Man's world, by contrast, randomization is a strategy the self can use on itself in order to expand freedom, foster creativity, and find new patches of food and entertainment. The uncertainty that comes with this may breed insecurity, but this can be resisted. Following Kant, E. R. Krüger (personal communication, ca. 1970) noted that we need to overcome what he called our inner pigdog in the interest of personal growth and that we do what we need to do even if we are worried about the consequences. Expressing the same sentiment in the film *A Dangerous Method* (Cronenberg, 2011), Otto Gross advises his therapist Carl Jung not to forget to stop for a drink when passing the oasis, transparently implying that Jung should submit to the charms of his patient Sabina Spielrein (Heuer, 2012). Trying to stand on firmer moral ground, conventional psychotherapy takes a similar tack when it recommends confrontation. Under some conditions, confrontation with a feared unknown is necessary for learning to occur and for fear to be mitigated (Meuret et al., 2012; see also Garfinkel, 1964, for an analysis of mildly non-normative behaviors in everyday life). In the context of behavioral research, experimenters may choose to expose themselves to alternate treatments and experiences before they subject their subjects to these (Roberts & Neuringer, 1998). Humans, like pigeons, can learn how to produce, without dice, sequences of behavior that are statistically indistinguishable from the truly random.

The act-different strategy, while liberating, can also signal resentment or *ressentiment* (Reginster, 2021). In a classic contribution to social psychology, Brehm (1966) introduced the concept of *reactance*. Reasoning that people seek to protect their freedom of action, they might choose options of low value as long as that choice signals to themselves and others that they are not slaves to the demands of others or the dictates of instrumental rationality. Dostoevsky raised reactance to a principle to live by. In *The Gambler* (1866/1887), he asserted that he'd rather roam the Kalmyk steppes and sleep in yurts than "bow to the German idol," by which he meant the prudential work ethic, thrift, and general joylessness (see Scheler, 1916, for a corroborating analysis). Though it might create a sense of freedom and self-determination, spite does not make a person unpredictable or unintelligible. Karl Popper (1957) talked of the "Oedipus effect" to refer to a paradoxical linkage between prediction and behavior. If the prophecy (i.e., the prediction) is that Oedipus will kill his father, and he vows not to do it, there is still a perfect, though negative, fit between prediction and intent. When intent and behavior are reactions to prediction, they are not independent of it (Sherman, 1980). This argument does not require the ultimate fulfillment of the prophecy (if that happens, we get an Oedipus *Complex*).

Conclusion

At the end of our foraging excursion into the tangled thicket of uncertainty, randomness, and freedom, we can ask what it is that people might want if they desire to feel less insecure. Would it not be nice to have one's freedom of the will assured while having no uncertainties, being faced with guileless opponents and trustworthy friends, and all the while being smarter than both? This, of course, is an idle dream, and one should beware of what one wishes for. A utopian state of psychological security leaves no room for learning, surprising victories (or defeats), or incentives to break new ground by breaking the old box. What we can say is that the interplay of the various forces we have considered is open to introspection and tractable in scientific study. Decisions as to how to apply these lessons rest with the individual. This is an existential freedom we would not want to take away.

Author Note. We thank George Cockcroft, a.k.a. Luke "The Dice Man" Rhinehart for showing us the joys of randomness in a deterministic world, Joe Forgas for adding nuance to our views on existential dread, and Klaus Fiedler for scattering our epistemic doubts when declaring that the findings 'had to be so.' Anna Cohenuram and Cliff Dutton provided helpful feedback on a draft of this chapter.

References

Berg, J., Dickhaut, J., & McCabe, K. (1995). Trust, reciprocity, and social-history. *Games and Economic Behavior, 10*(1), 122–142. https://doi.org/10.1006/game.1995.1027

Brehm, J. W. (1966). *A theory of psychological reactance.* New York: Academic Press.

Buhr, K., & Dugas, M. J. (2002). The intolerance of uncertainty scale: Psychometric properties of the English version. *Behavior Research and Therapy, 40*(8), 931–945. https://doi.org/10.1016/S0005-7967(01)00092-4

Cioran, E. (2012/1973). *The trouble with being born.* Arcade. Original published in French.

Clement, R. W., & Krueger, J. (1998). Liking persons versus liking groups: A dual-process hypothesis. *European Journal of Social Psychology, 28*(3), 457–469. https://doi.org/10.1002/(SICI)1099-0992(199805/06)28:3<457::AID-EJSP880>3.0.CO;2-T

Cronenberg, D. (2011). *A dangerous method.* https://www.imdb.com/title/tt1571222/

Dawes, R. M. (1988). *Rational choice in an uncertain world.* San Diego, CA: Harcourt Brace Jovanovich.

Denrell, J., & Le Mens, G. (2012). Social judgments from adaptive samples. In J. I. Krueger (Ed.), *Social judgment and decision making* (pp. 151–169). New York: Psychology Press.

Dixit, A. K., & Nalebuff, B. (2008). *The art of strategy: A game theorist's guide to success in business & life.* New York: Norton & Company.

Dostoevsky, F. (1866/1887). *The gambler.* London: Vizetelly & Co.

Evans, A. M., & Krueger, J. I. (2009). The psychology (and economics) of trust. *Social and Personality Psychology Compass: Intrapersonal Processes, 3*(6), 1003–1017. https://doi.org/10.1111/j.1751-9004.2009.00232.x

Evans, A. M., & Krueger, J. I. (2016). Bounded prospection in dilemmas of trust and reciprocity. *Review of General Psychology, 20*, 17–28. Doi.org/10.1037/gpr0000063

Evans, A. M., Ong, H. H., & Krueger, J. I. (2021). Social proximity and respect for norms in trust dilemmas. *Journal of Behavioral Decision Making, 34*(5), 657–668. https://doi.org/10.1002/bdm.2238

Fiedler, K., & Krueger, J. I. (2012). More than an artifact: Regression as a theoretical construct. In J. I. Krueger (Ed.). *Social judgment and decision making* (pp. 171–189). New York: Psychology Press.

Fisher, L. (2008). *Rock, paper, scissors: Game theory in everyday life*. New York: Basic Books.

Fiske, S. T., & Taylor, S. E. (1984). *Social cognition*. New York: Random House.

Garfinkel, H. (1964). Studies of the routine grounds of everyday activities. *Social Problems, 11*(3), 225–250. https://doi.org/798722

Genschow, O., & Lange, J. (2022). Belief in free will is related to internal attribution in self-perception. *Social Psychological and Personality Science*. https://doi.org/10.1177/19485506211057711

Gigerenzer, G. (2015). *Risk savvy: How to make good decisions*. New York: Penguin.

Gigerenzer, G. (2020). How to explain behavior? *Topics in Cognitive Science, 12*(4), 1363–1381. https://doi.org/10.1111/tops.12480

Gigerenzer (2022). *How to stay smart in a smart world: Why human intelligence still beats algorithms*. MIT Press.

Gold, N., & Colman, A. M. (2020). Team reasoning and the rational choice of payoff-dominant outcomes in games. *Topoi, 39*, 305–316. https://doi.org/10.1007/s11245-018-9575-z

Grüning, D. J., & Krueger, J. I. (2021). Strategic thinking: A random walk into the rabbit hole. *Collabra: Psychology, 7*(1), 24921. https://doi.org/10.1525/collabra.24921

Grüning, D. J., & Krueger, J. I. (2023). *Elements of free will belief: Self-enhancement, uncertainty intolerance, and self-other similarity*. Manuscript submitted for publication.

Grüning, D. J., & Salmen, K. (2021). Metacognition as monitoring and control of the cognition-environment fit: A lens model perspective. *The Brunswik Society Newsletter, 36*, 33–35.

Hertwig, R., & Engel, C. (2020) Deliberate ignorance: Choosing not to know. *Strüngmann Forum Reports, 29*, 257–287. Cambridge, MA: MIT Press.

Heuer, G. M. (2012). A most dangerous – and revolutionary – method: Sabina Spielrein, Carl Gustav Jung, Sigmund Freud, Otto Gross, and the birth of intersubjectivity. *Psychotherapy and Politics International, 10*(3), 261–278. https://doi.org/10.1002/ppi.1281

Karakowsky, L., Podolsky, M., & Elangovan, A. R. (2020). Signaling trustworthiness: The effect of leader humor on feedback-seeking behavior. *Journal of Social Psychology, 160*, 170–189. Doi: 10.1080/00224545.2019.1620161

Knight, F. (1921). *Risk, Uncertainty and Profit (vol. XXXI)*. Boston: Houghton Mifflin.

Krueger, J. I. (2007a). The flight from reasoning in psychology. *Behavioral and Brain Sciences, 30*(1), 32–33. https://doi.org/10.1017/S0140525X07000751

Krueger, J. I. (2007b). From social projection to social behaviour. *European Review of Social Psychology, 18*(1), 1–35. https://doi.org/10.1080/10463280701284645

Krueger, J. I. (2014). Heuristic game theory. *Decision, 1*, 59–61.

Krueger, J. I. (2022). Nietzsche's last will. Review of 'The will to nothingness. An essay on Nietzsche's On the genealogy of morality' by Bernard Reginster. *American Journal of Psychology, 135*, 359–362. https://psyarxiv.com/3w746/

Krueger, J. I., & Evans, A. M. (2013). Fiducia: Il dilemma sociale essenziale / Trust: The essential social dilemma. *In-Mind: Italy, 5*, 13–18. http://www.tonymevans.com/wp-content/uploads/2015/07/krueger-evans-2013.pdf

Krueger, J. I., Heck, P. R., & Athenstaedt, U. (2017). The search for the self. In T. Nelson (Ed.). *Getting grounded in social psychology: The essential literature for beginning researchers* (pp. 15–36). New York: Routledge.

Krueger, J. I., & Grüning, D. J. (2021). Psychological perversities and populism. In J. P. Forgas, W. D. Crano, & K. Fiedler (eds.). *The social psychology of populism: The tribal challenge to liberal democracy. The sydney symposium on social psychology, 22*, 125–142. Taylor & Francis.

Krueger, J. I., Hahn, U., Ellerbrock, D., Gächter, S., Hertwig, R., Kornhauser, L. A., Leuker, C., Szech, N., & Waldmann, M. R. (2020). Normative implications of deliberate ignorance. In R. Hertwig & C. Engel (Eds.) *Deliberate ignorance: Choosing not to know.* Strüngmann Forum Reports, 29, 257–287. Cambridge, MA: MIT Press.

Krueger, J. I., Heck, P. R., Evans, A. M., & DiDonato, T. E. (2020). Social game theory: Preferences, perceptions, and choices. *European Review of Social Psychology, 31*, 322–353.

Krueger, J. I., Massey, A. L., & DiDonato, T. E. (2008). A matter of trust: From social preferences to the strategic adherence to social norms. *Negotiation & Conflict Management Research, 1*, 31–52. Doi:10.1111/j.1750-4716.2007.00003.x

Krueger, J. I., Grüning, D. J., & Sundar, T. (2022). Power and sociability. In J. P. Forgas, W. D. Crano, & K. Fiedler (eds.), *The psychology of sociability. The Sydney Symposium on Social Psychology, 23*. New York: Routledge.

MacIntyre, A. C. (1957). Determinism. *Mind, 66*(261), 28–41.

Meuret, A. E., Wolitzky-Taylor, K. B., Twohig, M. P., & Craske, M. G. (2012). Coping skills and exposure therapy in panic disorder and agoraphobia: Latest advances and future directions. *Behavior Therapy, 43*(2), 271–284. https://doi.org/10.1016/j.beth.2011.08.002

Moore, D. A. (2020). *Perfectly confident: How to calibrate your decisions wisely.* New York: HarperCollins.

Morgan, S. L. (2014). *Counterfactuals and causal inference.* Cambridge: Cambridge University Press.

Murray, S., Murray, E., & Nadelhoffer, T. (2021). Piercing the smoke screen: Dualism, free will, and Christianity. *Journal of Cognition and Culture, 21*(1–2), 94–111. Doi: https://doi.org/10.1163/15685373-12340098

Nadelhoffer, T., Shepard, J., Nahmias, E., Sripada, C., & Thompson Ross, L. (2014). The free inventory: Measuring beliefs about agency and responsibility. *Consciousness and Cognition, 25*, 27–41. https://doi.org/10.1016/j.concog.2014.01.006

Pinker, S. (2021). *Rationality.* A. Lane.

Pleasants, N. (2019). Free will, determinism, and the "problem" of structure and agency in the social sciences. *Philosophy of the Social Sciences, 49*(1), 3–30. https://doi.org/10.1177/0048393118814952

Popper, K. R. (1957). *The poverty of historicism.* New York: Harper and Row.

Prager, J., Krueger, J. I., & Fiedler, K. (2018). Towards a deeper understanding of impression formation: New insights gained from a cognitive-ecological analysis. *Journal of Personality and Social Psychology, 115*(3), 379–397. https://doi.org/10.1037/pspa0000123

Reginster, B. (2021). *The will to nothingness.* New York: Oxford University Press.

Rhinehart, L. (1971). *The dice man.* New York: Overlook Press.

Roberts, S., & Neuringer, A. (1998). Self-experimentation. In K. A. Lattal & M. Perone (Eds.), *Handbook of research methods in human operant behavior* (pp. 619–655). New York: Plenum.

Roese, N. (1999). Counterfactual thinking and decision making. *Psychonomic Bulletin & Review, 6*(4), 570–578. https://doi.org/10.3758/BF03212965

Sarkissian, H., Chatterjee, A., De Brigard, F., Knobe, J., Nichols, S., & Sirker, S. (2010). Is belief in free will a cultural universal? *Mind & Language, 25*(3), 346–358. https://doi.org/10.1111/j.1468-0017.2010.01393.x

Seth, A. (2021). *Being you: A new science of consciousness*. London: Faber & Faber.

Scheler, M. (1917). *Über die Ursachen des Deutschenhasses*. Leipzig: Wolff.

Schelling, T. C. (1960). *The strategy of conflict*. Boston: Harvard University Press.

Schimmel, S. (1977). Free-will, guilt and self-control in rabbinic Judaism and contemporary psychology. *Judaism, 26*(4), 418–429.

Sherman, S. J. (1980). On the self-erasing nature of errors of prediction. *Journal of Personality and Social Psychology, 39*(2), 211–221. https://doi.org/10.1037/0022-3514.39.2.211

Stephens, D. W., Brown, J. S., & Ydenberg, R. C. (2007). *Foraging: Behavior and ecology*. Chicago: University of Chicago Press.

von Horváth, Ö. (1929/1978). Rund um den Kongreß. In *Gesammelte Werke / Bd. 3, Lyrik; Prosa; Romane*. Suhrkamp.

von Neumann, J., & Morgenstern, O. (1947). *Theory of games and economic behavior*. Princeton, NJ: Princeton University Press.

Westfal, M., Crusius, J., & Genschow, O. (2021). Imitation and interindividual differences: Belief in free will is not related to automatic imitation. *Acta Psychologica, 219*, 103374. https://doi.org/10.1016/j.actpsy.2021.103374

APPENDIX

Appendix A

Exhaustive presentation of the formula for posterior gain of certainty:

$$p(T) \times \frac{p(T) \times p(r|T)}{p(r)} + (1-p(T)) \times \frac{p(T) \times p(-r|T)}{1-p(r)} - p(T)$$

Averaging as an alternative weight of posterior certainties:

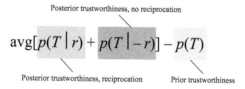

$$\text{avg}[p(T|r) + p(T|-r)] - p(T)$$

Posterior trustworthiness, no reciprocation
Posterior trustworthiness, reciprocation
Prior trustworthiness

Appendix B

Free Will Inventory (Nadelhoffer et al., 2014)

1. People always have the ability to do otherwise.
2. People always have free will.
3. How people's lives unfold is completely up to them.
4. People ultimately have complete control over their decisions and their actions.
5. People have free will even when their choices are completely limited by external circumstances.

6 People's choices and actions must happen precisely the way they do because of the laws of nature and the way things were in the distant past.
7 Human action can only be understood in terms of our souls and minds and not just in terms of our brains.
8 Each person has a non-physical essence that makes that person unique.
9 The human mind cannot simply be reduced to the brain.
10 The human mind is more than just a complicated biological machine.

Scale:

1 Completely disagree
2
3 Neither
4
5 Completely agree

Intolerance of Uncertainty Scale (Buhr & Dugas, 2002)

1 Unforeseen events upset me greatly.
2 It frustrates me not having all the information I need.
3 Uncertainty keeps me from living a full life.
4 One should always look ahead so as to avoid surprises.
5 A small unforeseen event can spoil everything, even with the best of planning.
6 When it's time to act, uncertainty paralyzes me.
7 When I am uncertain I can't function very well.
8 I always want to know what the future has in store for me.
9 I can't stand being taken by surprise.
10 The smallest doubt can stop me from acting.
11 I should be able to organize everything in advance.
12 I must get away from all uncertain situations.

Scale:

1 Not at all characteristic of me
2 A bit characteristic of me
3 Somewhat characteristic of me
4 Very characteristic of me
5 Entirely characteristic of me

10
SEEKING MORAL MEANING IN MISFORTUNE

Assigning Blame, Without Regard for Causation

Alan Page Fiske

DEPARTMENT OF ANTHROPOLOGY, UCLA, USA

Abstract

Faced with illness, suffering, death, famine, or disaster, in most of the world through most of history, people have consulted spirit mediums, oracles, and diviners who reveal who is morally responsible and what can be done to rectify the wrong: appease the offended spirits, ancestors, or deities – and/or identify the witch. Contrary to modern Western explanations of these practices, what is revealed has nothing to do with the material 'cause' of the misfortune. What people seek to know, and what is revealed, is who transgressed what social relationship with whom. In these practices, the biological or physical causation is typically not at issue in any respect, and is quite often not even considered relevant to moral responsibility. A person can be found to be at fault without anyone making the attribution that they had anything at all to do with the material facts of the case, such as the disease or accident. The suffering person or community may be found to be at fault, or they may be entirely innocent victims of wrongdoing. If the sufferer(s) are at fault, they need not have intended the transgression or even been aware that they committed any transgression; their suffering or death is regarded as just, regardless. In these crucial respects and others, the explanations that people seek and find can be said to be 'moral,' in that they concern transgressions of social relationships. But the morality in which people find meaning does not correspond to the modern Western academic concept of morality. It is notable that finding these 'moral' meanings does not make people feel secure, nor does it make them feel that the world is comprehensible, far less controllable. I illustrate these points from my two years of fieldwork among the Moosé of Burkina Faso, including my own consultations with diviners, my three experiences of becoming a diviner myself, and my participation in several funerals where the dead testified who was to blame for their deaths.

DOI: 10.4324/9781003317623-12

In a small West African village, after three days of funeral rituals, two men carry the deceased out of his hut on a litter, wrapped in a blanket and a mat, with only the soles of his feet visible (Fiske 1977–1979 fieldnotes and photographs). Before the deceased reluctantly departs toward his grave, an elder asks him, "Was there wrongdoing?" People want to know who is to blame for the death – not who caused it, but who is morally responsible. (The deceased himself could be at fault, but more often someone else is blamed.) People assume that there must be a transgression behind the death, so they want to know who did what wrong. The deceased knows, and he answers questions by quickly moving downward and toward the questioner to indicate 'yes,' or remaining immobile if his answer is 'no.' He indicates there was wrongdoing, so the elders ask follow-up questions about the nature of the transgression. Asked whether a witch was responsible for his death, the deceased gives an affirmative answer. To check this, another elder asks him the same question, which the deceased again answers affirmatively. So they ask him to identify the witch, which he then does by moving through the crowd right up to one woman. She immediately flees the village, ostracized forever, losing her children and all her property. There is no possible appeal. She returns to her home village; if she is of childbearing age, men will soon court her elders, who will give her in marriage to one of them.

When the man first became ill, he had consulted a diviner, without revealing why he sought the consultation. The diviner repeatedly tossed handfuls of cowrie shells on the ground, from which he read out the nature of the wrongdoing antecedent to the illness. The diviner specified what kind of chickens to sacrifice at what location, and the man soon carried out the prescribed sacrifices. But nonetheless he died. Within hours after his death the elders conducted a preliminary inquest to find out what transgression had been committed and who was to blame for the death – the victim, or someone else. To find out, they cut the throats of a series of chickens, asking a 'yes' or 'no' question of each in succession. If a chicken, after flopping around, came to rest on its breast, the answer was 'yes'; if it died on its back, the answer was 'no.' That inquest had indicated that a witch was responsible, but the elders did not ask the name of the witch.

I observed the testimony of the deceased at several funerals, consulted diviners on many occasions, and, with the guidance of three established diviners, established myself as a diviner. This was 42 years ago, and I've been trying to understand it all ever since.

When Things Go Wrong, *Someone* Must Have Done Something Wrong

One of the remarkable facts about the psychology and anthropology of misfortune is the consistency of response across all sorts of bad experiences. People seek moral meaning – the same sorts of moral meaning – when they, their family, or their community face

> Death
> Illness and pain
> Injury & disablement
> Miscarriage, stillbirth
> Infertility
> Damaging weather such as lightning, tornadoes, hurricanes, floods, tsunamis
> Volcanic eruptions and landslides
> Fires
> Drought and famine
> Pestilence and plague
> Loss of livestock or food stores
> Failures of fisheries or hunting grounds
> Setbacks and failures to attain important goals

In short, *anything* 'bad' evokes a search for moral meaning, no matter what the form of the negative event. When things go wrong, people want to know who did what wrong. It's the badness itself that invites the need for moral meaning; every sort of badness evokes the search for moral meaning. If people's moral accounts of misfortune were 'causal' in any ordinary material, mechanistic sense, then one would expect the contrary pattern: that within any given culture, at least, people would consistently explain each kind of misfortune as consequent to a distinct kind of transgression. In general, that's not the case. In some cultures, a few specific kinds of misfortune are attributed to particular sorts of transgressions, but that sort of linking is far and away the exception to the rule. For the most part, people do not believe they can identify the kind of transgression that has occurred from the kind of misfortune they are suffering.

The worse the event, the more sudden, the more unexpected, and in some sense the more dramatic it is, the more acute is the search for its moral meaning. Strikingly good events incite people to find a moral justification, but the asymmetry is huge: the need for a moral account is vastly greater for bad events than for good. No doubt this is due in part to the consistent tendency to give greater weight to negative events than positive ones (Rozin & Royzman 2001), but there is much more to it than that (see also von Hippel, this volume).

One might imagine that humans hold other persons accountable for misfortunes only when the biological, mechanical, or other known impersonal mechanisms cannot account for the event. But that is not what we observe. Even when, by local cultural standards, the impersonal causal mechanisms are well understood, people seek a moral account as well. The virologist with world-renown expertise in epidemiology and immunology still wants to know what her baby daughter could possibly have done to deserve her Ebola infection; the mother's comprehensive knowledge does not prevent her from raging at the unfairness of her baby being ill with this horrible disease. The West African man whose funeral we began with may have died of a snake bite or a fall from a tree, but his family's knowledge of these causes of mortality does not prevent their need to know who did what wrong. Moreover, in many or most cultures, assignment of blame for misfortune is not guided by beliefs about causal mechanisms, nor is there any sort of evidence about causal means. If a person died when struck by lightning, the search for the moral meaning of the death does not involve any consideration of any process by which the person who is blamed could control lightning. If a person dies when bitten by a snake known to have a lethal bite, typically there is no concern, no speculation, no evidence adduced for how the person held responsible caused the snake to be there and bite the person. In many or most cultures, when accusing a woman of being a witch morally responsible for the death of a person who died in a fall from a tree, no material evidence is presented to demonstrate that she had a means of making the deceased fall. Nor, when accused, does the woman defend herself by saying that she wasn't anywhere near the tree and there is no way she could have made the tree branch break.

Of course, much human suffering occurs at the hands of other humans who intend to do harm: people commit assault, rape, murder, arson, theft, and vandalism (Fiske & Rai 2014). Much other suffering results indirectly from neglect and social practices whose intent may not be harm per se but can be characterized as irresponsible, at the least (see also Forgas; Kreko, this volume). In these circumstances, blaming the perpetrator fits the core sense of what 'morality' means. Likewise, some harm comprises punishment, and so fits directly into our ordinary sense of morality. What I am concerned with in this essay are misfortunes where no human agency is evident, including those where there are no plausible grounds for imagining that human agency could possibly be involved, such as, say, volcanic eruptions or hurricanes. Why are people blamed when there is no manifest sign that humans caused the misfortune?

What Is a "Moral" Account?

Since my question is why people seek moral meaning in misfortune, it will help us to formulate this question more precisely if I clearly state what I mean by 'moral.' Morality concerns the congruence between action and the relational models that people are using to generate, understand, and evaluate their social coordination (Rai & Fiske 2011). Action is moral if it conforms to the cultural implementation of the relational model that people are using to coordinate. Action is immoral if it is incongruent with that model, as locally implemented. Hence, for example, it is morally correct to give each person exactly the same say in a group decision (for example, by voting) if they are making the decision in the framework of equality matching, but it is a moral violation for participants to claim equal say in the framework of an authority ranking relationship such as a military hierarchy.

Note that people often attribute a person's misfortune to social relationships between others. That is, X's suffering may be determined to be due to Y's transgression of a relationship with Z. For example, children may die because a parent transgresses a relationship with someone. The Moosé fear that all of the men in a lineage may die if one of them sleeps with the wife of another. In the Old Testament, Job's neighbors express the same idea – the suffering of his children is due to God's wrath toward Job. The children have no control and likely don't comprehend what's happening. Roughly speaking, we can call this a belief in 'collective responsibility,' or, perhaps better, 'collective consequence,' where anyone or everyone in a given group may suffer for the transgression of anyone else in that group. A person may also die if they themselves commit a transgression; for example, a witch may die if she tries to kill someone whose protective magic is too strong for her.

What do we see when we apply this conceptualization of morality as relationship regulation to the moral accounts that people seek and find for misfortune? We see that people attribute misfortune to the transgression of relationships among living persons, but also to violations of other social relationships. People often attribute misfortune to violations of their social relationships with nonhuman beings, either anthropomorphic or not. For example, one may have offended the spirit of a river, or the misfortune may have to do with the envy of the 'fairies.' People believe that many of their misfortunes are due to the transgression of relationships with ancestors, ghosts, or other deceased persons. And people attribute misfortunes to violation of social relationships with the nonmaterial aspects of living persons, i.e., with their spirits or with their witch aspect. Needless to say, in some (but not all) societies, people account for a great many of their misfortunes by inferring that they have angered deities of one sort or another. What I find notable about this is that in many cultural systems, one finds that these nonhuman beings play very little role in people's day-to-day lives *except* for accounting for misfortune. That is, in many systems, most people rarely have much to do with most of these beings except when attributing misfortunes to violations of human relationships with them. For example, one may find that people rarely mention and hardly ever initiate interaction with, say, ancestors, except when they attribute a misfortune to the anger of the ancestors, and hence make offerings to propitiate them. The rest of the time, people don't seem to think much about or be much concerned with these nonhuman beings. The exception is that people may periodically make offerings to them to show respect, please the beings, and generally sustain their commitments to the beings. That is, people take steps to avert misfortune by maintaining good relations with the nonhuman beings.

Reflecting on this, we may infer that in an important sense *these immaterial beings exist only to explain misfortune* – they have no other function. In short, in many cultures, the only major role that ancestors, spirits, and deities play in the lives of the living is to be credited with harming the living when the living transgress their relationships with such beings. If this is more or less correct, then why would people 'invent' such beings? More accurately, why would fear of and belief in such beings arise and be perpetuated by cultural evolution? Why would biological natural section make people susceptible to belief in and fear of supernatural beings? These are false beliefs, and they are very disruptive to communities, leading to near-universal distrust. Findings of responsibility may entail exile or execution, so determination of the moral meaning in misfortune is no small matter.

These are not simply personal fears and beliefs. In most cultures, they are socially shared, embedded in institutions, and condoned by religious and secular leaders. There are numerous appropriate practices for reaching a legitimate determination of fault: divination, oracles, spirit mediums, sacrificing chickens, witch ordeals, and so on. In these contexts, a person could not easily detach themselves from such ideas, emotions, and practices.

Yet even in cultures that have few of these practices or institutions, even people who do not explicitly believe in supernatural sanctions commonly seek moral meanings in misfortune. For example, when a child dies, religious people in the West do wonder why an all-powerful, all-knowing, perfectly good God could do that. Wondering about that – or being angry – implies that one feels that there is a moral meaning in misfortune. So does the fact that when their child dies, parents commonly feel not only grief, but guilt and anger. Furthermore, in the US, a great many law suits, rather than simply seeking compensation as such, seem to be aimed in whole or in part at assigning moral responsibility for a misfortune (Hensler 2003). More broadly, modern law is widely used as a means of creating justice when people suffer undeserved misfortunes (Friedman 1985). In the modern world, attributions of moral responsibility for collective suffering often take the form of conspiracy theories or populist blaming of elites (van den Bos; van Prooijen, this volume). These seem to get their traction from the feeling that bad things don't simply 'happen'; they must be due to the actions of malevolent people acting behind the scenes. We also see the proclivity to find moral meaning among modern victims of all sorts of trauma. "Often, survivors feel responsible for the death or injury of others, even when they had no real power or influence in the situation" (Murray, Pethania, & Medin 2021:1). Survivors may have the feeling that "other people dying instead of me means that I have done something wrong by surviving" or simply that the others did not deserve to die while I survived (Murray et al 2021:4). "Survivor guilt, however, often exists in the absence of a perception of responsibility. Survivors often know that there is nothing they could have done to prevent the death of another, but feel guilty nonetheless" (Murray et al 2021:4). Interviewing six patients in London who had survived a traumatic event in which others had died, Pethania & Brown (2018:2) found that

> All participants viewed their survival of the traumatic event as 'unfair' or fundamentally 'wrong' because others had died.... This was described alongside a persistent sense of guilt for surviving in the absence of any apparent wrongdoing.... Several participants were also left 'feeling responsible, for the death of others by surviving.

The Standard Account, and What's Wrong with It

Whenever I talk to social scientists or others about why people moralize misfortune, they always give pretty much the same account: People cannot bear to live in an incomprehensible, unpredictable, or uncontrollable world (see also Cooper & Pearce; and Kruglanski & Ellenberg, this volume). Suffering a misfortune threatens people's sense that the world is comprehensible, predictable, and controllable. It is crucial to have *some* account of what happened – any account – even if that account cannot be verified. So people attribute the misfortune to wrongdoing. This attribution is reassuring because it makes the world comprehensible, predictable, and controllable. (Just how a moral account does so is never quite specified.)

A variant of this account, complementary with it rather than mutually exclusive, is anthropomorphism. This perceptual account theorizes that humans

> interpret the world's ambiguities first as those possibilities that matter most. Such possibilities usually include living things and especially humans. Although this strategy leads to mistakes, it also leads to vital discoveries that outweigh them. We see shadows in alleys as persons and hear sounds as signals because if these interpretations are right they are invaluable, and if they are not, they are relatively harmless.
>
> *(Guthrie 1995: vi)*

There are a great many things wrong with this account, both logical and empirical. Logically, it's not apparent why adding one more link to the causal chain solves any of the three problems. The chain of causation cannot be traced back through infinitely many ramifying steps; what does it matter how far back one goes? Why is knowing, say, three links in the causal chain better than knowing only two? If one is the victim of wrongdoing, what is more comprehensible or predictable about the moral account than the causal account without that link? Does one know more about why and when other people commit transgressions against oneself, compared to what one knows about the causal account without that precursor link? Can one control when others harm one any better than one can control the causal processes absent that transgression? Even if the transgression is explained as the moral outcome of one's own transgression, can one typically understand, predict, and control when one transgresses? In any case, it is quite common for people in many cultural systems to explain X's misfortune as the moral repercussion of Y's transgression against Z. Third-party transgressions are neither controllable nor easy to predict, and may not be readily comprehensible. As the ultimate extension of this principle, Genesis says that all human suffering, including mortality itself, is due to God's wrath at Eve and Adam for disobeying him. This leaves humankind with no control over their misfortunes and little or no predictability. Calvinists who believe in predestination are in the same boat: their moral account gives them no control and no predictability (see also Krueger & Gruening, this volume). In societies whose moral code is focused on honor and shame, there is something analogous: a woman will be executed who has been raped – or might have been – though she may have been scrupulously careful not to put herself in jeopardy, and fiercely resisted. The redress of her patriline's honor is required regardless of her lack of control over the event that shamed her and her patriline.

The biggest logical problem with the standard account is that it fails the specificity test. Why is it that around the world and throughout history, people address the posited need for comprehensibility, predictability, and controllability with *moral* accounts, in particular? Wouldn't any other account work just as well, and perhaps better? Why don't people attribute misfortune to the shape of the clouds, the patterns of yellow flowers, the sounds of birds, or what they had to eat? The world would feel more comprehensible, predictable, and controllable if one believed that one's misfortunes were the result of the direction one gazed when first waking or precisely how one combed one's hair. The standard account never attempts to explain the universal predominance of *moral* accounts as such or the total absence of attributions of misfortunes to how one combed one's hair.

Empirically, the standard account makes a strong prediction that there should be a high negative correlation between the search for moral meanings in misfortune and the depth of a person's knowledge about the causal mechanisms leading to the misfortune, as well as with their expertise in controlling these mechanisms. I know of no evidence to support this prediction, and I doubt that it is true. However, it is one possible explanation for the apparent historical decline over the past couple of centuries in socially legitimated moral attributions about misfortune.

Evolutionary psychologists have offered theories to explain the disposition to believe in 'supernatural' *causes* of misfortune, and some of these theories focus on beliefs that misfortunes result from the victim's transgression of a social relationship (see also von Hippel, this volume). These theories generally posit that it is adaptively important to be hypervigilant to recognize without fail whenever one has committed what others perceive to be an offense. They argue that there are large fitness costs of failing to recognize instances when misfortune is a punishment or revenge by others who perceive that one has transgressed, but there are small costs of making the opposite error of thinking that one is perceived to have transgressed and hence been punished when in fact others do not perceive any transgression and are not in fact punishing. It seems to me that theories like this do not fit the ethnographic facts very well. First, because quite often people attribute misfortune to immoral actions of others against the sufferer. For example, the sufferer is perceived to be an innocent victim of a transgressive attack by a witch. Second, these theories fail to account for the common case in which people judge that the sufferer is the innocent victim of a transgression by one third party against another, where the sufferer was not involved in any relevant interaction with either party but nonetheless suffers in consequence. For example, a Moosé man of the chiefly clan dies because his wife had sex with another man. Or people die because the gods or ancestors destroy an entire community because of the sins of a few citizens. Third, these evolutionary theories do not account very well for instances in which a person's misfortune is attributed to the ancestors' or spirits' punishment for his having offended the ancestors or spirits by an act that did no harm to any living person. Fourth and finally, it is surely quite clear to any sentient mammal that no human agent could conceivably have caused her to fall from a tree, be struck by lightning, or get bitten by a snake in the remote bush; no sort of hypervigilance should lead to such absurd *causal* attributions. Yet it is precisely sudden deaths – with obvious causes – that especially evoke the search for moral meaning. Moosé call such an abrupt, dramatic deaths *ku miuugu*, 'red death,' and it is red death that they perceive as certainly indexing transgression by or against the victim. It does not seem to me that a cognitive account based on the relative costs of Type 1 versus Type 2 errors in moral attribution provides a plausible account of moral attributions for events like these that so plainly cannot be influenced by human volition, let alone human action.

The standard account posits that people are in some manner psychologically better off – less stressed, with fewer symptoms and negative sequelae – if they find a moral account of their misfortune that satisfies them. So far as I know, this basic assumption has not been tested (though I certainly don't know all of the relevant literature). But even if that were established, it leaves open the question of what proportion of people *seek* moral meaning in misfortune without *finding* a satisfactory meaning. In modern societies there are few or no legitimate institutions for establishing the true moral meaning of misfortune; there are no procedures for obtaining the testimony of the deceased; there are few diviners, spirit mediums, oracles, or witch-finders; and little legitimacy is accorded to the declarations of those who engage in those practices. In all likelihood, this means that in WEIRD societies, people often fail to find a satisfactory moral meaning, presumably leaving them worse off than if they had not sought one in the first place. The standard account does not address that issue.

The standard account is methodologically individualist, aiming to explain the search for moral meaning in the psychology of the individual, without considering her relationships, her community, and her culture. But of course there is wide cultural variation in specific moral rules and their relative importance, and what is a transgression in one society may be permissible or required in another. Moreover, moral psychology concerns relationship regulation (Rai & Fiske 2011). Hence when people seek moral meanings in misfortune, they examine their social relations to see what went wrong in what relationships. Moosé ask if a misfortune is a repercussion of offending the ancestors, of adultery within the lineage, of witches, and so on. Few people in WEIRD cultures look for moral meaning in those relationships. In many WEIRD cultures, people often regard it as being fair when things are even, such that everyone gets the same thing. In many non-WEIRD cultures, rank-ordered distributions and/or collectively shared resources are more prevalent standards of fairness. So we cannot understand what the search for moral meaning consists of without understanding the networks of relationships in which people live. As discussed above, in most cultures people do not seek moral meanings of misfortune alone, are obligated to use a legitimate means of *determining* the moral meanings of misfortunes, and are supported by social systems that punish, exile, or execute those held responsible. In other words, the most common case around the world is that a person is socially required to find the moral meaning of his misfortunes and those of his family, with the possible moral meanings culturally delineated and the selection of the correct one determined by an obligatory set of practices.

In any case, there is one thing I can say with confidence, based on my two years of participant observation among the Moosé and my fairly extensive reading of relevant ethnographies from around the world. (The classic is Evans Pritchard's wonderful 1937 ethnography; Hutton 2017 offers a broad review focused on fear of witches. On divination and oracles, see Johnston 2009, Devisch 2012, Boyer 2020, Loewe & Blacker 2021). People who live in worlds where everyone takes for granted that most misfortunes are explicable in terms of social transgressions do not find those worlds comfortingly comprehensible, positively predictable, or confidently controllable. On the contrary, they find the world frightening, they don't trust each other, and they live in constant uncertainty. That is, as sociocultural *systems,* the beliefs, fears, and practices that comprise the search for and determination of the moral meanings of misfortune definitely do not enhance psychological wellbeing. They make life stressful, even in between misfortunes.

Attributing Moral Responsibility to a Person Without Concern for How Or Whether She 'Caused' the Misfortune

The most fundamental problem with the standard account is that it is based on an unquestioned premise of modern Western moral philosophy and academic moral psychology: that people are morally responsible only for the effects of their actions. More specifically, this premise holds that a person is responsible for what they do when they either intend the consequences or at least know the consequences, or could reasonably have known them had they taken due care (see also Krueger & Gruening, this volume). Moral responsibility generally is extended to include a negligent failure to act when a prudent, reasonable person could have acted to prevent something bad happening. This 'causal' premise is so profoundly built into what we in the modern West mean by 'moral' that it is difficult to conceive of any other meaning. But suppose, without redefining that term, we posit that a transgression, violation, or wrongdoing may 'result' in a misfortune that it did not, in any mechanical or physical sense, 'cause.' Then it is perfectly intelligible and descriptively valid to say that people often seek to identify the wrongdoing that *resulted in* a given misfortune without implying that the wrongdoing was in any way a material cause of that misfortune.

If you can't escape your modern Western conceptualization of morality, you may object that it makes no sense to imagine transgressions that 'result in' misfortunes that they do not materially 'cause.' All I can say in answer to that objection is that in most cultures through most of history, people did and still do hold people responsible for misfortunes that their purported transgressions did not and often could not cause in any sense of the word. In cultural systems characterized by the relevant fears, beliefs, practices, and institutions, the search for and the finding of transgressions that resulted in one's misfortune generally is not concerned with determining the antecedent material, mechanical, physical causation of the misfortune. The processes of divination and so forth typically make no mention of any means or causal linkages to the misfortune. *Diviners, oracles, and spirit mediums seek and find only moral transgressions.* Conversely, people consulting oracles and spirit mediums often put questions to them, just as we saw in the Moosé death inquest and testimony of the deceased. *These questions invariably concern wrongdoing, with virtually no inquiry into the means or mechanisms of causation.* Furthermore, when accusing someone, say a witch, of moral responsibility for a death or other misfortune, means and material mechanisms are never mentioned, nor does the witch defend herself by arguing that she did not have the means to cause the misfortune.

It is crucial to emphasize that when I write about "causation," I am not referring to (much less limiting the account to) modern Western conceptions of causality, let alone specifically scientific explanations. When I write that in many or most cultures people attributing moral responsibility for misfortunes do not concern themselves with casual processes, I mean their own theories of causation. (Note that, like modern Western theories, some indigenous folk theories posit action at a distance, such that an agent may cause something without having any physical contact with the effects they cause.) Every culture has causal concepts, but most cultures nevertheless blame people for misfortunes without adducing any causal mechanism by which the accused could have brought about the misfortune. Even in cultures where people find physical evidence at autopsy showing that a person is a witch, they do not concern themselves with how this entity could cause misfortunes. The famous example is autopsy in which Azande find *mangu* in the body cavity that identifies a Zande woman as having been a witch, even though all acknowledge that the deceased may truly have never known she was a witch, much less ever intended to harm anyone (Evans-Pritchard 1937).

In sum, in most cultures for most of history, when people suffer a misfortune, most people cognize two processes that can be entirely distinct: the causal mechanism leading to the misfortune and the moral responsibility for it. These can intersect: a person might have poisoned the victim's food, or stolen the missing livestock. But *they need not intersect*. People think about separable causal and moral processes that may operate entirely independently of each other and are often perceived as unconnected – with no need to connect them.

How This Cognitive Predisposition Might Have Evolved

We can draw two conclusions from the ethnographic evidence. First, finding moral meaning in misfortune generally makes the world *less* predictable and *less* controllable than it would be otherwise. Second, moral meanings are not mechanistic causal explanations and typically are not linked to mechanistic causal explanations – the two forms of understanding operate in parallel, for the most part disconnected and independent of each other. So how can we explain why people around the world so often seek moral meanings in misfortune?

I begin with the premise that the limitations and biases of dispassionate cognition make it impossible to sustain cooperative social relationships without the support of social-relationally specific emotions and motives, including moral emotions and motives (Fiske 2002, 2010; Rai & Fiske 2011; Keltner & Haidt 1999). That idea can be extended to help explain the search for moral meaning in misfortune.

In order to conduct social relationships and be a trustworthy partner in them, it is essential that, from early childhood, a person reliably recognize punishment and retribution as such, connecting them to one's wrongdoing (or at least others' perception of one's wrongdoing). That is, a person must appreciate that the harm done to them is a signal that they must rectify a relationship, because it jeopardizes the relationship. This must include an appreciation that harm done to one's family members or community may be punishment for one's own 'individual' transgressions. It is important for a person to recognize not only retaliation by dyadic partners but also recognize third-person punishment. So I posit that *Homo sapiens* have evolved a sensitivity to punishment qua punishment, and not mere harm. In a relatively small, closed, and stable community, people may punish others covertly, in order to avoid the immediate risk and the long-term resentment or vengeance of those they punish. Hence those who suffer misfortunes with no evident human perpetrators may, in fact, have been punished for actual transgressions. And once punishment-recognition proclivities are established, it is likely that misattribution of misfortunes to transgressions is actually beneficial because it further supports cooperation. (For a related theory, see, for example, Norenzayan 2013, Norenzayan et al 2016; but for a review of evidence against this kind of thesis, see Bloom 2012.)

Moreover, it is certainly beneficial to act and speak so as to persuade others that misfortunes – everyone's misfortunes – result from transgression, because that belief makes *others* more reliable cooperators with oneself. Punishing purported perpetrators of transgressions against oneself and against others may be worth the cost if it persuades others to fear punishment for transgressions even when their transgressions might seem to be concealable. One's social relational partners will cooperate more reliably if they expect their violation of relationships to result in their suffering misfortune to themselves or those they care about, regardless of whether one conceals the transgression from human eyes. (Norenzayan, 2013, proposes this as an explanation for the rise of monotheistic religions.) So everyone should promulgate the expectation and fear that transgressions result in misfortune because everyone individually benefits if *everyone else* fears that. Among other implications is that failing to join in the condemnation and punishment of third-party 'transgressors' marks one as unmotivated by the belief that misfortunes are punishments; hence those who fail to join in punishing third-party transgressors will be seen as themselves less committed, less reliable partners. Moreover, depending on the particular cultural beliefs and fears, failure to join in punishment may suggest that one is a party to the transgression, or at least a supporter of the transgressor.

On top of this is another factor. In most systems it is the head of the household or some senior man who consults diviners, shamans, or other misfortune interpreters on behalf of his dependents. If most people believe that misfortunes have moral meanings, then when a dependent suffers a misfortune, a responsible and devout man should consult a diviner. Failure to do so is neglect – it is a failure to protect and care for dependents. So if senior men perceive that others interpret misfortunes as morally meaningful, they will act as if they do, too.

In short and in sum, regardless of whether one is actually afraid of and believes that misfortunes are morally meaningful, it is advantageous for others to think that one does. Thus in the limiting case, even if no one actually feared or believed in supernatural sanctions, it is conceivable that everyone might believe that everyone else does. Then it would be dangerous to be outed as a doubter. This is not as far-fetched a possibility as it may seem at first glance: pluralistic ignorance about moral standards may be fairly common (Prentice & Miller 1996; Vandello, Cohen, & Ransom 2008; Bjerring, Hansen, & Pedersen 2014; Sargent & Newman 2021). That is, most people may believe that most others in their community hold moral standards that, in fact, few or none of them actually hold.

It is important to recognize that what I am proposing we consider is an account at the level of *individual* inclusive fitness (fitness of the individual's genes, carried in herself and her kin). I am not suggesting that this disposition might have evolved through biological group selection, which is apparently rare, if it ever occurs (Soltis, Boyd, & Richerson 1995). This predisposition to find moral meaning in misfortune can clearly evolve through normal selection at the level of individual phenotypes.

So *Homo sapiens* may have evolved the proclivity to believe that misfortunes generally can be attributed to wrongdoing, and, even when they doubt that, to act as if they believed it so as to impel others to believe it.

Author's Note: I am grateful to Joseph Forgas, Bill Crano, and Klaus Fiedler for the invitation to participate in the Visegrad conference where these papers were presented, to the participants for their insightful comments, and to Klaus and Joe for valuable comments on my draft chapter. My thanks also go to Geir Overskeid for his comments on a recent draft.

References

Bjerring, Jens Christian, Hansen, Jens Ulrik, & Pedersen, Nikolaj Jang Lee Linding 2014. On the rationality of pluralistic ignorance. *Synthese* 191:2445–2470.

Bloom, Paul 2012. Religion, morality, evolution. *Annual Review of Psychology* 63:179–199.

Boyer, Pascal 2020. Why divination? Evolved psychology and strategic interaction in the production of truth. *Current Anthropology* 61:100–123.

Devisch, René 2012. Divination in Africa. In Elias Kifon Bongmba, Ed., *The Wiley-Blackwell Companion to African Religions* (pp: 79–96). Malden, MA: Wiley & Sons.

Evans-Pritchard, Edward Evan 1937. *Witchcraft, Oracles and Magic among the Azande*. Oxford: Clarendon Press.

Fiske, Alan Page 2002. Moral emotions provide the self-control needed to sustain social relationships. *Self and Identity* 1:169–175.

Fiske, Alan Page 2010. Dispassionate heuristic rationality fails to sustain social relationships. In Andrea W. Mates, Lisa Mikesell, & Michael Sean Smith, Eds., *Language, Interaction and Frontotemporal Dementia: Reverse Engineering the Social Brain* (pp. 199–241). Oakville, KY: Equinox.

Fiske, Alan Page, & Tage Shakti Rai 2014. *Virtuous Violence: Hurting and Killing to Create, Sustain, End, and Honor Social Relationships*. Cambridge: Cambridge University Press.

Friedman, Lawrence M. 1985. *Total Justice*. New York: Russell Sage Foundation.

Guthrie, Stewart Elliott 1993. *Faces in the Clouds: A New Theory of Religion*. Oxford: Oxford University Press.

Hensler, Deborah R. 2003. Money talks: Searching for justice through compensation for personal injury and death. *DePaul Law Review* 53:417–456.

Hutson, Sadie P., Joanne M. Hall, & Frankie L. Pack 2015. Survivor guilt. *Advances in Nursing Science* 38:20–33.
Hutton, Ronald 2017. *The Witch: A History of Fear, from Ancient Times to the Present.* New Haven: Yale University Press.
Johnston, Sarah Iles 2009. *Ancient Greek Divination.* Oxford: John Wiley & Sons.
Keltner, Dacher & Jonathan Haidt 1999. Social functions of emotions at four levels of analysis. *Cognition & Emotion* 13:505–521.
Kushner, Harold S. 1981. *When Bad Things Happen to Good People.* New York: Anchor.
Loewe, Michael, & Carmen Blacker, Eds. 2021. *Divination and Oracles.* Abingdon: Routledge.
Nisbett, Richard E., & Dov Cohen 2018. *Culture of Honor: The Psychology of Violence in the South.* New York: Routledge.
Norenzayan, Ara 2013. *Big Gods: How Religion Transformed Cooperation and Conflict.* Princeton: Princeton University Press.
Norenzayan, Ara, Azim F. Shariff, Will M. Gervais, Aiyana K. Willard, Rita A. McNamara, Edward Slingerland, & Joseph Henrich 2016. The cultural evolution of prosocial religions. *Behavioral and Brain Sciences* 39 E1. doi:10.1017/S0140525X14001356.
Pethania, Y., Murray, H., & Brown, D. 2018. Living a life that should not be lived: A qualitative analysis of the experience of survivor guilt. *Journal of Traumatic Stress Disorder and Treatment* 7. DOI: 10.4172/2324–8947.1000183
Piaget, Jean 1965. *The Moral Judgment of the Child.* (Marjorie Gabain, translator.) New York: Free Press. (Original work published 1932.)
Prentice, Deborah A., & Dale T. Miller 1996. Pluralistic ignorance and the perpetuation of social norms by unwitting actors. *Advances in Experimental Social Psychology* 28:161–209.
Rai, Tage Shakti, & Alan Page Fiske 2011. Moral psychology is relationship regulation: Moral motives for unity, hierarchy, equality, and proportionality. *Psychological Review* 118: 57–75.
Rozin, Paul, & Edward B. Royzman 2001. Negativity bias, negativity dominance, and contagion. *Personality and Social Psychology Review* 5:296–320.
Sargent, Rikki H., & Leonard S. Newman 2021. Pluralistic ignorance research in psychology: A scoping review of topic and method variation and directions for future research. *Review of General Psychology* 25:163–184.
Soltis, Joseph, Robert Boyd, & Peter J. Richerson 1995. Can group-functional behaviors evolve by cultural group selection?: An empirical test. *Current Anthropology* 36:473–494.
Vandello, Joseph A., Dov Cohen, & Sean Ransom 2008. US Southern and Northern differences in perceptions of norms about aggression: Mechanisms for the perpetuation of a culture of honor. *Journal of Cross-Cultural Psychology* 39:162–177.

PART III
The Role of Insecurity in Social Relationships

11
ATTACHMENT SECURITY AND COPING WITH EXISTENTIAL CONCERNS

Studying Security Dynamics in Dyadic, Group, Sociopolitical, and Spiritual/Religious Relationships

Mario Mikulincer

REICHMAN UNIVERSITY (IDC) HERZLIYA, ISRAEL

Phillip R. Shaver

UNIVERSITY OF CALIFORNIA, DAVIS, USA

Abstract
Attachment theory has become one of the most influential theories in psychology, generating an explosion of empirical research and myriad clinical and educational applications. At the heart of attachment theory is a conception of what we are calling *security dynamics*, a set of psychological processes involved in the search for social sources of safety and security in the context of threats and challenges and the consequences of being comforted, appreciated, supported, and encouraged by others for emotional regulation and goal pursuit. In the current chapter, we focus on the role security dynamics play in shaping the ways we think, experience, and cope with existential concerns of uncertainty, mortality, and meaninglessness. Specifically, we present ideas and evidence that people tend to search for the proximity and support from close relationship partners, social groups and organizations, and supernatural figures when facing existential threats and challenges. Moreover, we review evidence that feeling safe and secure regarding the support of these protective figures in times of need (attachment security) contributes to effective management of existential concerns and adaptive functioning of the social mind/brain.

Attachment theory (Bowlby, 1982) is one of the leading frameworks for understanding emotion regulation, and personal and social adjustment in general, and responses to threats and challenges (see also Arriaga & Kumashiro, this

DOI: 10.4324/9781003317623-14

volume). At the heart of attachment theory is a conception of what we are calling *security dynamics*, a set of psychological processes involved in the search for social sources of security in the context of threats and challenges and the consequences of security attainment for emotional stability and personal growth. Stated simply, actual or anticipated dangers, threats, and challenges motivate us to seek the proximity and support of competent and benevolent others and rely on them for restoring emotional balance and pursuing personal goals. These reactions may well have evolutionary origins, as human survival in our ancestral environment depended on close association and collaboration with others (see also von Hippel; Hirschberger, this volume). Being well cared for by these figures, a person feels appreciated, relaxed, and able to return optimistically to other pursuits. Failure to attain a sense of security leaves distress unresolved and interferes with effective emotion regulation and goal pursuit.

In the current chapter, we focus on the role security dynamics play in shaping the ways we think, experience, and cope with existential concerns of uncertainty, mortality, and meaninglessness (see Cooper & Pearce; Forgas; Kreko; Kruglanski & Ellenberg; Pyszczynski & Sundby; and Hirt et al., this volume, for other ways of coping with these existential concerns). Specifically, we present ideas and evidence that people cope with existential concerns by seeking proximity and support from close relationship partners, the social groups and organizations they belong to and interact with, and God and its earthly representatives (see also Hirschberger; Murray & Lamarche; and van den Bos, this volume). In addition, we show that people who feel safe and secure regarding the support of these protective figures in times of need (i.e., who hold a solid sense of attachment security) tend to effectively manage existential concerns and maintain emotional composure and an optimistic outlook during distressing circumstances.

Overview of Adult Attachment Theory

The Search for Social Sources of Protection and Support

According to attachment theory (Bowlby, 1982), human beings are born with a biologically evolved psychobiological system (*attachment behavioral system*) that motivates them to seek proximity to supportive others (*attachment figures*) in times of need as a way of reducing anxiety and obtaining protection from threats. In Bowlby's (1982) view, attachment figures are special individuals or social entities that are perceived as competent and benevolent and as a potential source of protection and support when needed. People turn to attachment figures for two main provisions: a physical and emotional *safe haven* (i.e., which alleviates fears and distress) and a *secure base* from which to explore, learn, and thrive in a confident manner (Ainsworth, 1991). The goal of proximity seeking is twofold – to feel appreciated, protected, and comforted by an attachment figure when threatened, and to feel that one's strivings for competence and autonomy are approved, supported, and empowered by this figure.

Having a safe haven and secure base over time creates a sense of *attachment security* (confidence that one is worthy and lovable and that others will be protective and supportive when needed). This sense is not just a momentary feeling or the absence of a feeling (e.g., the lack of fear, threat, or insecurity). It is also a matter of cognition (expectations, assumptions, and beliefs) and action tendencies, many of which are automatic and not fully conscious. The sense of attachment security shares with related constructs of personal security, safety, and predictability the belief that the world is safe, but it goes beyond these constructs by emphasizing the importance of a warm and inviting social source of protection and support that sustains hope, courage, and optimism even in distressing circumstances.

Bowlby (1982, 1988) believed that the search for a safe haven and secure base, which is critical for survival during infancy, continues to operate throughout life, as indicated by adults' needs for affection, support, and encouragement when facing threats or challenges (Zeifman & Hazan, 2016). However, proximity-seeking bids may operate somewhat differently at different ages, despite the need for a safe haven and a secure base being important across the lifespan. For example, the identity and type of targeted attachment figures tend to change with development. During infancy, primary caregivers (usually parents or parent substitutes) typically occupy the role of attachment figures (Ainsworth, 1991). However, during adolescence and adulthood, there is extensive evidence that additional relationship partners also become potential attachment figures, including close friends and romantic partners (Zeifman & Hazan, 2016).

Beyond these relationship partners, a wide variety of people can occupy the role of strong and wise caregivers in particular contexts. These figures include teachers, coaches, and mentors in educational and recreational settings; managers in work settings; officers in the military; therapists in clinical settings; health workers in medical settings,; clergy in religion-related settings; and leaders in social and political organizations. Research indicates that any of these people, depending on circumstances, can be perceived as a potential source, in context, of a safe haven and secure base (e.g., Mallinckrodt et al., 2009; Mayseless & Pooper, 2019; Sabol & Pianta, 2012). There is also evidence that other figures and social entities can be recruited as potential attachment figures, such as groups, social institutions, and God (e.g., Granqvist, 2020; see also Hirschberger; Hogg & Gaffney; Murray & Lamarche; and van den Bos, this volume).

The tactics people use for obtaining a safe haven and secure base tend to differ at different ages. In infancy, the tactics are largely innate (e.g., crying when frightened, reaching out to be picked up and held). As a child develops and enters more complex social relationships, attachment behaviors become increasingly flexible, context-sensitive, and skillful (e.g., communicating needs and feelings coherently). In adolescence and adulthood, these tactics are expanded to include many other methods of establishing contact with close relationship partners and asking for protection and support (e.g., phoning or FaceTiming, sending an e-mail or text message). They also include calling upon soothing, encouraging mental representations of attachment figures (e.g., memories, visual and auditory images), and restoring symbolic proximity to these sources of felt security.

At the cognitive level, proximity seeking also includes a wide variety of cognitive tactics aimed at increasing feelings of emotional connection and relatedness with the targeted attachment figure. For example, people can increase their identification with the attachment figure (e.g., self-other overlap, in-group identification, tribalism), emphasize consensus with others in beliefs and values, or heighten their trust on attachment figure's power and benevolence (see Forgas; Kreko; Pyszczynski; and van den Bos, this volume). In this way, others' strengths and resources are experienced, to some extent, as one's own (a process that Aron et al. 1991, called *inclusion of the other in the self*), thereby strengthening confidence that one can count on these resources and strengths for attaining a safe haven and secure base.

Attachment-Figure Responsiveness, Felt Security, and Insecure Forms of Attachment

Bowlby (1973) also discussed important individual differences in the extent to which a person possesses a stable sense of attachment security. In his view, these individual differences are rooted in reactions of attachment figures, over extended periods of time, to one's bids for protection and support in times of need and in the incorporation of these reactions into mental representations of self and others (*internal working models*). Interactions with attachment figures who are sensitive and responsive to one's bids for protection and support foster a sense of attachment security and contribute to positive working models of self and others. When a person's attachment figures are not reliably available and supportive, however, his or her sense of lovability and worth is shaky or absent, others' benevolence is in doubt, and the person becomes less confident in dealing with threats and challenges (Bowlby, 1973).

Pursuing these theoretical ideas in adulthood, many researchers have focused on a person's *attachment orientation*, a systematic pattern of relational expectations, emotions, and behaviors that results from a particular history of interactions with attachment figures (Fraley & Shaver, 2000). These orientations can be conceptualized as regions in a continuous two-dimensional space (e.g., Brennan et al., 1998). One dimension, *attachment-related avoidance*, reflects the extent to which a person distrusts others' benevolence and defensively strives to maintain independence and emotional distance from relationship partners. The other dimension, *attachment anxiety*, reflects the extent to which a person worries that others will not be responsive in times of need and is preoccupied with and intrusive in close relationships. The two dimensions can be measured with reliable and valid self-report scales and are associated in theoretically predictable ways with many aspects of personal well-being and relationship quality (see Mikulincer & Shaver, 2016).

We have proposed that a person's location in the two-dimensional space defined by attachment anxiety and avoidance reflects both the person's sense of attachment security and the ways in which he or she deals with threats and stressors (Mikulincer & Shaver, 2016). People who score low on these dimensions are generally secure and tend to employ constructive and effective affect-regulation strategies. In contrast, people who score high on either attachment anxiety or avoidance or both (a condition called fearful avoidance) suffer from attachment insecurities and tend to hyperactivate or deactivate attachment needs and behaviors in an effort to cope with threats (Arriaga & Kumashiro, this volume; Cassidy & Kobak, 1988).

People who score high on attachment anxiety rely on hyperactivating strategies – energetic attempts to achieve protection and support combined with a lack of confidence that these resources will be provided and feelings of sadness or anger when they are in fact not provided. These reactions occur in relationships in which an attachment figure is sometimes responsive but unreliably so, placing the needy person on a partial reinforcement schedule that rewards exaggeration and persistence in proximity-seeking attempts because these efforts sometimes succeed (Ainsworth et al., 1978). In contrast, people who score high on avoidant attachment tend to use deactivating strategies: Trying not to seek proximity to others when threatened, denying attachment needs, and avoiding closeness and interdependence in relationships. These strategies develop in relationships with attachment figures who disapprove of and punish frequent bids for closeness and expressions of need (Ainsworth et al., 1978).

In short, each attachment strategy has a major regulatory goal (insisting on proximity to an attachment figure or on self-reliance), which goes along with particular cognitive and affective processes that facilitate goal attainment. These strategies affect the formation and maintenance of close relationships (Arriaga & Kumashiro, this volume). Moreover, the strategies affect the ways in which people experience and cope with threatening events, including existential threats – the focus of the following sections of this chapter.

Attachment Orientations and Coping with Existential Concerns

We propose that the attachment system was "designed," or selected, by evolution as a regulatory device for dealing with all kinds of stressors and threats, including existential concerns about annihilation or death, which Bowlby (1982) discussed in relation to the threat of predation in early humans' environments of adaptation (see also von Hippel, this volume). According to several social psychology theories, life events or circumstances that remind us about the fragility of our existence, our biological finitude, or our inability to predict and control the course of one's life are a major source of existential anxiety (about death, uncertainty, or lack of control) and a powerful trigger of psychological defenses (Forgas; Krueger & Gruening; and Pyszczynski & Sundby, this volume). From an attachment perspective, these existential concerns might automatically energize a person's attempts to attain care, protection, and safety. This means that a sense of attachment security should be an effective psychological defense against existential concerns, because it restores a person's sense of value and continuity and renders other symbolic defenses less necessary. In contrast, a lack of responsive attachment figures may cause people to rely on other forms of defense against existential concerns.

Existential Concerns and the Search for Social Sources of Protection and Support

In a study of the mental accessibility of attachment-related representations, Mikulincer et al. (2000) found that even preconscious reminders of death can automatically activate attachment-relevant mental representations. They (Mikulincer et al., 2000, Study 3) exposed participants to the word "death" or a neutral word for 22 milliseconds and found that words related to attachment security (e.g., love, hug, closeness) became more available for processing (as indicated by faster reaction times in a lexical decision task) following the death prime. The word "death" had no effect on the mental access of attachment-unrelated positive words.

There is also evidence that conscious death reminders cause a person to think of seeking proximity to a close other (Pyszczynski & Sundby, this volume). For example, as compared to a neutral condition, experimentally induced death reminders tend to heighten desire for affectionate touch and preference for sitting close to other people in a group discussion (Koole et al., 2014; Wisman & Koole, 2003) and to strengthen the tendency to overestimate the extent to which one's beliefs are shared by others – false consensus effect (McGregor et al., 2005). This tendency to heighten proximity to protect others when facing existential threats is targeted not only to friends, romantic partners, and spouses but also to social groups, sociopolitical entities (e.g., government), and spiritual figures (e.g., God).

Dyadic relationships. Terror management studies revealed that death awareness heightens proximity seeking to romantic partners (see Mikulincer et al., 2003; Plusnin et al., 2018; Pyszczynski & Sundby, this volume, for reviews). In a pioneering study, Florian et al. (2002) randomly assigned participants to write about either their own death (the mortality salience condition) or a neutral topic (watching TV) and then assessed participants' *personal* commitment to their romantic partner as well as their *moral* commitment to the institution of marriage. As compared to the neutral condition, mortality salience increased personal commitment but had no effect on moral commitment. Florian et al. (2002) concluded that death reminders increase the desire to stay close to a romantic partner in the present but not the adherence to cultural norms concerning marriage.

Subsequent studies have extended this line of research to other relational motives, cognitions, and behaviors. In general, their findings indicate that people facing existential threats express more interest in initiating a romantic relationship and heightened desire for intimacy in such a relationship (e.g., Hoppe et al., 2018; Silveira et al., 2014). Moreover, as compared to participants in a control condition, participants exposed to death reminders are more likely to make compromises in their mate selection criteria in order to secure a long-term mate (e.g., Hirschberger et al., 2002). Research also indicates that death reminders increase participants' desire for emotional intimacy with a romantic partner even after this partner has disapproved of or criticized them, implying that death awareness makes people willing to pay the price of diminished self-esteem to maintain relational closeness (Hirschberger et al., 2003).

Group relationships. When facing existential threats, people can turn for protection and support to the social group they belong (Fiske; Hirschberger, this volume). This defensive proximity seeking can be manifested in heightened emotional connection to, and identification with, the group, and the resulting tendencies of accepting, valuing, and endorsing the goals, beliefs, values, and prototypical traits and behaviors of the group (Hogg & Gaffney, this volume). These kinds of proximity-seeking bids might also result in positive attitudes toward other group members and a preference for in-group members over outgroup members.

In a series of surveys, a dramatic increase in patriotism (love of country and in-group solidarity) and identification with national symbols (e.g., flag and anthem) was found among American adults immediately after the 9/11 terrorist attacks (e.g., Li & Brewer, 2004). A similar increase in patriotism was reported recently by Sibley et al. (2020) among New Zealanders immediately following the COVID-19 outbreak (compared to a matched pre-pandemic group). Heightened in-group identification has also been noted following exposure to physical threats (e.g., potential virus contagion; Bélanger et al., 2013), economic and ecological threats (e.g., Barth et al., 2018; Uhl et al., 2018), and threats to one's sense of control and agency (Fritsche et al., 2013). For example, experimental reminders of lack of personal control over life events have been found to increase in-group identification, mainly when groups are framed as strong and competent (Stollberg et al., 2015). These reminders have also been found to increase positive appraisals of in-groups (Fritsche et al., 2013) and conformity with salient in-group (but not out-group) norms (Stollberg et al., 2017).

Terror management studies provide some of the best evidence that existential concerns (i.e., death reminders) activate proximity-seeking inclinations in relation to a group (Pyszczynski & Sundby, this volume). For example, Castano et al. (2002) found that, as compared to Italian participants in a control condition, Italians exposed to death reminders displayed stronger identification with Italians, perceived themselves to share more common characteristics with Italians, and held more positive attitudes toward Italians (vs. Germans). Moreover, mortality salience has been found to increase self-group overlap, causing people to conform to an in-group's beliefs and overestimate their similarity to in-group members in both psychological traits and opinions (Pyszczynski et al., 1996, Renkema et al., 2008; Watanabe & Karasawa, 2012).

Mortality salience also increases positive attitudes toward in-group members. In one of the early terror management studies, Greenberg et al. (1990) found that Christians who were reminded of their mortality gave more positive evaluations of Christians but more negative evaluations of Jews. This finding was conceptually replicated when in-groups and out-groups were arbitrarily formed in the context of a lab experiment (using the minimal group paradigm; Harmon-Jones et al., 1996). Moreover, as compared to a control condition, participants exposed to death reminders tend to report more interest in interacting with an in-group member than an out-group member (e.g., Frischlich et al., 2015).

Sociopolitical relationships. There is also some evidence supporting the idea that existential concerns increase citizens' tendency to rely on social institutions (e.g., government, army) for protection and support (van den Bos; van Prooijen, this volume). Longitudinal surveys conducted in the US, Switzerland, and New Zealand indicate that people expressed greater trust in the government during a pandemic (H1N1, COVID-19) than in a pre-pandemic assessment (Bangerter et al., 2012; Quinn et al., 2013; Sibley et al., 2020). Experimentally induced threats to one's sense of continued existence and control also seem to heighten reliance on governmental services and foster more positive attitudes toward social organizations. For example, Kay et al. (2008) found that, as compared with a control condition, a threat to personal control (requiring participants to recall uncontrollable events) increased explicit requests for help from the government. Prusova and Gulevich (2019) found that a mortality salience manipulation caused Russian students to give the government more responsibility and control over their lives, and Jonas et al. (2011) reported that participants exposed to death reminders were more likely to endorse aspects of their workplace culture than participants in a control condition.

The search for protection and support in times of need may also be directed at social or political leaders (see Forgas; and Kreko, this volume). Popper and Mayseless (2003) proposed that the desire for a strong leader tends to arise in times of personal or collective crisis, trauma, or uncertainty. During demanding and challenging periods, people wish to feel close to and rely upon a leader who can protect them. This tendency was first documented in 1970 by John Mueller, a political scientist who found that sudden international crises during the

Cold War led Americans to rally behind their leaders and increase their trust in American presidents. This "rally behind the leader" tendency was also found in laboratory studies: As compared to a control condition, American participants who were reminded of the 9/11 attacks (by watching a short video clip) expressed more favorable attitudes toward President Bush (Lambert et al., 2011).

Religious/spiritual relationships. Believers turn to God when distressed and tend to strengthen their emotional connection to this omnipotent and omnibenevolent attachment-like figure by praying or heightening their endorsement of religious beliefs and engagement in religious practices (Granqvist, 2020; Koenig et al., 2012). Indeed, studies conducted with patients diagnosed with a serious physical illness have documented more frequent attendance at religious services and a greater tendency to pray for God's protection and support during the illness (e.g., Keefe et al., 2001). In addition, a study of weekly variations in Google search in 16 nations over a period of 12 years revealed that a larger than usual weekly Google search volume for life-threatening illnesses (e.g., cancer, diabetes, hypertension) predicted increases in searches for religious content (e.g., God, Jesus, prayer) during the following week (Pelham et al., 2018). This effect evaporated when examining searches for a non-life-threatening illness (e.g., sore throat).

Increases in religiosity have also been found during and following natural and man-made disasters. For example, Sibley and Bulbulia (2012) found an increase in religious faith among people who were directly affected by the 2011 Christchurch earthquake in New Zealand, and Davis et al. (2019) reported that Louisiana flood survivors who were directly affected by the flooding were especially prone to view God as a source of protection, comfort, and care. Using a large worldwide sample ($N = 86,272$), Du and Chi (2016) found that people living in areas that suffered from more wars or violent conflicts and were more worried about wars tended to be more religious, as assessed by religious practices, religious identity, and belief in God. However, in this case, one can equally argue that heightened religiosity and tribalism might be the cause and not the consequence of intergroup conflicts and wars (Forgas, this volume).

Heightened emotional connection with God can also be observed following experimental inductions of existential threats (making salient one's lack of control or one's mortality) in the laboratory. Kay et al. (2008) found, for example, that, as compared to a control condition, participants who were asked to recall an uncontrollable personal event were more likely to say they believed in God's existence and power. In addition, terror management studies have found stronger endorsement of religious beliefs following a mortality salience manipulation than following neutral inductions among Christians, Muslims, and Agnostics (e.g., Vail et al., 2012).

The dark side of attachment-system activation. Although the search for protection and support from one's social or religious group, institutions, or leaders might, under ideal conditions, result in the provision of a safe haven and secure base and a renewal of the sense of attachment security, it can create two important societal problems under less optimal conditions. First, this search can contribute to the rise of autocratic leaders or dictators who promise to restore order and security at the price of eroding personal freedom and trampling on democratic values and practices (see Forgas; and Kreko, this volume). As Fromm noted in his 1941 book *Escape from Freedom*, fear and insecurity often lead to "a readiness to accept any ideology and any leader if only he offers a political structure and symbols which allegedly give meaning and order to an individual's life" (p. 124). In times of collective crises, people are likely to be particularly susceptible to the influence of and more prone to rely upon charismatic and confident-seeming leaders who proclaim that they have the power to rid the world of distress and chaos. Therefore, when attachment-system activation is directed at political leaders, it can inadvertently contribute to the rise of dictators and the erosion of democracy.

This effect was illustrated by Hertzber (1940), who analyzed the rise of 35 dictators and found that all of them took power in times of collective crises (e.g., Adolf Hitler during a depression following Germany's defeat in World War I). In social psychological research, Gelfand (2018) reported results from surveys she conducted before presidential elections in the US and France showing that people who were more fearful about various social threats (e.g., illegal immigration, unemployment, crime, terrorism) were more supportive of autocratic candidates who offered simple solutions to these complex problems (Donald Trump, Marine Le Pen).

However, these findings do not mean that every collective crisis necessarily results in the rise of a dictator, because more democratic leaders have also attained office during times of upheaval (e.g., Franklin Roosevelt, Winston Churchill). In fact, a solid sense of attachment security might counter this trend and encourage reliance on leaders who not only provide a safe haven during a collective crisis but also respect and support citizens' autonomy and individuality. Attachment-secure people are so confident in their lovability and value and have developed such a strong sense of mastery and autonomy that they might be able to disengage from and rebel against a leader who dismisses or frustrates their needs for freedom and autonomous growth. For example, a person who is securely attached to democratic institutions and confidently relies on them in times of need might be able to fight actively against leaders who promise to restore order and safety at the price of destroying these valued sociopolitical sources of felt security. In our opinion, "rallying behind the leader" in times of need does not necessarily require *escaping from freedom* and endorsing a dictatorship. These possible destructive results of attachment-like behavior within sociopolitical relationships might depend on a person's sense of attachment-felt security and might be particularly characteristic of attachment-insecure people who either anxiously intensify their dependence on strong leaders or search for social tightness and restoration of the social status quo.

The second problem is that manipulative leaders can take advantage of people's need for security and deliberately emphasize the imminence and severity of external threats as a way to increase people's reliance on their leadership (see Forgas; and Kreko, this volume). In a recent article, *Creating Fear and Insecurity for Political Goals*, Bar-Tal (2020) provided examples of authoritarian leaders intentionally attempting to evoke fear and insecurity among their followers as a method of increasing dependence, conformity, and obedience. From our attachment-theory perspective, messages that emphasize external threats can be viewed as activating followers' proximity-seeking bids, which can be targeted toward a leader who promises to eliminate the threats and restore security. In other words, autocratic leadership involves a cynical manipulation of security dynamics in order to amplify followers' search for protection from a leader, resulting in heightened dependence on the leader and increased compliance with his prescriptions and demands.

A similar dangerous dynamic can be observed in religious/spiritual relationships. Placing one's problems in God's hands can put emotionally troubled people at risk of being exploited by manipulative and deceptive religious leaders who promise redemption and inner peace at the price of giving up their individuality and autonomy. As a result, believers can be entrapped in a self-exacerbating cycle of dependence, self-dissolution, fanatical adherence to a faith, and blind merger with the religious group, leader, and cause. This process might be particularly characteristic of attachment-anxious believers, who may be overly dependent on their religious faith and leaders (Granqvist, 2020). However, this dark side of attachment to God might be avoided by a solid sense of attachment security, which sustains and promotes more mature forms of spirituality, curious exploration of others' religious beliefs, and the ability to disengage from religious/spiritual groups, leaders, and institutions that threaten one's individuality and autonomy (Mikulincer & Shaver, 2016).

Attachment-Related Differences in Managing Existential Concerns

From an attachment perspective, individual differences in the sense of attachment security are highly relevant for explaining the ways people experience and manage existential concerns. In this context, several studies have found that attachment orientations moderate the effects of mortality salience. For example, Mikulincer and Florian (2000) and Mikulincer et al. (1990) found that attachment security is associated with lower levels of death-related thoughts and fear of death measured by self-report scales, projective tests (narrative responses to pictures from the Thematic Apperception Test (TAT)), and cognitive tasks (completion of death-related word fragments). In contrast, attachment anxiety is associated with a heightened fear of death as measured by both self-reports and TAT responses and with greater accessibility of death-related thoughts even when no death reminder is present. Attachment-related avoidance is related to a lower self-reported fear of death but with a higher level of death-related thoughts and anxiety in TAT responses. That is, avoidant individuals tend to suppress death concerns and exhibit dissociation between their conscious and unconscious thoughts about death.

Attachment-related differences have also been found in people's construal of death anxiety (Florian & Mikulincer, 1998; Mikulincer et al., 1990). Anxiously attached people tend to attribute this fear to the loss of social identity after death (e.g., "People will forget me"), whereas avoidant people tend to attribute it to the unknown nature of the hereafter (e.g., "uncertainty about what to expect"). These findings are compatible with secondary attachment strategies. Anxious people hyperactivate worries about rejection and abandonment, viewing death as yet another relational setting in which they can be abandoned or forgotten. Avoidant people work to sustain self-reliance and strong personal control, which leads to fear of the uncertain and unknown aspects of death – threats to perceived control.

Secure and insecure people differ in the way they manage concerns related to death. Although endorsement and validation of one's cultural worldview have been considered the normative defense against existential threats, there is evidence that this response is more characteristic of insecure than secure individuals. For example, experimentally induced death reminders produced more severe judgments and punishments of moral transgressors, greater willingness to die for a political cause, and more support for a conservative presidential candidate only among insecurely attached people (Caspi-Berkowitz et al., 2019; Mikulincer & Florian, 2000; Weise et al., 2008). Securely attached people were less affected by death reminders. Moreover, the experimental priming of attachment security buffered the effects of mortality salience on increased support for violent measures against terrorists (Weise et al., 2008) and increased support for the war in Iraq and harsh foreign policy toward North Korea (Gillath & Hart, 2010).

Some of the studies reveal ways in which secure people react to death reminders. Mikulincer and Florian (2000) found that secure people reacted to mortality salience with an increased sense of symbolic immortality – a constructive, transformational strategy that, while not solving the unsolvable problem of death, leads people to invest in their children's care and to engage in creative, growth-oriented activities whose products live on after death. Secure people also reacted to mortality salience with a heightened desire for intimacy in close relationships (Mikulincer & Florian, 2000), heightened reliance on others in times of need (Cox et al., 2008), and a greater willingness to engage in social interactions (Taubman Ben-Ari et al., 2002). In addition, Yaakobi et al. (2014) found that parenthood can serve as a buffer against mortality salience mainly among more secure people (those scoring relatively low on the avoidance dimension).

Caspi-Berkowitz et al. (2019) also found that secure people reacted to death reminders by strengthening their desire to care for others. In her study, people read hypothetical scenarios in which a relationship partner was in danger of death; the participants were then asked about their willingness to endanger their own life to save their partner's life. Secure people reacted to death reminders with a heightened willingness to sacrifice themselves. Insecure people were generally averse to self-sacrifice and reacted to death reminders with less willingness to save others' lives. It's notable that insecure individuals, who seem more ready than secure ones to die for their cultural worldviews, are more reluctant to sacrifice themselves for a particular other person.

People differing in attachment security also differ in their perceptions of life's meaning and ways of coping with the threat of meaninglessness. For example, feelings of closeness and social support (which are core aspects of the sense of attachment security) are associated with a heightened sense of life's meaning (e.g., Hicks & King, 2009; Steger et al., 2008). Similarly, Lambert et al. (2010) reported that perceived closeness to family members and support from them was associated with greater meaning in life among young adults, even when self-esteem, feelings of autonomy and competence, and social desirability were statistically controlled. Moreover, implicit priming of relational closeness increased the perception of life's meaning when participants were in a bad mood (Hicks & King, 2009). In contrast, experimental manipulations of rejection, social exclusion, and loneliness (which are related to attachment insecurity) reduce people's sense that life is meaningful (e.g., Hicks et al., 2010; Stillman et al., 2009; Williams, 2007).

Concluding Remarks

In this chapter, we have reviewed the evidence that when people encounter distressing threats and challenges, they tend to seek proximity to social sources of protection and support. This tendency is not only targeted at close relationship partners but also at groups, sociopolitical entities, and supernatural figures that people perceive as competent and benevolent and a potential source of safety and security. As shown throughout the chapter, many mental and social processes studied by personality and social psychologists working outside the attachment framework, including many within this volume, can also be conceptualized as instances of attachment-like behaviors with regard to groups, organizations, or supernatural figures.

We also attempted to highlight the important role that the sense of attachment security plays in coping with existential concerns. Although existential threats are obviously real and of great consequence, it would be a mistake to conclude that human beings are insufficiently equipped to deal with them or cannot do so without erecting psychologically distorting and socially damaging defenses. A host of studies show that people who have developed dispositional attachment security deal effectively with the fact of mortality and the need for meaning, certainty, and control. Moreover, they deal with these threats while remaining relatively open, optimistic, internally integrated, and well connected socially. We had space here to focus on only a few examples, but there are other relevant and important studies of attachment security and honesty, authenticity, and creativity (e.g., Gillath et al., 2010; Mikulincer et al., 2011). Overall, a coherent body of research indicates that people who are treated well by others, beginning early in life, find life engaging, enjoyable, and meaningful. In this way, the current chapter helps tie together many phenomena and findings that highlight both our natural tendency to rely on others to cope with stress (social allostasis) and the crucial role that the sense of attachment security plays in sustaining effective emotional regulation and goal pursuit.

References

Ainsworth, M. D. S. (1991). Attachment and other affectional bonds across the life cycle. In C. M. Parkes, J. Stevenson-Hinde, & P. Marris (Eds.), *Attachment across the life cycle* (pp. 33–51). New York: Routledge.

Ainsworth, M. D. S., Blehar, M. C., Waters, E., & Wall, S. (1978). *Patterns of attachment: Assessed in the Strange Situation and at home*. Hillsdale, NJ: Erlbaum.

Aron, A., Aron, E. N., Tudor, M., & Nelson, G. (1991). Close relationships as including other in the self. *Journal of Personality and Social Psychology, 60,* 241–253.

Bangerter, A., Krings, F., Mouton, A., Gilles, I., Green, E. G. T., & Clémence, A. (2012). Longitudinal investigation of public trust in institutions relative to the 2009 H1N1 pandemic in Switzerland. *PLoS One, 7,* Article e49806.

Bar-Tal, D. (2020). Creating fear and insecurity for political goals. *International Perspectives in Psychology: Research, Practice, Consultation, 9,* 5–17.

Barth, M., Masson, T., Fritsche, I., & Ziemer, C.-T. (2018). Closing ranks: Ingroup norm conformity as a subtle response to threatening climate change. *Group Processes & Intergroup Relations, 21,* 497–512.

Bélanger, J. J., Faber, T., & Gelfand, M. J. (2013). Supersize my identity: When thoughts of contracting swine flu boost one's patriotic identity. *Journal of Applied Social Psychology, 43,* E153–E155.

Bowlby, J. (1973). *Attachment and loss: Vol. 2. Separation: Anxiety and anger*. New York: Basic Books.

Bowlby, J. (1982). *Attachment and loss: Vol. 1. Attachment* (2nd ed.). New York: Basic Books.

Bowlby, J. (1988). *A secure base: Clinical applications of attachment theory*. London: Routledge.

Brennan, K. A., Clark, C. L., & Shaver, P. R. (1998). Self-report measurement of adult attachment: An integrative overview. In J. A. Simpson & W. S. Rholes (Eds.), *Attachment theory and close relationships* (pp. 46–76). New York: Guilford Press.

Caspi-Berkowitz, N., Mikulincer, M., Hirschberger, G., Ein-Dor, T., & Shaver, P. R. (2019). To die for a cause but not for a companion: Attachment-related variations in the terror management function of self-sacrifice. *Journal of Personality and Social Psychology, 117,* 1105–1126.

Cassidy, J., & Kobak, R. R. (1988). Avoidance and its relationship with other defensive processes. In J. Belsky & T. Nezworski (Eds.), *Clinical implications of attachment* (pp. 300–323). Hillsdale, NJ: Erlbaum.

Castano, E., Yzerbyt, V., Paladino, M.-P., & Sacchi, S. (2002). I belong, therefore, I exist: Ingroup identification, ingroup entitativity, and ingroup bias. *Personality and Social Psychology Bulletin, 28*, 135–143.

Cox, C. R., Arndt, J., Pyszczynski, T., Greenberg, J., Abdollahi, A, & Solomon, S. (2008). Terror management and adults' attachment to their parents: The safe haven remains. *Journal of Personality and Social Psychology, 94*, 696–717.

Davis, E. B., Kimball, C. N., Aten, J. D., Hamilton, C., Andrews, B., ..., & Chung, J. (2019). Faith in the wake of disaster: A longitudinal qualitative study of religious attachment following a catastrophic flood. *Psychological Trauma: Theory, Research, Practice, and Policy, 11*, 578–587.

Du, H., & Chi, P. (2016). War, worries, and religiousness. *Social Psychological and Personality Science, 7*, 444–451.

Florian, V., & Mikulincer, M. (1998). Symbolic immortality and the management of the terror of death: The moderating role of attachment style. *Journal of Personality and Social Psychology, 74*, 725–734.

Florian, V., Mikulincer, M., & Hirschberger, G. (2002). The anxiety buffering function of close relationships: Evidence that relationship commitment acts as a terror management mechanism. *Journal of Personality and Social Psychology, 82*, 527–542.

Fraley, R. C., & Shaver, P. R. (2000). Adult romantic attachment: Theoretical developments, emerging controversies, and unanswered questions. *Review of General Psychology, 4*, 132–154.

Frischlich, L., Rieger, D., Dratsch, T., & Bente, G. (2015). Meet Joe Black? The effects of mortality salience and similarity on the desire to date in-group versus out-group members online. *Journal of Social and Personal Relationships, 32*, 509–528.

Fritsche, I., Jonas, E., Ablasser, C., Beyer, M., Kuban, J., Manger, A.-M., & Schultz, M. (2013). The power of we: Evidence for group-based control. *Journal of Experimental Social Psychology, 49*, 19–32.

Fromm, E. (1941). *Escape from freedom.* New York: Farrar & Rinehart.

Gelfand, M. (2018). *Rule makers, rule breaker: How tight and loose cultures wire our world.* New York: Scribner.

Gillath, O., & Hart, J. (2010). The effects of psychological security and insecurity on political attitudes and leadership preferences. *European Journal of Social Psychology, 40*, 122–134.

Gillath, O., Sesko, A. K., Shaver, P. R., & Chun, D. S. (2010). Attachment, authenticity, and honesty: Dispositional and experimentally induced security can reduce self- and other-deception. *Journal of Personality and Social Psychology, 98*, 841–855.

Granqvist, P. (2020). *Attachment in religion and spirituality.* New York: Guilford Press.

Greenberg, J., Pyszczynski, T., Solomon, S., Rosenblatt, A., Veeder, M., Kirkland, S., & Lyon, D. (1990). Evidence for terror management theory II: The effects of mortality salience on reactions to those who threaten or bolster the cultural worldview. *Journal of Personality and Social Psychology, 58*, 308–318.

Harmon-Jones, E., Greenberg, J., Solomon, S., & Simon, L. (1996). The effects of mortality salience on intergroup bias between minimal groups. *European Journal of Social Psychology, 26*, 677–681.

Hertzber, J. O. (1940). Crises and dictatorships. *American Sociological Review, 5*, 157–160.

Hicks, J. A., & King, L. A. (2009). Positive mood and social relatedness as information about meaning in life. *Journal of Positive Psychology, 4,* 471–482.

Hicks, J. A., Schlegel, R. J., & King, L. A. (2010). Social threats, happiness, and the dynamics of meaning in life judgments. *Personality and Social Psychology Bulletin, 36,* 1305–1317.

Hirschberger, G., Florian, V., & Mikulincer, M. (2002). The anxiety buffering function of close relationships: Mortality salience effects on the readiness to compromise mate selection standards. *European Journal of Social Psychology, 32,* 609–625.

Hirschberger, G., Florian, V., & Mikulincer, M. (2003). Strivings for romantic intimacy following partner complaint or partner criticism: A terror management perspective. *Journal of Social and Personal Relationships, 20,* 675–687.

Hogg, M. A. (2007). Uncertainty-identity theory. In M. P. Zanna (Ed.), *Advances in experimental social psychology* (Vol. 39, pp. 69–126). New York: Academic Press.

Hoppe, A., Fritsche, I., & Koranyi, N. (2018). Romantic love versus reproduction opportunities: Disentangling the contributions of different anxiety buffers under conditions of existential threat. *European Journal of Social Psychology, 48,* 269–284.

Jonas, E., Kauffeld, S., Sullivan, D., & Fritsche, I. (2011). Dedicate your life to the company! A terror management perspective on organizations. *Journal of Applied Social Psychology, 41,* 2858–2882.

Kay, A. C., Gaucher, D., Napier, J. L., Callan, M. J., & Laurin, K. (2008). God and the government: Testing a compensatory control mechanism for the support of external systems. *Journal of Personality and Social Psychology, 95,* 18–35.

Keefe, F. J., Affleck, G., Lefebvre, J., Underwood, L., Caldwell, D. S., Drew, J., Egert, J., Gibson, J., & Pargament, K. (2001). Living with rheumatoid arthritis: The role of daily spirituality and daily religious and spiritual coping. *The Journal of Pain, 2,* 101–110.

Koenig, H. G., King, D. E., & Carson, V. B. (2012). *Handbook of religion and health* (2nd ed.). New York: Oxford University Press.

Koole, S. L., Sin, M. T. A., & Schneider, I. K. (2014). Embodied terror management: Interpersonal touch alleviates existential concerns among individuals with low self-esteem. *Psychological Science, 25,* 30–37.

Lambert, N. M., Stillman, T. F., Baumeister, R. F., Fincham, F. D., Hicks, J. A., & Graham, S. M. (2010). Family as a salient source of meaning in young adulthood. *Journal of Positive Psychology, 5,* 367–375.

Lambert, A. J., Schott, J. P., & Scherer, L. (2011). Threat, politics, and attitudes: Toward a greater understanding of rally-'round-the-flag effects. *Current Directions in Psychological Science, 20,* 343–348.

Li, Q., & Brewer, M. B. (2004). What does it mean to be an American? Patriotism, nationalism, and American identity after 9/11. *Political Psychology, 25,* 727–739.

Mallinckrodt, B., Daly, K., & Wang, C.-C. D. C. (2009). An attachment approach to adult psychotherapy. In J. H. Obegi & E. Berant (Eds.), *Attachment theory and research in clinical work with adults* (pp. 234–268). New York: Guilford Press.

Mayseless, O., & Popper, M. (2019). Attachment and leadership: Review and new insights. *Current Opinion in Psychology, 25,* 157–161.

McGregor, I., Nail, P. R., Marigold, D. C., & Kang, S.-J. (2005). Defensive pride and consensus: Strength in imaginary numbers. *Journal of Personality and Social Psychology, 89,* 978–996.

Mikulincer, M., Birnbaum, G., Woddis, D., & Nachmias, O. (2000). Stress and accessibility of proximity-related thoughts: Exploring the normative and intraindividual components of attachment theory. *Journal of Personality and Social Psychology, 78*, 509–523.

Mikulincer, M., & Florian, V. (2000). Exploring individual differences in reactions to mortality salience: Does attachment style regulate terror management mechanisms? *Journal of Personality and Social Psychology, 79*, 260–273.

Mikulincer, M., Florian, V., & Hirschberger, G. (2003). The existential function of close relationships: Introducing death into the science of love. *Personality and Social Psychology Review, 7*, 20–40.

Mikulincer, M., Florian, V., & Tolmacz, R. (1990). Attachment styles and fear of personal death: A case study of affect regulation. *Journal of Personality and Social Psychology, 58*, 273–280.

Mikulincer, M., & Shaver, P. R. (2016). *Attachment in adulthood: Structure, dynamics, and change* (2nd edition). New York: Guilford Press.

Mikulincer, M., Shaver, P. R., & Rom, E. (2011). The effects of implicit and explicit security priming on creative problem solving. *Cognition and Emotion, 25*, 519–531.

Mueller, J. E. (1970). Presidential popularity from Truman to Johnson. *American Political Science Review, 64*, 18–34.

Pelham, B. W., Shimizu, M., Arndt, J., Carvallo, M., Solomon, S., & Greenberg, J. (2018). Searching for God: Illness-related mortality threats and religious search Vol. in Google in 16 nations. *Personality and Social Psychology Bulletin, 44*, 290–303.

Plusnin, N., Pepping, C. A., & Kashima, E. S. (2018). The role of close relationships in terror management: A systematic review and research agenda. *Personality and Social Psychology Review, 22*, 307–346.

Popper, M., & Mayseless, O. (2003). Back to basics: Applying a parenting perspective to transformational leadership. *Leadership Quarterly, 14*, 41–65.

Prusova, I., & Gulevich, O. (2019). *The effect of mortality salience on attitudes toward national outgroups.* Higher School of Economics Research Paper No. WP BRP 105/PSY/2019.

Pyszczynski, T., Wicklund, R. A., Floresku, S., Koch, H., Gauch, G., Solomon, S., & Greenberg, J. (1996). Whistling in the dark: Exaggerated consensus estimates in response to incidental reminders of mortality. *Psychological Science, 7*, 332–336.

Quinn, S. C., Parmer, J., Freimuth, V. S., Hilyard, K. M., Musa, D., & Kim, K. H. (2013). Exploring communication, trust in government, and vaccination intention later in the 2009 H1N1 pandemic: results of a national survey. *Biosecurity and Bioterrorism, 11*, 96–106.

Renkema, L. J., Stapel, D. A., & Van Yperen, N. W. (2008). Go with the flow: Conforming to others in the face of existential threat. *European Journal of Social Psychology, 38*, 747–756.

Sabol, T. J., & Pianta, R. C. (2012). Recent trends in research on teacher–child relationships. *Attachment & Human Development, 14*, 213–231.

Sibley, C. G., & Bulbulia, J. (2012). Faith after an earthquake: A longitudinal study of religion and perceived health before and after the 2011 Christchurch New Zealand Earthquake. *PLoS One, 7*, Article e49648.

Sibley, C. G., Greaves, L. M., Satherley, N., Wilson, M. S., Overall, N. C., …, & Barlow, F. K. (2020). Effects of the COVID-19 pandemic and nationwide lockdown on trust, attitudes toward government, and well-being. *American Psychologist, 75*, 618–630.

Silveira, S., Graupmann, V., Agthe, M., Gutyrchik, E., Blautzik, J., ..., & Hennig-Fast, K. (2014). Existential neuroscience: Effects of mortality salience on the neurocognitive processing of attractive opposite-sex faces. *Social Cognitive and Affective Neuroscience, 9*, 1601–1607.

Steger, M. F., Kashdan, T. B., Sullivan, B. A., & Lorentz, D. (2008). Understanding the search for meaning in life: Personality, cognitive style, and the dynamic between seeking and experiencing meaning. *Journal of Research in Personality, 42*, 660–678.

Stillman, T. S., Baumeister, R. F., Lambert, N. M., Crescioni, A. W., DeWall, C. N., & Fincham, F. D. (2009). Alone and without purpose: Life loses meaning following social exclusion. *Journal of Experimental Social Psychology, 45*, 686–694.

Stollberg, J., Fritsche, I., & Bäcker, A. (2015). Striving for group agency: Threat to personal control increases the attractiveness of agentic groups. *Frontiers in Psychology, 6*, Article 649.

Stollberg, J., Fritsche, I., & Jonas, E. (2017). The groupy shift: Conformity to liberal in-group norms as a group-based response to threatened personal control. *Social Cognition, 35*, 374–394.

Stouffer, S. A. (1949). *The American Soldier: Vol. 2. Combat and Its Aftermath.* Princeton, NJ: Princeton University Press.

Taubman - Ben-Ari, O., Findler, L., & Mikulincer, M. (2002). The effects of mortality salience on relationship strivings and beliefs: The moderating role of attachment style. *British Journal of Social Psychology, 41*, 419–441.

Uhl, I., Klackl, J., Hansen, N., & Jonas, E. (2018). Undesirable effects of threatening climate change information: A cross-cultural study. *Group Processes & Intergroup Relations, 21*, 513–529.

Vail, K. E. III, Arndt, J., & Abdollahi, A. (2012). Exploring the existential function of religion and supernatural agent beliefs among Christians, Muslims, Atheists, and Agnostics. *Personality and Social Psychology Bulletin, 38*, 1288–1300.

Watanabe, T., & Karasawa, K. (2012). Self-ingroup overlap in the face of mortality salience. *Japanese Journal of Experimental Social Psychology, 52*, 25–34.

Weise, D. R., Pyszczynski, T., Cox, C. R., Arndt, J., Greenberg, J., & Solomon, S. (2008). Interpersonal politics: The role of terror management and attachment processes in shaping political preferences. *Psychological Science, 19*, 148–155.

Williams, K. D. (2007). Ostracism. *Annual Review of Psychology, 58*, 425–452.

Wisman, A., & Koole, S. L. (2003). Hiding in the crowd: Can mortality salience promote affiliation with others who oppose one's worldview. *Journal of Personality and Social Psychology, 84*, 511–527.

Yaakobi, E., Mikulincer, M., & Shaver, P. R. (2014). Parenthood as a terror management mechanism: The moderating role of attachment orientations. *Personality and Social Psychology Bulletin, 40*, 762–774.

Zeifman, D., & Hazan, C. (2016). Pair bonds as attachments: Mounting evidence in support of Bowlby's hypothesis. In J. Cassidy & P. R. Shaver (Eds.), *Handbook of attachment: Theory, research, and clinical applications* (3rd ed., pp. 416–434). New York: Guilford Press.

12
BEYOND DYADIC INTERDEPENDENCE
Romantic Relationships in an Uncertain Social World

Sandra L. Murray

UNIVERSITY AT BUFFALO, THE STATE UNIVERSITY OF NEW YORK, USA

Veronica M. Lamarche

UNIVERSITY OF ESSEX, UK

Abstract

This chapter broadly examines how individuals sustain social connections with others in the face of uncertainty about the safety of depending on them. It proposes that romantic relationships not only exist as part of a dyadic interdependent network of close *personal* bonds (e.g., children, parents, and friends) but also extended *collective* bonds (e.g., neighbors, health systems, and governments). Consequently, experiences that engender uncertainty about the safety of depending on specific relationships in this network put the overall safety of social connection itself in question. This chapter reviews recent research that suggests romantic relationships play dual roles in lending safety and certainty to personal and collective social connections. As a *source* of uncertainty, being surprised or taken aback by the behavior of romantic partners motivates individuals to find greater safety in the *collective* relationships they share with others, perhaps seeing public-health officials as especially wise or a Prime Minister as an especially adept caregiver of the nation. As an antidote to *uncertainty*, however, questioning the wisdom or being taken aback by the behavior of the *collective* motivates individuals to impose greater safety on the *personal* relationships they share with others, seeing family members as especially nurturant and friends and extended family as especially welcoming. The chapter will conclude by outlining the commonalities between situations that create uncertainty about the safety of social connection and pointing to avenues for future research.

DOI: 10.4324/9781003317623-15

In his 1967 Christmas Sermon on Peace, Martin Luther King Jr. asked his audience:

> Did you ever stop to think that you can't leave for your job in the morning without being dependent upon most of the world? Before you finish eating breakfast in the morning, you've depended on more than half the world. This is the way our universe is structured. It is its interrelated quality...

Humans exist socially connected to others from birth to death (Baumeister & Leary, 1995). People are not only connected to those they hold dear, such as a spouse, child, or cherished friend (Murray et al., 2006), but to individuals and entities they might never meet, such as a reclusive neighbor, a political pundit spewing conspiracy theories, a deadlocked Congress, or a divisive Prime Minister (Holt-Lundstadt, 2018; see also Mikulincer & Shaver, this volume). Although the ways in which our fates are entwined with others' may not always be obvious, our fates are dependent on the actions of other individuals and collectives nonetheless.

On the upside, being able to count on the support of a romantic partner, the goodwill of friends, the communality of neighbors, or the foresight of political leaders can strengthen immune responses (Cohen et al., 2015), attenuate physical pain (Master et al., 2009), alleviate death anxiety (Cox & Arndt, 2012; Plusnin et al., 2018), facilitate personal goal pursuits (Feeney & Collins, 2015; Fitzsimons et al., 2015), make potential foes appear less physically intimidating in stature (Fessler & Holbrook, 2013), and lessen endemic government distrust (Holt-Lundstadt, 2018; Hudson, 2006; Lamarche, 2020; Murray, Lamarche et al., 2021). On the downside, the potential to be disappointed or harmed by the selfishness of romantic partners, the disloyalty of friends, the carelessness of neighbors, or the fecklessness of political leaders creates tremendous vulnerability. People can be left unsure of their identity after a romantic breakup (Slotter et al., 2010), nursing a wounded ego after soliciting advice from a friend (Leary et al., 1995), becoming infected with a life-threatening virus after sharing coffee with a convivial neighbor (Bai et al., 2020), being stressed and distraught over a Presidential election (Blanton et al., 2012), or struggling financially after unexpected government cuts to valued social welfare programs (Hudson, 2006).

The personal and collective ties we share with others thus offer the potential for benefit as well as harm, reducing as well as increasing experiences of insecurity. Depending on the exigencies of the situation, a spouse might be supportive or critical, a friend might be congratulatory or jealous, or a president might be accommodating or obstructive (Murray et al., 2006). Recognizing this duality, Baumeister and Leary (1995) concluded that people are fundamentally motivated to belong – to feel included in *secure or safe* social connections where others protect and care for them rather than hurt or exploit them.

Reflecting the importance of this motivation for human survival, people are equipped with regulatory systems for minimizing the risks of social connection (Kenrick et al., 2010; see also Fiedler & McCaughey, this volume). For instance, the *behavioral-immune* system motivates people to avoid others when they are potentially infected with contagious diseases (Miller & Maner, 2011; 2012; Murray & Schaller, 2006), whereas the *risk-regulation* system motivates people to avoid intimates who might reject or ostracize them (Cameron & Granger, 2019; Murray et al., 2002).

However, in depending on others, people risk exposing themselves to more than physical or emotional harm. They also risk having their very understanding of reality challenged by the actions of others (Heine et al., 2006; Jonas et al., 2014; Murray et al., 2017; see also Cooper & Pearce; Krueger & Gruening, this volume). For instance, a Liberal might be bewildered by her spouse's unexpected tirade against critical race theory, a conservative might be nonplussed by a president's renewed imposition of a public-health mandate, a father might be taken aback by a teenager's inexplicable meltdown over a seemingly minor criticism, and a friend's deep dive into conspiratorial thinking may leave one mystified (see also Jussim, Finkelstein & Stevens; van Prooijen, this volume).

In this chapter, we examine how individuals cope with existential threats to the safety of social connection — those everyday experiences that suggest that people might not understand others or the reality others inhabit as well as they thought. We examine how existential uncertainty (i.e., experiences that cannot be anticipated or explained) can motivate people to feel secure or safe in the world around them rather than insecure or unsafe. We first describe the theoretical underpinnings of the social-safety system, the defensive system that restores perceived security or safety to social connection in the face of uncertainty. In the second part of this chapter, we describe the current state of empirical research supporting the model. In concluding the chapter, we discuss how the social-safety's operation ultimately affects feelings of safety and security in the relational world and point to directions for future research.

The Social-Safety System

People live immersed in multiple layers of social connection. People not only share social connections with those they *personally* know (or love or loathe), but they also share social connections with those they will never personally know but nonetheless depend. Specifically, people share collective ties to others across *personal* (Feeney & Collins, 2015; Mikulincer & Shaver, 2003, this volume) and *sociopolitical* relational worlds (Hudson, 2006).

Personal relational worlds involve *close* others, such as a spouse, sibling, parent, child, in-law, friend, or valued coworker, that people can choose to nurture and value to a greater or lesser degree (Feeney & Collins, 2015). However, *sociopolitical* relational worlds involve relationships with *non-close* others, such as an employer, a teacher, a fellow citizen, a Congress, or a President, that living as part of an organized society foists on people (Hudson, 2006). Despite dissimilarities in familiarity and volition, these relational worlds nonetheless share a defining feature: One's security and fate depend on the actions of others (Kelley, 1979), a type of solidarity Durkheim (1997) described as organic as opposed to mechanical.

For instance, depending on a spouse's advice and comfort can result in hurtful criticisms or reassuring praise; cultivating a teen's excitement for a family trip can court excitement or sullen indifference; and disclosing a secret to a friend can result in greater closeness or betrayed confidences. Similarly, depending on an employer for family leave can result in vacation days spent on a beach or taking one's children to the doctor, trusting local governments to provide clean water can result in safe or tainted water supplies, and relying on fellow community members to vote sensibly can elect experienced politicians or Q-Anon followers to Congress.

Fortunately, people have some power to keep themselves safe from being hurt by the actions of others. However, to exercise this self-protective power, people need to be able to reliably anticipate how others are likely to behave and adjust their own behavior accordingly (Kelley, 1979; Tooby & Cosmides, 1996). For instance, Arya can better safeguard herself against being criticized, maligned, or misled by correctly anticipating when her spouse is motivated to be supportive (vs. critical), her friends are motivated to be congratulatory (vs. jealous), or a president is motivated to be honest (vs. duplicitous). In such situations, correctly anticipating the motivations and/or behavior of a family member, employer, or president provides Arya with reassuring evidence that she understands the reality they inhabit (Higgins et al., 2021). This understanding then allows her to adjust her behavior toward others by seeking advice when her spouse is motivated to be supportive, sharing good news when her friend is motivated to be congratulatory, and being judiciously disbelieving when a president is likely to be duplicitous.

However, people often err when they try to forecast others' behavior because they naively assume that others perceive the same reality they do (Griffin & Ross, 1990; Peetz et al., 2022). This results in individuals, entities, or institutions behaving in ways that violate personal (e.g., "My spouse is a feminist just like me"), historic (e.g., "I've never seen my son enjoy vegetables"), and/or normative (e.g., "Presidents should be prudent") expectations. For instance, people overestimate how positively and negatively others are likely to feel (Pollmann & Finkenauer, 2009), leaving them vulnerable to being bewildered by a spouse's ennui after a promotion or unsettled by an employer's nonchalance in the face of poor earnings. People also misjudge how much gratitude acts of kindness will elicit in others (Kuma & Epley, 2018), leaving them suspicious of a friend's glee over a small favor, puzzled when trading partners reject concessions, or flummoxed when Congress passes bipartisan bills.

Finding Safety in the Face of the Unexpected

With our colleagues, we developed the model of the social-safety system illustrated in Figure 12.1. The system defends people against the potential threat that unexpected and thus uncertainty-evoking behavior poses to the collective safety of social connection. In outlining the daily operation of this system, we first explain when unexpected behavior poses a greater existential threat to the safety of social connection. We then explain the perceptual/cognitive defenses that restore the perception of collective safety to social connection in the face of such threats to certainty (Murray, Lamarche et al., 2021).

Triggering a safety threat. Reflecting its roots in interdependence theory (Kelley, 1979), the model contends that people are *more* strongly motivated to feel safe in social connection when their personal outcomes are *more* (vs. less) tied to the actions of others (Murray et al., 2006; Murray, Lamarche et al., 2021). Outcome-dependence varies by *situation* because situations vary in their features (Kelley et al., 2003). For instance, Arya's outcomes are more dependent on her husband's actions when she needs a favor than when she does not. Arya's outcomes are also more dependent on the actions of their neighbors when COVID-19 cases are rising (vs. falling) in her community (Murray, Seery et al., 2021). And her outcomes are more dependent on the actions of the populace when votes are being cast in more (vs. less) consequential elections (Blanton et al., 2012; Stanton et al., 2009; Trawalter et al., 2011).

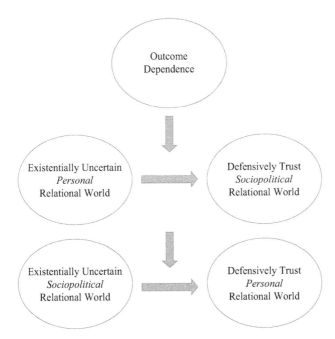

FIGURE 12.1 The social-safety system.

Outcome-dependence also varies by *person* because people differ in the extent to which they are more (vs. less) vulnerable to being harmed by the actions of others. According to evolutionary theorists, people evolved to detect and seek out intimates they can reliably depend upon – intimates who see them as indispensable or special – because loyal alliances afford protection against harm (Tooby & Cosmides, 1996; see also von Hippel & Merakovsky; Hogg & Gaffney; Hirschberger, this volume). In modern life, people rely heavily on romantic partners to provide such protection (Arriaga et al., 2018; Feeney & Collins, 2015; Finkel et al., 2014; Szepsenwol & Simpson, 2019; see also Arriaga & Kumashiro, and Mikulincer & Shaver, this volume). This reliance leaves people who are *less* certain they can trust their romantic partner *more* vulnerable to harm than people who are *more* certain they can trust their romantic partner. For instance, people who are *less* certain they can trust their romantic partners are *more* readily hurt by their partner's transgressions as compared to people who are certain they can trust their partner (Murray et al., 2003). People who are uncertain they can trust their romantic partner are also more readily agitated by the thought of physical pain or human mortality, suggesting they generally feel more vulnerable to the actions of others (Cox & Arndt, 2012; Plusnin et al., 2015).

The model further contends that outcome-dependence is not uniformly threatening. Instead, whether outcome-dependence feels more or less safe depends on existential certainty, which varies *situationally*. According to the model, when a spouse, friend, community, government institution, or president behaves *expectedly*, it provides reassuring evidence that one understands their reality. The resulting state of existential *certainty* makes it easier to believe in one's power to keep oneself safe from harm in such situations (Heine et al., 2006; Jonas et al., 2014; Kay et al., 2015; Murray et al., 2017). Conversely, when a significant person behaves *unexpectedly*, it provides troubling evidence that reality is not consensually perceived, putting one's own understanding of reality into question. The resulting state of existential *uncertainty and insecurity* undermines one's sense of self and identity (see for example van den Bos, and Hogg & Gaffney in this issue), and makes it harder to believe in one's power to keep oneself safe from harm in such situations (Murray et al., 2018).

For instance, a spouse's unexpected behavior poses a greater threat to the safety of social connection when people are counting on their spouse for a specific sacrifice than when they are not. A neighbor's unexpected behavior also poses a greater threat to the collective safety of social connection when *rising* COVID-19 cases make people more dependent on their neighbors to keep them safe from infection than when falling COVID-19 cases make them less dependent. Similarly, a president's unexpected behavior poses a greater existential threat to the collective safety of social connection when being uncertain of a romantic partner's trustworthiness makes people more vulnerable to others. In highly outcome-dependent situations, experiencing *more* unexpected behavior makes salient the existentially troubling possibility that one might not understand others or the reality they inhabit well enough to keep oneself safe.

Indeed, the model assumes that unexpected behavior poses *generalized* rather than localized threats to the safety of social connection in a given relational world. That is, when Arya's spouse behaves unexpectedly, Arya does more than wonder whether she really understands him. She might also wonder, however fleetingly, whether she actually understands anyone close to her. Why would existential uncertainty generalize in this way? Relationships within a given relational world are interconnected, both experientially and cognitively. Experiences depending on a spouse are bound up in experiences depending on children, in-laws, and family friends (Holt-Lundstadt, 2018; Mikulincer & Shaver, 2000). Similarly, experiences depending on fellow community members to behave in socially or legally prescribed ways are bound up in experiences depending on local, state, and federal officials and institutions (Anderson, 2010; Hudson, 2006). Because the experiences people have with others share these past and present interconnections, unexpected behavior on the part of one inhabitant of a given relational world can put the motivations of *other* inhabitants in that relational world into question as well, threatening the safety of that relational world *as a whole*.[1]

Defending against a safety threat. To recap the "threat" side of Figure 12.1, ongoing experiences threaten feelings of *collective* safety in social connection when people are (1) *highly* outcome-dependent (i.e., highly vulnerable to others) and (2) *unable* to anticipate the behavior of the inhabitants of a specific relational world (i.e., highly existentially uncertain). The "defense" side of Figure 12.1 describes how people alleviate the resulting threat to their experience of safety.

Much as pain motivates reflexively withdrawing from its source, we contend that being highly outcome-dependent on the unexpectedly behaving inhabitants of one relational world motivates people to psychologically escape this now riskier relational world (Cavallo et al., 2009; Murray et al., 2006). People typically escape threatening experiences by adopting the beliefs that can most readily and compellingly restore feelings of equanimity (Heine et al., 2006). When unexpected behavior alerts people to the possibility that they might *not* understand one's relational world as well as they thought, the model contends that people can most readily convince themselves that they are still safe *nonetheless* by imposing caring and well-meaning intentions on the inhabitants of the *alternate*, more perceptually *pliable*, relational world (Zunda, 1990). Consistent with this logic, people can escape anxieties about being rejected by others by seeing greater acceptance in the tabula rasa afforded by *new* acquaintances (Maner et al., 2007; Richman & Leary, 2009; Williams et al., 2000). They also escape anxiety about death by turning to those close relationship partners who best afford safety (Plusnin et al., 2018; Young et al., 2015; see also Pyszczynski & Sundby, this volume).

Operating in conjunction in daily life, the social-safety system's dual defenses link the experience of a threat to the safety of social connection in one relational world to its associated *defense* for making collective social connections feel safer nonetheless. Specifically, *personal-to-sociopolitical* threat-defense links the anxiety that one might understand the inhabitants of one's *personal* relational world to the defensive inclination to impose understandable and benevolent motivations on the inhabitants of one's *sociopolitical* relational world. Conversely, *sociopolitical-to-personal* threat-defense links the anxiety that one might not understand the inhabitants of one's sociopolitical relational world to the defensive inclination to impose understandable and benevolent motivations on the inhabitants of one's *personal* relational world. We detail and illustrate each defense in turn.

Personal-to-sociopolitical threat-defense. Through this threat-defense, people counteract the existential uncertainty posed by high outcome-dependence on unexpectedly behaving *intimates* by defensively perceiving greater reason to trust in their *sociopolitical* relational worlds. For instance, being taken aback by a Liberal spouse's complaints about critical race theory or a friend gushing over for a small favor might motivate people to defensively perceive their neighbors, legislators, or president as being *more* capable of taking care of them.

Sociopolitical-to-personal threat-defense. Through this threat-defense, people counteract the existential uncertainty posed by high outcome-dependence on *unexpectedly behaving sociopolitical* relational worlds by defensively perceiving greater reason to trust in their *personal* relational worlds. For instance, being taken aback by a president tweeting about the girth of his "nuclear button" or good employment numbers triggering stock selloffs might motivate people to defensively perceive their family members and friends as being *more* capable of taking care of them. Consistent with this logic, people acutely threatened by national economic uncertainty report more trust in the good intentions of people they know (Navarro-Carrillo et al., 2018), and people primed to think their country is doing poorly report more trust in their partner and in the institution of marriage (Day et al., 2011).

Empirical Examples of the Social-Safety System in Operation

In our research on the social-safety system, we rely on naturalistic daily diary and longitudinal study designs so that we can capitalize on unexpected events as they happen in the *real* world. Accordingly, we report findings drawn from two daily diary studies conducted during the initial years of the Trump administration (Murray, Lamarche et al., 2021), an eight-week longitudinal study of the 2018 U.S. Midterm election (Murray, Lamarche et al., 2021; Murray, Seery et al., 2021), and a three-week longitudinal study of the initial months of the COVID-19 pandemic in the United States (Murray, Seery et al., 2021). In all of these studies, we required participants to be involved in live-in romantic relationships. Given this constraint, most were married for more than a decade and had an average of two children still at home. In the daily diary studies, participants provided online daily reports each day for nine–ten consecutive days. In the midterm election study, participants provided once-weekly online reports for each of the six weeks preceding the 2018 U.S. midterm election, the day after the midterm election, and one week after the election. In the COVID-19 study, participants provided reports every other day for three weeks. We provide the evidence for *personal-to-sociopolitical* and *sociopolitical-to-personal* threat-defense in turn.

Personal-to-Sociopolitical Threat-Defense

When people exhibit this threat-defense, they counteract the existential uncertainty posed by depending on unexpectedly behaving intimates (i.e., the *threat*) by perceiving greater reason to trust in *sociopolitical* relational worlds (i.e., the defense). We examined these dynamics in one daily diary study and the 2018 midterm election study. In these studies, we expected people to defensively trust *more* in the *sociopolitical* relational world when they were (1) *highly outcome-dependent* and (2) existentially confused by the *unexpected* behavior of their intimates.

The daily diary study. In this study, we operationalized *outcome-dependence* through differences between *people* – that is, through individual differences in expressions of romantic trust (i.e., "I can trust my romantic partner completely"; "I can always count on my romantic partner to be responsive to my needs and feelings"; "My romantic partner is always there for me"). As in prior research, we expected participants who were *less* certain they could trust their romantic partners to be more outcome-dependent or vulnerable to being harmed by others (Murray et al., 2006), sensitizing them to the threat posed by unexpected behavior. We operationalized intimates' *unexpected behavior* through daily "yes" responses to six items that asked participants to indicate whether their romantic partner or child(ren) had said or done anything "out-of-the-ordinary", "they did not expect", or anything that "did not make sense" that day. We operationalized *trust in the sociopolitical relational world* through daily expressions of faith that the Trump administration was a trustworthy steward of the nation, which we captured through the daily value participants personally placed on political conservatism and their trust that the (Republican) federal government was doing a good job (Murray, Lamarche et al., 2021).

Because the data had a nested structure (i.e., day within person), we used multilevel analyses to test our hypotheses. These analyses revealed that participants who were less certain they could trust their romantic partner (i.e., *high* outcome-dependence) reported *greater* faith that Republican government's actions secured the country's welfare on the days after family members behaved *more* unexpectedly than usual (i.e., *high* existential threat to safety) as compared to days their family members behaved *less* unexpectedly (i.e., *low* existential threat to safety). However, people who were already certain they could trust their romantic partner evidenced no such compensatory effect (i.e., *low* outcome-dependence).

The 2018 midterm election study. As we did in the daily diary study, we indexed *outcome-dependence* through individual differences in expressions of romantic trust. However, the midterm election study also allowed us to index outcome-dependence *situationally*. With each passing week during the 2018 midterm election season, the public became increasingly aware of the eventual electoral result – Democrats gaining control of the House and Republicans retaining the Senate. Because neither party gained unilateral control of Congress on election day, we expected *not* knowing whether one's preferred party would ultimately wield enough power over the country's governance *in the future* to make the risks of having one's fate tied to the votes cast by fellow community members more salient to partisans of both stripes. Therefore, we expected participants to be more keenly aware of their dependence on the collective populace to cast the "right" votes *after* the election than before the election.

We operationalized the daily *unexpected behavior* of intimates through "yes" responses to eight items that asked participants to separately indicate whether their romantic partner or child had said or done anything "out-of-the-ordinary" anything "they did not expect" or anything that "did not make sense" and whether *they themselves* had any thoughts or feelings about their romantic partner or children "they did not expect to have" that day. We operationalized *trust in the sociopolitical relational world* through sympathy for the brand of Republicanism President Trump routinely Tweeted. Namely, each week, we asked participants to report how much they personally (1) distrusted the media (e.g., "The mainstream media cannot be trusted"), (2) distrusted progressivism (e.g., "American society needs to be radically restructured", reversed, "The structure of American society needs to change", reversed), and (3) favored economic (e.g., "fiscal responsibility", "business") over social conservatism (e.g., "the family", "religion"). We targeted the relative priority of economic over social conservatism as a metric of sympathy for President Trump because perceiving such an economic bias in his policies made Trump more sympathetic and appealing to swing voters in the 2016 federal election (Silver, 2019).

Multilevel analyses revealed the expected operation of the social-safety system. After the results of the election were known (i.e., *high* outcome-dependence), participants who were less certain they could trust their romantic partner (i.e., *high* outcome-dependence) reported significantly *greater* faith that the Republican government's actions secured the country's welfare on the days after family members behaved *more* unexpectedly than usual (i.e., high existential threat to safety) as compared to days their family members behaved *less* unexpectedly (i.e., low existential threat to safety). However, these compensatory effects were not evident well in advance of the election (i.e., *low* outcome-dependence), nor were they evident when people were certain they could trust their romantic partner (i.e., *low* outcome-dependence).

Of course, it is relatively easy for participants who are motivated to find safety in the *sociopolitical* relational world to profess greater trust in the Republican-led government. To provide an even more telling test of the social-safety system's power, we asked participants to report how they voted. Even though voting is strongly partisan in nature, these data nevertheless revealed a shift in voting preferences, but only when people needed to find *greater* safety in the *sociopolitical* world to defend against unexpected behavior in their *personal* relational world. Specifically, participants who were *less* certain they could trust their romantic partner (i.e., *high* outcome-dependence) were *more* likely to cast votes for Republican candidates when their family members behaved *more* unexpectedly in the five weeks preceding the midterm (i.e., *high* existential uncertainty), as compared to less certain participants whose family members behaved *less* unexpectedly in the five weeks preceding the midterm (i.e., *low* existential uncertainty). In other words, the defensive need to trust and find safety in the Trump administration in the face of unexpectedly behaving family members predicted voting in Republicans to support it.

In sum, findings from both studies suggest that participants defensively professed *greater* trust in the safety of their *sociopolitical* relational worlds when they were (1) *highly outcome-dependent* and (2) existentially confused by the unexpected behavior of their intimates.

Sociopolitical-to-Personal Threat-Defense

When people engage in this threat-defense, they counteract the existential uncertainty posed by *high* outcome-dependence on *unexpectedly behaving* sociopolitical relational worlds (i.e., the *threat*) by perceiving greater reason to trust in their intimates (i.e., the *defense*). We examined this threat-defense in the daily diary studies, the 2018 midterm election study, and our three-week study of the initial period of the COVID-19 pandemic in the United States. We expected people to defensively trust *more* in the *personal* relational world when they felt (1) *highly outcome-dependent*, but (2) existentially confused by the *unexpected* behavior of *sociopolitical* agents, whether the behavior of government leaders or fellow members of the U.S. community.

The daily diary studies. As we did before, we operationalized *outcome-dependence* through differences between *people* – that is, through individual expressions of romantic trust. We then turned to two *real-world* indicators to operationalize *unexpected behavior* on the part of government leaders in the *sociopolitical* relational world. Specifically, we categorized the *expectedness* of *government* behavior each day through (1) the Chicago Board Options Exchange Volatility Index, an economic marker of political instability,[2] and (2) the Google searches of people living in the same zip code as the participant, assuming that Google searches that day would capture existential uncertainty about unexpected government behavior. For instance, the news cycle in the first study included Trump tweeting threats of nuclear war with North Korea and impugning the FBI director. So, we tracked how often people in the same zip code as the participant searched for "uncertainty", "North Korea", "Trump lies", "Comey", and "terrorism," assuming that more frequent searches would capture greater collective existential anxiety about these presumably unexpected behaviors. We operationalized *trust in personal relational worlds* through perceptions of love and acceptance from immediate family members (e.g., "my partner made me feel especially loved"; "my child expressed love and affection toward me") and the daily quality of these family relationships (from "terrible" to "terrific").

Multilevel analyses revealed that participants who were *less* certain they could trust their romantic partner (i.e., *high* outcome-dependence) reported greater trust or faith in the benevolent motivations on days after the sociopolitical world behaved *more* unexpectedly, as captured by spikes in the VIX or Google-search activity (i.e., *high* existential threat to safety), as compared to days when sociopolitical powers behaved *less* unexpectedly (i.e., low existential threat to safety). However, people who were already certain they could trust their romantic partner evidenced no such compensatory effect (i.e., *low* outcome-dependence). In sum, participants defensively professed *greater* trust in the safety of their *personal* relational worlds when they were (1) *highly outcome-dependent* and (2) existentially confused by *unexpected* government behavior.

The midterm election and pandemic studies. The midterm election and pandemic studies allowed us to operationalize both outcome-dependence and expectedness *situationally*. In the midterm study, we again operationalized *outcome-dependence* through knowledge of the election's results in the midterm study. As we explained earlier, we expected participants to feel *more* keenly dependent on fellow community members to vote the "right way" after the election resulted in a divided Congress (i.e., *high* outcome-dependence) than before (i.e., *low* outcome-dependence). In the pandemic study, we operationalized *outcome-dependence* through the total number of COVID-19 cases in the United States. We expected participants to feel *more* keenly dependent on their fellow community members to keep them safe from infection on days when the total number of U.S. cases increased *more* than usual (i.e., *high* outcome-dependence), as compared to *less* than usual (i.e., *low* outcome-dependence).

We operationalized expectedness through the *consistency* between one's personal view of President Trump's stewardship and popular consensus. We did this by tracking the percent of social media posts mentioning the president that were *negative* on each assessment day. We then used political partisanship to define (1) days when U.S. citizens *more often* posted negative comments mentioning Donald Trump as *more* unexpected for conservatives, given the positive reality of his stewardship most conservatives perceived, and (2) days when U.S.

citizens *less often* posted negative comments mentioning Donald Trump as *more* unexpected for liberals, given the *negative* reality of his stewardship most *liberals* perceived. We operationalized *trust in personal relational worlds* through participants reporting (1) the daily quality of their relationships with their romantic partners and children (from "terrible" to "terrific") and (2) daily doubt and conflict in these familial relationships, which we reversed, such that higher scores captured greater daily happiness, and thus safety, within these family relationship bonds.

Multilevel analyses revealed that on days when the election results were known (i.e., *high* outcome-dependence), *liberals* reported greater family relationship happiness on days when popular sentiment toward President Trump was *less negative* than usual (i.e., *high* existential uncertainty for liberals) than *more negative* than usual (i.e., *low* existential uncertainty for liberals). However, *conservatives* reported greater family relationship happiness on days when popular sentiment toward President Trump was *more negative* than usual (i.e., *high* existential uncertainty for conservatives) than *less negative* than usual (i.e., *low* existential uncertainty for conservatives). Similarly, on days when COVID-19 cases spread *more* rapidly (i.e., *high* outcome-dependence), *liberals* reported greater family relationship happiness on days when popular sentiment toward President Trump was *less negative* than usual (i.e., *high* existential uncertainty for liberals) than *more negative* than usual (i.e., *low* existential uncertainty for liberals). However, *conservatives* reported greater family relationship happiness on days when popular sentiment toward President Trump was *more negative* than usual (i.e., *high* existential uncertainty for conservatives) than *less negative* than usual (i.e., *low* existential uncertainty for conservatives). No such compensatory effects emerged for liberals or conservatives in either study on days when the election results were unknown or U.S. COVID-19 cases spread less rapidly (i.e., *low* outcome-dependence).

In sum, we found evidence for *sociopolitical-to-personal* threat-defense in all four studies. Participants defensively professed *greater* trust in the safety of their *personal* relational worlds on days when they were highly *outcome-dependent* (i.e., vulnerable to others) and existentially confused by government leaders and community members behaving unexpectedly.

Looking Forward

The findings that we have reviewed in this chapter suggest that people look to *alternate* relational worlds for safety when they feel more insecure and need to count on others (i.e., *high* outcome-dependence) in a given relational world but cannot anticipate or foresee how others are likely to behave (i.e., *high* existential uncertainty). For instance, when Arya questions her spouse's trustworthiness or fears a rapidly circulating virus, *not* being able to anticipate the behavior of her president or fellow community members threatens the collective safety of social connection, motivating her to perceive greater evidence of safety in her family relationships (i.e., *sociopolitical-to-personal* threat-defense). However, *not* being able to anticipate the behavior of her spouse or children motivates her to perceive greater evidence of safety in her relationship with her government or fellow community members (i.e., *personal-to-sociopolitical* threat-defense).

Of course, the evidence that people engage in these defenses in daily life does not address the question of whether the defense "works." While we do not yet have a definitive answer, the initial evidence points to their effectiveness. For instance, we took advantage of a longitudinal study of newlywed couples to examine whether newlyweds who were *more* likely to find greater reason to be happy in their marriage on days when the VIX spiked fared better over time than newlyweds who were *less* likely to evidence this defense. For people who married *less* than completely certain they could trust their new spouse, being *more* likely to find happiness in their marriage on days when the VIX spiked was protective. Namely, being able to find greater safety in their marriage on days when the world behaved more unpredictably helped strengthen their trust in their partner and the security of their marriage over time (Murray, Lamarche et al., 2021).

Nevertheless, there is still more that we need to understand in delineating how people find a sense of safety in the collective relationships that surround them (see also Arriaga & Kumashiro; Crano & Hohman; Kruglanski & Ellenberg; this volume). For instance, future research might broaden the collective relational world to include group or out-group members (Hogg & Gaffney, this volume). On days when children or romantic partners behave acutely and unexpectedly, perhaps people impose greater safety on the *sociopolitical* world by believing more in stereotypes that depict in-group members as warm and disbelieving stereotypes that depict out-group members as hostile. They might also profess greater than usual faith in the importance of religion in their lives. Future research should also examine the social-safety system's operation in different political contexts. The studies we presented were conducted in the U.S. during the Trump administration, which confounds the motivation to believe in the prevailing sociopolitical powers with conservative (Republican) thinking. In this context, people had to believe in right-wing ideology and policy (no matter how personally foreign they normally found it) to impose safety on the sociopolitical world. In a Democratic-led administration, people might instead need to believe in more left-wing ideology and policy (no matter how personally foreign they find it) to impose safety on the sociopolitical world. Additionally, the presented studies were done in a bipartisan context where group status (e.g., Republican vs. Democrat) is relatively homogenous and unambiguously defined and offers reasonably limited opportunities for polarization (e.g., liberals have no left-of-Democrat parties to identify with). However, uncertainty can also push people to reject moderate establishments and embrace both extreme left- and right-wing ideologies (see, for example, van Prooijen in this issue). Thus, in sociopolitical contexts with multiple competing parties or multi-party governing systems, such as coalition governments, people may instead impose safety by polarizing their ideologies and policies in a direction *consistent* with their personal views if those parties represent entitative groups (see, for example, Hogg & Gaffney in this issue).

Conclusion

There is no escaping the solidarity of life. Indeed, the way people think about the ties they share with those closest to them is intricately bound up with the way they think about the collective ties they share with people and institutions they will never meet. In recognizing that interdependence extends beyond the dyad, the proposed model of the social-safety system sheds new light on how people sustain a sense of safety in their collective ties to others, allowing them to better reap the potential benefits of these relationships.

Notes

1 We assume that unexpected events can generate existential uncertainty regardless of whether they are positive, negative, or neutral (Kruglanski & Ellenberg, this volume).
2 Derived from the behavior of financial traders, the VIX is a daily economic indicator that tracks uncertainty in national and international events by forecasting greater volatility in the stock market over the next 30 days (Bloom, 2014).

References

Arriaga, X. B., & Kumashiro, M. Adult attachment insecurity during the COVID pandemic: Heightened insecurity and its undoing. In J. P. Forgas, W. D. Crano, & K. Fiedler (Eds.), *The psychology of insecurity: Seeking certainty where none can be had.* New York: Routledge.
Arriaga, X. B., Kumashiro, M., Simpson, J. A., & Overall, N. C. (2018). Revising working models across time: Relationship situations that enhance attachment security. *Personality and Social Psychology Review, 22,* 71–96.
Bai, Y., Yao, L., & Wei, T. (2020). Presumed asymptomatic carrier transmission of COVID-19. *JAMA, 323,* 1406–1407.
Baumeister, R. F., & Leary, M. R. (1995). The need to belong: Desire for interpersonal attachments as a fundamental human motivation. *Psychological Bulletin, 117,* 497–529.
Blanton, H., Strauts, E., & Perez, M. (2012). Partisan identification as a predictor of cortisol response to election news. *Political Communication, 29,* 447–460.
Bloom, N. (2014). Fluctuations in uncertainty. *Journal of Economic Perspectives, 28,* 153–176.
Cameron, J. J., & Granger, S. (2019). Does self-esteem have an interpersonal imprint beyond self-reports? A meta-analysis of self-esteem and objective interpersonal indicators. *Personality and Social Psychology Review, 23,* 73–102.

Cavallo, J., Fitzsimons, G. M., & Holmes, J. G. (2009). Taking chances in the face of threat: Romantic risk regulation and approach motivation. *Personality and Social Psychology Bulletin, 35,* 737–751.

Cohen, S., Janicki-Deverts, D., Turner, R. B., & Doyle, W. J. (2015). Does hugging provide stress-buffering social support? A study of susceptibility to upper respiratory infection and illness. *Psychological Science, 26,* 135–147.

Cox, C. R., & Arndt, J. (2012). How sweet it is to be loved by you: The role of perceived regard in terror management in relationships. *Journal of Personality and Social Psychology, 102,* 616–632.

Day, M. V., Kay, A., C., Holmes, J. G., & Napier, J. L. (2011). System justification and the defense of committed relationship ideology. *Journal of Personality and Social Psychology, 101,* 291–306.

Durkheim, Emile (1997). *The division of labour in society.* Trans. W. D. Halls, intro. Lewis A. Coser (pp. 39, 60, 108). New York: Free Press.

Feeney, B. C., & Collins, N. L. (2015). A new look at social support: A theoretical perspective on thriving through relationships. *Personality and Social Psychology Review, 19,* 113–147.

Fessler, D. M. T., & Holbrook, C. (2013). Friends shrink foes: The presence of comrades decreases the envisioned physical formidability of an opponent. *Psychological Science, 24,* 797–802.

Finkel, E. J., Hui, C. M., Carswell, K. L., & Larson, G. M. (2014). The suffocation of marriage: Climbing Mount Maslow without enough oxygen. *Psychological Inquiry, 25,* 1–41.

Fitzsimons, G. M., Finkel, E., J., & van Dellen, M. R. (2015). Transactive goal dynamics. *Psychological Review, 122,* 648–673.

Griffin, D. W., & Ross, L. (1991). Subjective construal, social inference and human misunderstanding. In M. P. Zanna (Ed.), *Advances in experimental social psychology* (Vol. 24, pp. 319–359). CA: Academic Press.

Heine, S. J., Proulx, T., & Vohs, V. (2006). The meaning maintenance model: On the coherence of social motivations. *Personality and Social Psychology Review, 10,* 88–110.

Higgins, E. T., Rossignac-Milon, M., & Echterhoff, G. (2021). Shared reality: From sharing-is-believing to merging minds. *Current Directions in Psychological Science, 30,* 103–110.

Hogg, M. A., & Gaffney, A. M. Social identity dynamics in the face of overwhelming uncertainty. In J. P. Forgas, W. D. Crano, & K. Fiedler (Eds.), *The psychology of insecurity: Seeking certainty where none can be had.* New York: Routledge.

Holt-Lunstad, J. (2018). Why social relationships are important for physical health: A systems approach to understanding and modifying risk and protection. *Annual Review of Psychology, 69,* 437–458.

Hudson, J. (2006). Institutional trust and subjective well-being across the EU. *KYKLOS, 59,* 43–62.

Jonas, E., McGregor, I., Klackl, J., Agroskin, D., Fritsche, I., Holbrook, C., Nash, K., Proulx, T., & Quirin, M. (2014). Threat and defense: From anxiety to approach. In J. M. Olson & M. P. Zanna (Eds.), *Advances in experimental social psychology* (pp. 219–286). Amsterdam: Elsevier Inc.

Kelley, H. H. (1979). *Personal relationships: Their structures and processes.* NY: Psychology Press.

Kelley, H. H., Holmes, J. G., Kerr, N. L., Reis, H. T., Rusbult, C. E., & Van Lange, P. A. M. (2003). *An atlas of interpersonal situations*. Cambridge: Cambridge University Press.

Kenrick, D. T., Griskevicius, V., Neuberg, S. L., & Schaller, M. (2010). Renovating the pyramid of needs: Contemporary extensions built upon ancient foundations. *Perspectives on Psychological Science, 5*(3), 292–314.

Kumar, A., & Epley, N. (2018). Undervaluing gratitude: Expressers misunderstand the consequences of showing appreciation. *Psychological Science, 29*, 1423–1435.

Kunda, Z. (1990). The case for motivated reasoning. *Psychological Bulletin, 108*(3), 480–498.

Lamarche, V. M. (2020). Socially connected and COVID-19 prepared: The influence of sociorelational safety on perceived importance of COVID-19 precautions and trust in government responses. *Social Psychological Bulletin, 15*(4), 1–25. https://doi.org/10.32872/spb.4409

Leary, M. R., Tambor, E. S., Terdal, S. K., & Downs, D. L. (1995). Self-esteem as an interpersonal monitor: The sociometer hypothesis. *Journal of Personality and Social Psychology, 68*, 518–530.

Maner, J. K., DeWall, C. N., Baumeister, R. F., & Schaller, M. (2007). Does social exclusion motivate interpersonal reconnection? Resolving the "porcupine problem." *Journal of Personality and Social Psychology, 92*, 42–55.

Master, S. L., Eisenberger, N. I., Taylor, S. E., Naliboff, B. D., Shirinyan, D., & Lieberman, M. D. (2009). A picture's worth: Partner photographs reduce experimentally induced pain. *Psychological Science, 20*, 1316–1318.

Mikulincer, M. Attachment security and coping with existential concerns: Studying security dynamics in dyadic, group, sociopolitical, and spiritual/religious relationships. In J. P. Forgas, W. D. Crano, & K. Fiedler (Eds.), *The psychology of insecurity: Seeking certainty where none can be had*. New York: Routledge.

Mikulincer, M., & Shaver, P. R. (2003). The attachment behavioral system in adulthood: Activation, psychodynamics, and interpersonal processes. In M. Zanna (Ed.), *Advances in experimental social psychology* (Vol. 35, pp. 52–153). New York: Academic Press.

Miller, S. L., & Maner, J. K. (2011). Sick body, vigilant mind: The biological immune system activates the behavioral immune system. *Psychological Science, 22*, 1467–1471.

Miller, S., L., & Maner, J. K. (2012). Over-perceiving disease cues: The basic cognition of the behavioral immune system. *Journal of Personality and Social Psychology, 102*, 1198–1213.

Murray, S. L., Holmes, J. G., & Collins, N. L. (2006). Optimizing assurance: The risk regulation system in relationships. *Psychological Bulletin, 132*, 641–666.

Murray, S. L., Lamarche, V., Gomillion, S., Seery, M. D., & Kondrak, C. (2017). In defense of commitment: The curative power of violated expectations in relationships. *Journal of Personality and Social Psychology, 13*, 627–729.

Murray, S. L., Lamarche, V., & Seery, M. D. (2018). Romantic relationships as shared reality defense. *Current Opinion in Psychology, 23*, 34–37.

Murray, S. L., Lamarche, V., Seery, M. D., Jung, H. Y., Griffin, D. W., & Brinkman, C. (2021). The social-safety system: Fortifying relationships in the face of the unforeseeable. *Journal of Personality and Social Psychology, 120*, 99–130.

Murray, S. L., Rose, P., Bellavia, G., Holmes, J., & Kusche, A. (2002). When rejection stings: How self-esteem constrains relationship-enhancement processes. *Journal of Personality and Social Psychology, 83*, 556–573.

Murray, D. R., & Schaller, M. (2016). The behavioral immune system: Implications for social cognition, social interaction, and social influence. In *Advances in experimental social psychology* (Vol. 53, pp. 75–129). Cambridge, MA: Academic Press.

Murray, S. L., Seery, M. D., Lamarche, V., Jung, H. Y., Saltsman, T. L., Griffin, D. W., Dubois, D., Xia, J., Ward, D. E., & McNulty, J. (2021). Looking for safety in all the right places: When threatening political reality strengthens family relationship bonds. *Social Psychological and Personality Science, 12*, 1193–1202.

Navarro-Carrillo, G., Valor-Segura, I., Lozano, L. M., & Moya, M. (2018). Do economic crises always undermine trust in others? The case of generalized, interpersonal, and in-group trust. *Frontiers in Psychology, 9*. doi: 10.3389/fpsyg.2018.01955

Peetz, J., Shimizu, J. P., & Royle, C. (2022). Projecting current feelings into the past and future: Better current relationship quality reduces negative retrospective bias and increases positive forecasting bias. *Journal of Social and Personal Relationships*, 02654075221084280.

Pitts, S., Wilson, J. P., & Hugenberg, K. (2014). When one is ostracized, others loom: Social rejection makes other people appear closer. *Social Psychological and Personality Science, 5*, 550–557.

Plusnin, N., Pepping, C. A., & Kashima, E. S. (2018). The role of close relationships in terror management: A systematic review and research agenda. *Personality and Social Psychology Review, 22*, 307–346.

Pollmann, M. M. H., & Finkenauer, C. (2009). Empathic forecasting: How do we predict other people's feelings? *Cognition and Emotion, 23*, 978–1001.

Richman, L. S., & Leary, M. R. (2009). Reactions to discrimination, stigmatization, ostracism, and other forms of interpersonal rejection: A multi-motive model. *Psychological Review, 116*, 365–383.

Slotter, E. B., Gardner, W. L., & Finkel, E. J. (2010). Who am I without you? The influence of romantic breakup on the self-concept. *Personality and Social Psychology Bulletin, 36*, 147–160.

Stanton, S. J., LaBar, K. S., Saini, E. K., Kuhn, C. M., & Beehner, J. C. (2010). Stressful politics: Voters' cortisol responses to the outcome of the 2008 United States Presidential election. *Psychoneuroendocrinology, 35*, 768–774.

Szepsenwol, O., & Simpson, J. A. (2019). Attachment within life history theory: An evolutionary perspective on individual differences in attachment. *Current Opinion in Psychology, 25*, 65–70.

Tooby, J., & Cosmides, L. (1996). Friendship and the banker's paradox: Other pathways to the evolution of adaptations for altruism. In W. G. Runciman, J. M. Smith, & R. I. M. Dunbar (Eds.), *Evolution of social behaviour patterns in primates and man* (pp. 119–143). Oxford: Oxford University Press.

Trawalter, S., Chung, V. S., DeSantis, A. S., Simon, C. D., & Adam, E. K. (2012). Physiological stress responses to the 2008 U. S. presidential election: The role of policy preferences and social dominance orientation. *Group Processes and Intergroup Relations, 15*, 333–345.

van den Bos, K. Trust in social institutions: The role of informational and personal uncertainty. In J. P. Forgas, W. D. Crano, & K. Fiedler (Eds.), *The psychology of insecurity: Seeking certainty where none can be had*. New York: Routledge.

van Prooijen, J.-W. Feelings of insecurity as drive of anti-establishment sentiments. In J. P. Forgas, W. D. Crano, & K. Fiedler (Eds.), *The psychology of insecurity: Seeking certainty where none can be had*. New York: Routledge.

Wilson, T. D., & Gilbert, D. T. (2003). Affective forecasting. In M. P. Zanna (Ed.), *Advances in experimental social psychology* (Vol. 35, pp. 345–411). San Diego, CA: Elsevier Academic Press.

Young, S. G., Slepian, M. L., & Sacco, D. F. (2015). Sensitivity to perceived facial trustworthiness is increased by activating self-protection motives. *Social Psychological and Personality Science, 6*, 607–613.

13
ADULT ATTACHMENT INSECURITY DURING THE COVID PANDEMIC

Heightened Insecurity and Its Undoing

Ximena B. Arriaga

PURDUE UNIVERSITY, USA

Madoka Kumashiro

GOLDSMITHS, UNIVERSITY OF LONDON, UK

Abstract

The COVID-19 pandemic has highlighted the importance of social connections. Attachment theory provides a compelling framework for understanding how pandemic conditions have affected people and their most important relationships. The pandemic has created chronic uncertainties and stressors, which are precisely the conditions that trigger attachment insecurity. Chronically activated insecurities, in turn, strain important close relationships. However, pandemic conditions have also provided opportunities to foster greater security. This chapter provides an up-to-date review of research on how early pandemic conditions affected individuals and relationships and then discusses the undoing of insecurity through two processes: (1) when a person feels insecure, relationship partners can adopt effective strategies to manage the person's insecure feelings, and (2) lasting decreases in insecurity occur when new experiences contradict the mental representations that underlie insecurity. This chapter emphasizes how relationships provide an important context in which people may flourish or languish under conditions of great uncertainty.

People strive to feel confident and secure about their abilities to pursue what they desire, manage problems effectively, and prepare for what the future holds. The absence of such security is aversive but has evolved to signal the need to protect oneself (Simpson et al., 2008; see also Fiedler & McCaughey; von Hippel & Merakovsky, this volume). For many people, the 2020 onset of the COVID-19 pandemic evoked personal insecurity by creating unabated risks to health and well-being, threats to financial security, and disruptions of social

DOI: 10.4324/9781003317623-16

ties. Pandemic conditions robbed people of agency and control in carrying on with life as usual and created chronic uncertainty about when and how the future would return to "normal". Most people intuitively understood threats to financial insecurity or risks of disease and health problems. However, many people did not anticipate or fully grasp how deeply they would be affected by changes in their social connections and relationships.

This chapter provides an attachment theory account of the impact of pandemic conditions on adult romantic involvements. Why focus on adult romantic involvements? Relationships fundamentally affect a person's "outcomes" – the subjective daily experience of rewarding, neutral, or costly moments. People feel ready to navigate daily events when they attain validation and support from their closest relationships (even imperceptible support; Bolger et al., 2001). When people do not attain validation and support, they languish in life's moments with insecurity or even detachment, without necessarily recognizing that the source of their suboptimal state traces back to problems in their closest relationships (see Mikulincer and Shaver's chapter in this volume on attachment security beyond awareness). Consider the person who is self-reliant out of necessity because of earlier experiences in which others were unavailable or unreliable for care or support, versus the person who is self-reliant because they always felt loved and supported in ways that afforded authentic confidence. These tendencies develop through consequential experiences with others. For many adults, romantic involvements are precisely the bond that sets the course for navigating other contexts.

This volume includes different ways of conceptualizing uncertainty. Attachment anxiety shares features with: (1) personal uncertainty, insofar as both involve subjective discomfort when feeling uncertain about oneself (see Van den Bos, this volume); (2) existential uncertainty as reflected by activated concerns about the future (see Murray & Lemarche; Pyszczynbsky & Sundby, this volume); and (3) existential threats (see van Prooijen, this volume, and Hirschberger, this volume). *Momentary attachment insecurity* triggers a threat response. People deploy defenses that afford the most protection in the moment, which may be, for example, either constructive versus maladaptive emotion regulation, appropriate reliance on others versus inappropriate overreliance or underreliance, and genuine confidence in tackling problems versus inflated overconfidence or desperate underconfidence. *Chronic attachment insecurity* reflects generalized tendencies and strategies that a person has devised to manage similar threats. A later section differentiates different dimensions of insecurity (attachment anxiety and attachment avoidance).

This chapter examines psychological changes during the pandemic using an attachment theory framework. First, it reviews recent research on the psychological effects of pandemic conditions. Second, it highlights attachment concepts that are relevant for understanding the effects of pandemic conditions. Third, it describes a novel framework to understand the basic processes through which insecurity declines in romantic involvements (Arriaga et al., 2018). The chapter ends with conclusions and directions for future research.

Impact of Pandemic Conditions

During the early stages of the pandemic, it is difficult to imagine anyone who did not experience moments of concern for their own or others' safety and well-being. People worldwide confronted unexpected uncertainty about whether they and close others would contract COVID-19, what the symptoms and prognosis could be, and the extent to which pandemic conditions would strain social ties, finances, long-term health, or lifestyle.

Consider the many ways in which early pandemic conditions disrupted, strained, or severed close connections with others. Elderly people in care facilities were forbidden from leaving their rooms for their own protection; marriages were strained with the never-ending juggling acts of child-care and work demands; people living with aggressive individuals could not escape easily; students were relegated to computer-mediated interactions (often with "video off") or to in-person interactions that were stymied by face masks and physical distance; and nearly everyone in the US knew of someone who died of COVID-19.

Indeed, mental health problems spiked during the initial month of lockdowns (for reviews, see Aknin et al., 2021; Robinson et al., 2022; Wirkner et al., 2021). Interestingly, these reviews and longitudinal studies also show that after a few months, mental health levels began to improve. Psychological well-being (life satisfaction, meaning in life, happiness, and anxiety) rebounded close to pre-pandemic levels but still lagged, especially whenever a new variant emerged and renewed uncertainty/social restrictions (Lewis et al., 2022). Longer-term effects of the pandemic manifest as "languishing" – a "void between depression and flourishing" that causes subtle reductions in concentration, lowered motivation, and the absence of well-being (Grant, 2021).

Although well-being rebounded close to pre-pandemic levels after initial lockdowns, certain groups continued to struggle: those who were female, young, and living alone or with young children; those in conflictual relationships; and those experiencing direct negative health effects of COVID-19 or economic hardship (Aknin et al., 2021; Robinson et al., 2022; Wirkner et al., 2021). Each of these high-risk groups represents people who would have benefited from effective social support. For example, women often do more of the household work and care for others, which probably became more burdensome during pandemic conditions; people living alone became lonely during lockdown periods; and people experiencing health or economic hardships needed help. The relatively worse outcomes for these groups suggest harm from having insufficient support from others. Supportive relationships provide protection in uncertain or difficult times, whereas distressed relationships provide little protection. In fact, single people reported better mental health than people in conflict-ridden relationships (Pieh et al., 2020), and people who experienced stress from living with family members sought social support elsewhere by reactivating dormant relationships (Yang et al., 2021).

This recent body of research suggests that good relationships serve as a "social vaccine" when social restrictions and uncertainty cause personal harm (e.g., poor mental health, loneliness), but this is not true for everyone. There is more to attaining relational protective features than merely being in a relationship.

Beyond changes in individual mental health, how did relationships fare during pandemic conditions? Those who sought to initiate intimate relationships early in the pandemic learned that social restrictions could suspend their romantic lives indefinitely. On average, established relationships fared well. Findings from a recent random sample indicated that more people were extremely satisfied with their relationship in 2021 than in 2022 or before the pandemic, and a similar pattern emerged in believing their partner is extremely important (Lewandowski & Murray, 2022). Another large-scale panel survey indicated similar pandemic boosts to relationships (Sanders, 2021).

However, outcomes varied depending on cohabitation status. Couples who lived apart declined in relationship satisfaction, frequency of joint activities, and sexual activity, whereas those living together increased in satisfaction (unless they argued a lot before the pandemic), frequency of conversations, and joint activities (Vigl et al., 2022; for an exception, see Schmid et al., 2021). Living with a partner predicted being less lonely early in the pandemic (Ray, 2021).

More generally, the quality of a relationship prior to the pandemic predicted the quality during the pandemic. In a study that compared couples before and after the pandemic started (prospectively), couples in high-quality relationships were more likely to experience increases in their relationship satisfaction by avoiding conflict and attributing instances of poor partner behavior to the pandemic rather than to negative partner traits (Williamson, 2020; see also Fleming & Franzese, 2021). In another study, pandemic strains were linked to increased conflict, lower intimacy, and less sexual behavior for only one-third of the couples (Luetke et al., 2020). Couples that reported elevated negative emotions during the pandemic also reported greater relational tension (Goodboy et al., 2021). Some studies reported that intimate partner violence increased during the pandemic (see Candel & Jitaru, 2021), but this may have been the case for couples that already reported violence.

Pandemic conditions thus seemed to have had an amplifying effect on the quality of relationships. Relationships that were functioning well prior to the pandemic tended to improve, whereas those that were vulnerable to strain prior to the pandemic declined in quality (e.g., Romeo et al., 2021).

Based on this recent body of research, several conclusions can be drawn about the effects of early pandemic conditions on the individual and relationship functioning. First, the early stages of the pandemic had varying effects, ranging from mild uncertainty to extreme distress. Second, these conditions strained some relationships but strengthened others. Third, well-functioning relationships protected against personal harm (e.g., poor mental health, loneliness). The next section discusses how consequential new moments, such as those experienced in the pandemic, can predict a person's momentary psychological state, resilience, and longer-term outcomes.

An Attachment Analysis of Pandemic Experiences

Attachment theory provides an understanding of how people respond to uncertain and stressful moments (Bowlby, 1982). Several interpersonal theories (e.g., attachment theory, interdependence theory) help to understand which social experiences and moments will cause people to revise the way they see themselves, others, or their relationships. These include moments that are unexpected, uncertain, stressful, or in any other way have enduring effects on mental representations of oneself and others.

Working Models and Attachment Orientations

Attachment orientations develop and change when interactions afford information about whether another person will be caring and supportive in times of distress and will inspire confidence to pursue goals or tackle problems (Feeney, 2004; Mikulincer & Shaver, 2016). These meaningful moments become encoded into elaborate mental representations with memories, beliefs, and emotions that interpret those moments, which then create expectations and scripts about the future (Bretherton & Munholland, 2008; Cassidy & Kobak, 1988). These mental representations form, endure, and change through a history of experiences with others.

Attachment theorists take particular interest in mental representations, or "internal working models", of one's self and close others (e.g., partners, friends, parents, co-workers). People feel secure when they have positive experiences with trusting others, being loved and valued by others, and having a safety net when feeling distressed. Chronically activating secure working models of self and others forms the basis of a secure attachment orientation. Chronically secure individuals exhibit high self-esteem and self-efficacy; comfort with closeness, intimacy, and trust; positive expectations when managing difficult situations; constructive responses to relationship conflicts and effective regulation of negative emotions; and high-quality relationships (Mikulincer & Shaver, 2003; Salvatore et al., 2011; Simpson et al., 1992).

Others have experiences that undermine trust and develop into insecure attachment orientations, including attachment anxiety (elevated concerns about how one is regarded by partners) or attachment avoidance (discomfort with closeness and dependence). A person's most salient experiences may feature unreliable love or moments of abandonment. This causes concerns about being important to others and fears of others losing interest in them. When they confront a problem, anxiously-attached individuals have elevated needs for reassurance combined with doubts about their ability to attain sufficient support. They generally have a negative working model of the self (e.g., that the self is not worthy or competent) and an ambivalent model of others (e.g., desire for closeness but disappointment in others' care).

Some individuals have salient past experiences with neglect or harm by close others – emotionally painful moments that reinforce the belief that others cannot be trusted – which develops into excessive self-reliance and avoidant attachment orientation. Avoidantly-attached individuals develop relatively negative models of others, becoming averse to relying on others and annoyed when others must rely on them. Their self-reliance extends to ignoring others when developing their own self-concept, which can lead to an inflated (albeit fragile) model of the self.

Pandemic conditions have been sufficiently stressful to activate momentary insecurity even among chronically secure individuals. We next examine attachment-relevant responses during the pandemic and the implications for individual and relationship outcomes.

Pandemic Outcomes of Insecure Individuals

Pandemic conditions made many people feel anxious, which predicted worse individual outcomes. In a study assessing daily states, individuals who reported pandemic-related anxiety and stress experienced greater loneliness and a desire for interaction, but also more interpersonal conflict and lower communication quality with others, which in turn predicted lower optimism about the pandemic (Merolla et al., 2021). These reactions are precisely how anxiously-attached individuals respond to stress, and the consequences are negative. Chronically anxiously-attached individuals exhibited worse coping and mental health outcomes during pandemic conditions relative to others (Kural & Kovacs, 2021; Mazza et al., 2021; Moccia et al., 2020; Vowels et al., 2022). Among individuals who lost a spouse or close family member to COVID-19, attachment anxiety predicted longer and maladaptive grief resolution (Katzman & Papouchis, 2022).

Interestingly, anxiously-attached individuals often acted as "sentinels" of COVID danger by warning others to practice safe hygiene habits (e.g., washing hands; Lozano & Fraley, 2021). Their negative emotions predicted both greater use of COVID protections and greater struggles to remain socially distant (Gruneau Brulin et al., 2022; Von Mohr et al., 2021).

Research so far has not revealed a robust pattern of outcomes for avoidantly-attached individuals. While some studies report that an attachment avoidance orientation did not predict mental health outcomes (e.g., Vowels et al., 2022) or bereavement-related grief (Katzman & Papouchis, 2022), one study reported that "discomfort with closeness" even served as a protective factor in coping with social isolation during the initial lockdown (Moccia et al., 2020). There was some evidence that relative to others, avoidantly-attached individuals had less effective coping strategies during the pandemic, which had indirect negative effects on their resilience levels (Kural & Kovacs, 2021). Avoidantly-attached individuals suppress their negative emotions, which may have obscured signals of distress; as they downplay adversity, they may have acclimated more quickly to the "new normal".

Chronic insecurities predicted worse relationship outcomes. For example, couples who had less contact with each other became less satisfied with their relationship during the early months of the pandemic, but this was especially true for insecure individuals (Overall, Pietromonaco, & Simpson, 2022; Vigl et al., 2022). Anxiously-attached individuals felt angry or hurt if a partner was unwilling to ignore lockdown restrictions to be together (Gruneau Brulin et al., 2022). Anxiously-attached individuals often use guilt to get compliance from a partner (Jamahaya et al., 2017); when this is done repeatedly, partners resent being made to feel guilty and become less satisfied with their relationship (Overall et al., 2014). Anxiously-attached individuals also reported more severe relationship problems during the pandemic, especially when they were experiencing stress, and their partners reported lower relationship quality (Overall, Chang, et al., 2022).

Other vulnerabilities have been specific to attachment avoidance. For example, people who tried to carry on with their personal goals felt less supported by their partner to the extent that the partner was avoidantly-attached (Vowels & Carnelley, 2021).

Why Insecurities Strain Relationships

Why do chronically insecure individuals have worse relationship outcomes? Chronic attachment anxiety strains relationships. Anxiously-attached individuals perceive more interpersonal problems than others (Gere et al., 2013); require more care and reassurance, which exhausts partners over time (Lemay & Dudley, 2011); and overinterpret negative interactions with a partner as reflecting major issues (Campbell et al., 2005), which creates new issues with which the partner must contend. Anxiously-attached individuals often use guilt to get compliance from a partner (Jamahaya et al., 2017); when this is done repeatedly, partners resent being made to feel guilty and become less satisfied with their relationship (Overall et al., 2014). These tendencies cause tension and conflict with relationship partners.

Chronic attachment avoidance also strains relationships (Overall et al., 2022). Avoidant individuals prioritize their own needs over others' and lack empathy; they often react with anger or hostility to others' requests; and they struggle to feel and communicate love, care, or closeness (Mikulincer & Shaver, 2016). Independence is more important to them than intimacy or commitment (Ren et al., 2017). This explains why their relationships have lower warmth, support, closeness, and commitment (Li & Chan, 2012). Partners grow tired of hoping for mutual love and high regard, and they eventually give up on intimacy (Overall & Lemay, 2015).

Benefits of Relationships in Pandemic Conditions: An ASEM Perspective

Pandemic conditions were ripe for frequently activating insecurities. The disruption of social connections and support likely amplified attachment anxiety concerns, and isolation reinforced avoidant responses of self-distancing and self-reliance. But earlier, we described how some relationships provided a "social vaccine" for coping with pandemic conditions. How do people protect their relationships from the negative effects of activated insecurities? Through what processes might people even become more secure?

These questions are addressed by the Attachment Security Enhancement Model (ASEM), displayed in Figure 13.1 (Arriaga et al., 2018; Arriaga & Kumashiro, 2019; Kumashiro & Arriaga, 2021). One part of the ASEM posits specific types of partner behavior that mitigate a person's insecurity (safe or soft strategies). Why is it important for one or both partners to manage insecurities effectively? This prevents relationship tension. It also allows for novel moments that can contradict and revise the mental models that have sustained insecurity (Arriaga et al., 2018). The second part of the ASEM posits specific processes that target insecure mental models. People encounter new situations that, if sufficiently salient, revise insecure working models and redirect chronic insecurities.

Three features of the ASEM are worth highlighting. First, reducing insecurity involves more than merely evoking positive sentiments; addressing momentary insecurities and revising working models may occur in unique ways, as described below. Second, reducing insecurity is not merely about providing reassurance; the process of reducing anxiety versus avoidance may be

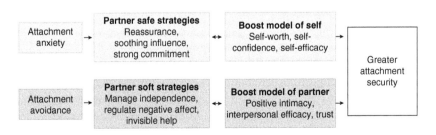

FIGURE 13.1 The Attachment Security Enhancement Model.

quite distinct. Third, the ASEM provides general principles of mitigating insecurity and revising insecure working models rather than specific, narrow, or proscriptive suggestions. Couples may do quite different things that nonetheless satisfy the same function or process.

The ASEM proposes: (1) targeted partner behaviors and strategies that mitigate momentary insecurity, given that insecurity erodes relationships when left unabated; and (2) the specific models that must change for lasting declines in insecurity. When a person experiences attachment anxiety, partners manage it effectively by showing they are a reliable source of safety and support, i.e., safe strategies; longer-term declines in anxiety occur when a partner genuinely is proud of an anxious person's contributions or accomplishments (i.e., a partner amplifies other's personal efficacy or qualities). When a person experiences attachment avoidance, partners manage this effectively by navigating interdependence (mutual influence and reliance) without making the realities of interdependence obvious, i.e., soft strategies; longer-term declines in avoidance occur when an avoidant person forms positive associations with relationships or feels interpersonal efficacy (fun moments, positive experiences providing or receiving support). See Arriaga et al., (2018).

Reducing Attachment Anxiety

The ASEM proposes that moments in which a person experiences attachment anxiety are effectively managed through a partner's use of *safe strategies*, which convey that a partner is a reliable source of safety and support in times of need. Specifically, partners exhibit the use of safe strategies when they do any of the following (see Figure 13.1):

1. Provide reassurance and encouragement that a problem can be contained.
2. Soothe anxious or negative emotions and defuse spiraling drama by speaking with a calm voice or using affectionate touch (Jakubiak & Feeney, 2016; Kim et al., 2018).
3. Convey a strong commitment and reassure the other that the relationship is on stable ground (Lemay & Dudley, 2011; Tran & Simpson, 2009).

Recent research supports the idea that safe strategies are the preferred strategies for anxiously-attached individuals (Fuentes & Jakubiak, 2022).

Relationship quality generally declined for people who were experiencing pandemic-related psychological distress, but *not* if they perceived that their partner understood the difficult emotions they were experiencing, helped them reframe a difficult issue, or provided direct assistance (Randall et al., 2022). This is consistent with the idea that moments of active distress and anxiety do not harm a relationship when partners regulate negative emotions and provide the needed reassurance. More generally, people want to know that their hardships are of concern to a partner and that the partner is responsive to their needs (Banford Willing et al., 2021; Rice et al., 2020).

Does attachment anxiety decline over the long term from safe-strategy support? Anxiously-attached individuals certainly may feel reduced momentary anxieties when a partner is soothing, reassuring, and visibly committed. Repeated exposure to safe strategies reinforces the idea that a partner can be trusted, which fosters greater security in a person's working model of relationships. However, a partner's use of safe strategies alone may not be sufficient to revise a person's insecure model of self. Reassurance from a partner may be necessary to reduce attachment anxiety, but alone, it may not be sufficient (Jakubiak et al., 2022).

The second part of the ASEM posits that specific revisions to working models lead to greater security over time (Arriaga et al., 2018). For anxiously-attached individuals, a key process concerns revisions to an insecure model of self (see Figure 13.1). Relative to others, anxiously-attached individuals doubt their own abilities, efficacy, and self-worth (see Mikulincer & Shaver, 2016, Table 6.1).

Attachment anxiety declines over time when individuals gain confidence that what they do is meaningful and valued, which is different from getting emotional reassurance and soft-strategy behavior from a partner. These distinct processes were apparent in a longitudinal study. People became less anxiously-attached when they believed that their personal goals mattered to their partner. In fact, partner validation predicted declines in attachment anxiety more than did feeling trust in a partner. However, low trust correlated concurrently with attachment anxiety, whereas low validation and attachment anxiety were not concurrently linked (Arriaga et al., 2014). These time-dependent effects suggest different momentary versus long term processes. Trust indicates safety and reassurance (soft strategies that mitigate momentary attachment anxiety), whereas believing that one's personal goals are valued by a partner improves one's model of self.

Another longitudinal study examined couples transitioning to parenthood (Arriaga et al., 2021). Concurrently, couples felt less anxiously-attached to the extent that they perceived proximal and sensitive reassurance from their partner, which reflects a partner's use of soft strategies as needed. However, longitudinally, anxious attachment declined over time from gaining confidence and self-efficacy in the new role of parenting, whereas reassurance from a partner did not predict longitudinal declines in attachment anxiety. In other research, gratitude from a partner predicted declines in attachment anxiety (Park et al., 2019), perhaps because receiving gratitude affirms what one is doing (i.e., it boosts the working model of self).

None of these studies were conducted during the pandemic, which has robbed many of feeling of personal control and efficacy (Ritchie et al., 2021). Still, pandemic conditions conceivably created opportunities for people to feel valued for their personal efforts and contributions, which should boost a person's model of self. Some people have developed "pandemic projects" (e.g., gardening, exercising, or other hobbies; Lades et al., 2020) that could translate into personal accomplishments. Personal successes, when amplified as reflecting a person's competence and achievement, can function to contradict an insecure model of self.

Reducing Attachment Avoidance

The ASEM posits specific insecurity-mitigating and model-revision processes that reduce avoidance as being distinct from those that reduce attachment anxiety. The first part of the ASEM suggests that when a person experiences attachment avoidance, partners manage this effectively by navigating interdependence (mutual influence and reliance) without making the realities of interdependence obvious. These are *soft strategies* (see Figure 13.1; Arriaga et al., 2018; Farrell et al., 2016; Simpson & Overall, 2014), as when partners:

1 Make requests in ways that preempt an avoidant person's hostility or anger. Avoidant individuals often interpret a partner's comments as criticism and become defensive. Partners prevent an avoidant person's hostility by first building goodwill before making a request through respect and acknowledgment of what the person already does (giving "due credit").
2 Regulate interactions to avoid interpersonal tension and negative emotions. A partner's emotionality, manipulative behavior, complaints, and sulking immediately trigger an avoidant person's distancing behavior. Partners mitigate avoidant reactions with direct, reasonable, and unemotional requests.
3 Provide help without *the appearance* of help (Howland & Simpson, 2010), given that avoidant individuals react strongly to being pitied or feeling indebted or obliged to others.

Recent research supports the idea that soft strategies, more so than safe partner behaviors, are the preferred strategies for avoidantly-attached individuals (Fuentes & Jakubiak, 2022).

Pandemic conditions included lockdowns forcing people to stay home. Avoidantly-attached individuals who were apart from their partner could regulate how much time they spent connecting online with their partner to attain an optimal balance of closeness and distance and to dodge tense discussions. Avoidant individuals who lived with their partner had to make many psychological and relational adjustments. Under normal conditions, they have the option of seeking "me time" to keep their avoidant tendencies at bay. Without the option of independent time, a partner's use of soft strategies may have been particularly predictive of mitigating relational tension. One example of using soft strategies would involve attributing a relationship problem to the pandemic rather than to hostile or distancing behavior by an avoidant partner (Williamson, 2020). Partners who generally agreed on or had similarly experienced life concerns – those who felt a "shared reality" during the pandemic – were better at being responsive to each other, which mitigated declines in relationship satisfaction (Enestrom & Lydon, 2021).

The second part of the ASEM suggests that avoidantly-attached individuals become more secure through experiences that contradict and revise their (relatively) negative working models of others (see Figure 13.1; Rholes et al., 2021). Relative to others, avoidantly-attached individuals harbor negative beliefs about trust, closeness, and dependence (Mikulincer & Shaver, 2016). In romantic involvements, the ASEM posits that avoidantly-attached individuals become more secure over time when their working model of romantic relationships incorporates trust and positive sentiments about relationships (Arriaga et al., 2018). In a series of studies combining different methods, positive relationship experiences during non-distressing moments reduced attachment avoidance but not attachment anxiety, providing direct evidence of targeted processes (model-specific revisions) in reducing insecurity (Bayraktaroglu et al., 2022).

Some pandemic conditions may have led to new and positive mental representations of relationships as couples sought positive time together through new routines and activities (Lillie et al., 2021), which can reduce avoidance (Stanton et al., 2017). Pandemic conditions may also have provided opportunities to experience unexpected benefits from helping others. When an avoidant individual unexpectedly provides help and enjoys the experience, this disrupts the idea that helping is aversive (see Rholes et al., 2021). Savvy partners of avoidant individuals who understand this generate helping situations that they know will be experienced in a positive way, such as an easy chore that provides supplies to people who then convey their gratitude or babysitting children while enjoying fun games together.

Of course, chronically insecure individuals may rely on their insecure lens in these new situations; their insecure working models direct how they interpret these new situations, react emotionally, and respond behaviorally. This is why relationships provide a powerful context for enhancing security: Partners have a critical role in directing the lasting effect of model-revising moments by underscoring and amplifying security-enhancing messages. With help from a partner, anxiously-attached individuals learn that they may be competent and valued after all, and avoidantly-attached individuals learn that relationships feel good and are worthy of trust.

Summary, Conclusions, and Future Directions

The pandemic has provided an unfortunate natural global experiment to observe and study reactions to the unprecedented levels of insecurity and uncertainty in recent history. In addition to the fear of death and acute economic hardship experienced by many, people experienced major disruptions in their daily routines and draconian social restrictions. As new COVID variants emerged without a good playbook for how to manage uncertainty, the lasting impact of the pandemic on relationships and well-being has yet to be fully understood.

Research on the early conditions of the pandemic suggests that close relationships played a key role in predicting outcomes. In fact, the pandemic appears to have amplified existing strengths and vulnerabilities in the relationships. Many couples were forced to spend more time together, sometimes cutting themselves off from their other social ties, while others spent more time apart than desired. People in bad relationships became more aware of their bad relationship, and people in satisfying relationships strengthened their bonds.

These conditions of uncertainty and stress amplify insecurities. Anxiously-attached individuals experienced strain in their relationships when they could not be with their partner or benefit from partner support. The effects on avoidant individuals were less robust. However, one can speculate that these conditions strained their relationships as well if they spent too little or too much time with their partner, who experienced their own stressors and need for support. Attachment concepts have been extremely useful for understanding how people navigate these conditions of uncertainty and strain (e.g., Overall et al., 2021; Vowels & Carnelley, 2021).

The ASEM provides a generative framework for understanding the specific interactions that mitigate the effects of stressors on relationships and the specific situations that revise insecure working models (Arriaga et al., 2018). The pandemic created situations that were novel, salient, and potentially stressful, which are conditions suited for revising working models. Of course, some partners are more likely than others to be able to foster an optimal environment for security-enhancing processes (Eller & Simpson, 2020). Some individuals do not have the psychological bandwidth to mitigate tensions or coordinate moments that improve specific working models. Recent research has documented that people become more secure when greater attachment security is something they desire and seek (Hudson et al., 2020).

More research is needed on pandemic outcomes because publications lag well behind when specific conditions reflected in the data occurred (e.g., what stage of lockdown the sample was in, how far into the pandemic). Many of the existing publications reflect an observational stage and need to pivot to the stage of systematic study. Some notable exceptions include systematic reviews and longitudinal research that compared well-being and relational quality before and after the social restrictions were imposed (e.g., Wirkner et al., 2021); these reviews have been effective in unearthing how different groups of people were affected by the pandemic and what kinds of conditions (e.g., financial hardship and limited space) may be influencing relational outcomes. Going forward, it is important to have more avenues for making longitudinal observations across a diverse range of people and groups.

New research also should adopt a wide perspective on possible outcomes of the pandemic. Due to severe social restrictions and lockdowns imposed by many governments around the world, the pandemic may have even caused a paradigm shift in how we approach our lives, work, and relationships. Combined with great improvements to technology, the pandemic has made viable new and more diverse ways of having relationships, as well as new approaches toward flexible working arrangements that can inadvertently affect close relationships. Due to improved online communication, it has become more viable to maintain or develop long-distance relationships, and partners can choose how much communication to have with each other even when apart from each other. This may enable people to better maintain an optimal level of balance between their personal and relational needs, which can in turn positively affect personal and relational well-being (Kumashiro et al., 2008).

Moreover, the pandemic may have changed expectations for many people in what they seek in their close relationships, and it is vital to have more systematic studies on differences among couples whose relationships formed during the pandemic, couples who experienced higher vs lower quality relationships during the pandemic, and couples whose relationships ended during the pandemic. Finally, future research needs to consider that these critical personal and relational outcomes cannot be studied in isolation and that future research needs to consider interdependent processes – characteristics of all interaction partners and the situation (e.g., Arriaga, 2013; Righetti et al., 2020) – that contribute to shaping critical outcomes and personality processes. It is especially important to consider attachment orientations of both partners and to examine if new situations imposed by the pandemic helped revise critical working models of self and others that may have helped individuals become more or less secure in their relationships.

In conclusion, the last two years have seen an unprecedented level of disruption in much of the world in recent history. Much remains unknown about what lies ahead, but what seems clear is that close relationships can both serve as a "social vaccine" to inoculate against the potentially devastating impact of pandemic-related insecurity on mental health and well-being but also worsen such outcomes. Somewhat surprisingly, many relationships seem to have even flourished during the pandemic. Attachment theory offers a useful framework for understanding who may flourish or languish under these conditions of great uncertainty. Close relationships can provide a level of security and certainty in such an uncertain world, serving as both a safe haven from the unknown and a secure base from which to venture into the new unknown reality. We offer insights into how people may be able to develop greater security in their relationships and in themselves, which will likely facilitate smoother sailing through the rough waters of the next uncertain phase of the pandemic.

Author Note: Preparation of this chapter was supported by funding from the National Science Foundation (BCS-1531226) and the Economic and Social Research Council (No. ES/N013182/1).

References

Aknin, L. B., De Neve, J. E., Dunn, E. W., Fancourt, D. E., Goldberg, E., Helliwell, J. F.,... & Ben Amor, Y. (online, 19 January, 2021). Mental health during the first year of the COVID-19 pandemic: A review and recommendations for moving forward. *Perspectives on Psychological Science*. https://doi.org/10.1177/17456916211029964

Arriaga, X. B. (2013). An interdependence theory analysis of close relationships. In J. A. Simpson & L. Campbell (Eds.), *The Oxford handbook of close relationships* (pp. 39–65). Oxford, England: Oxford University Press.

Arriaga, X. B., Eller, J., & Kumashiro, M., Rholes, W. S., & Simpson, J. A. (2021). Self-efficacy and declines over time in attachment anxiety during the transition to parenthood. *Social Psychological and Personality and Science, 12*, 658–666. doi/10.1177/1948550620933411

Arriaga, X. B., & Kumashiro, M. (2019). Walking a security tightrope: Relationship-induced changes in attachment security. *Current Opinion in Psychology, 25*, 121–126. doi.org/10.1016/j.copsyc.2018.04.016

Arriaga, X. B., Kumashiro, M., Finkel, E. J., VanderDrift, L. E., & Luchies, L. B. (2014). Filling the void: Bolstering attachment security in committed relationships. *Social Psychological and Personality Science, 5*, 398–405. doi:10.1177/1948550613509287

Arriaga, X. B., Kumashiro, M., Simpson, J. A., & Overall, N. C. (2018). Revising working models across time: Relationship situations that enhance attachment security. *Personality and Social Psychology Review, 22*, 71–96. doi: 10.1177/1088868317705257

Banford Witting, A., Busby, D. M., & Rellaford, S. R. (online, 22 December, 2021). Longitudinal anxiety in couples during a global pandemic: Considering loss, attachment behaviors, and trauma coping self-efficacy. *Family Process*. https://doi.org/10.1111/famp.12742

Bayraktaroglu, D., Gunaydin, G., Selcuk, E., Besken, M., & Karakitapoglu-Aygun, Z. (2022). The role of positive relationship events in romantic attachment avoidance. *Journal of Personality and Social Psychology*. Advance online publication. https://doi.org/10.1037/pspi0000406

Bolger, N., Zuckerman, A., & Kessler, R. C. (2000). Invisible support and adjustment to stress. *Journal of Personality and Social Psychology, 79*, 953–961.

Bowlby, J. (1982/1969). *Attachment and loss, Vol. 1: Attachment (2nd ed.)*. New York: Basic Books.

Bretherton, I., & Munholland, K. A. (2008). Internal working models in attachment relationships: Elaborating a central construct in attachment theory. In J. Cassidy, P. R. Shaver, J. Cassidy, P. R. Shaver (Eds.), *Handbook of attachment: Theory, research, and clinical applications (2nd ed.)* (pp. 102–127). New York, NY: Guilford Press.

Campbell, L., Simpson, J. A., Boldry, J., & Kashy, D. A. (2005). Perceptions of conflict and support in romantic relationships: The role of attachment anxiety. *Journal of Personality and Social Psychology, 88*, 510–531.

Candel, O. S., & Jitaru, M. (2021). COVID-19 and romantic relationships. *Encyclopedia, 1*, 1038–1046. https://doi.org/10.3390/encyclopedia1040079

Cassidy, J., & Kobak, R. R. (1988). Avoidance and its relationship with other defensive processes. In J. Belsky & T. Nezworski (Eds.), *Clinical implications of attachment* (pp. 300–323). Hillsdale, NJ: Erlbaum.

Eller, J., & Simpson, J. A. (2020). Theoretical boundary conditions of partner buffering in romantic relationships. *International Journal of Environmental Research and Public Health, 17,* 6880. https://doi.org/10.3390/ijerph17186880

Enestrom, M. C., & Lydon, J. E. (2021). Relationship satisfaction in the time of COVID-19: The role of shared reality in perceiving partner support for front-line health-care workers. *Journal of Social and Personal Relationships, 38,* 2330–2349. https://doi.org/10.1177/02654075211020127

Farrell, A. K., Simpson, J. A., Overall, N. C., & Shallcross, S. L. (2016). Buffering the responses of avoidantly attached romantic partners in strain test situations. *Journal of Family Psychology, 30,* 580–591. http://dx.doi.org/10.1037/fam0000186

Feeney, B. C. (2004). A secure base: Responsive support of goal strivings and exploration in adult intimate relationships. *Journal of Personality and Social Psychology, 87,* 631–648.

Fleming, C. J. E., & Franzese, A. T. (2021). Should I stay or should I go? Evaluating intimate relationship outcomes during the 2020 pandemic shutdown. *Couple and Family Psychology: Research and Practice, 10,* 158–167. https://doi.org/10.1037/cfp0000169

Fuentes, J. D., & Jakubiak, B. J. (2022, February 23–25). *Attachment insecurity predicts preferences for and effectiveness of attachment-matched support.* Society of Personality and Social Psychology Convention, San Francisco, CA.

Gere, J., MacDonald, G., Joel, S., Spielmann, S. S., & Impett, E. A. (2013). The independent contributions of social reward and threat perceptions to romantic commitment. *Journal of Personality and Social Psychology, 105,* 961–977.

Goodboy, A. K., Dillow, M. R., Knoster, K. C., & Howard, H. A. (2021). Relational turbulence from the COVID-19 pandemic: Within-subjects mediation by romantic partner interdependence. *Journal of Social and Personal Relationships, 38*(6), 1800–1818.

Grant, A. (2021, April 19). There's a name for the blah you're feeling: It's called languishing. *New York Times.* https://www.nytimes.com/2021/04/19/well/mind/covid-mental-health-languishing.html

Gruneau Brulin, J., Shaver, P. R., Mikulincer, M., & Granqvist, P. (2022). Attachment in the time of COVID-19: Insecure attachment orientations are associated with defiance of authorities' guidelines during the pandemic. *Journal of Social and Personal Relationship, 39,* 2528–2548. https://doi.org/10.1177/02654075221082602

Howland, M., & Simpson, J. A. (2010). Getting under the radar: A dyadic view of invisible support. *Psychological Science, 21,* 1878–1885.

Hudson, N. W., Fraley, R. C., Chopik, W. J., & Briley, D. A. (2020). Change goals robustly predict trait growth: A mega-analysis of a dozen intensive longitudinal studies examining volitional change. *Social Psychological and Personality Science, 11,* 723–732. https://doi.org/10.1177/1948550619878423

Jakubiak, B. K., & Feeney, B. C., (2016). A sense of security: Touch promotes state attachment security. *Social Psychological and Personality Science, 7,* 745–753.

Jakubiak, B. K., Fuentes, J. D., & Feeney, B. C. (2022, February 23–25). *Reducing attachment anxiety is a balancing act.* Society of Personality and Social Psychology Convention, San Francisco, CA.

Jayamaha, S. D., Girme, Y. U., & Overall, N. C. (2017). When attachment anxiety impedes support provision: The role of feeling unvalued and unappreciated. *Journal of Family Psychology, 31,* 181–191.

Katzman, W., & Papouchis, N. (online, 18 February, 2022). Grief Responses during the COVID-19 Pandemic: Differences in Attachment and Emotion Regulation. *Journal of Loss and Trauma.* https://doi.org/10.1080/15325024.2022.2040154

Kim, K. J., Feeney, B. C., & Jakubiak, B. K. (2018). Touch reduces romantic jealousy in the anxiously attached. *Journal of Social and Personal Relationships, 35,* 1019–1041.

Kumashiro, M., & Arriaga, X. B. (2020). Attachment security enhancement model: Bolstering attachment securing through close relationships. In B. A. Mattingly, K. P. McIntyre, & G. W. Lewandowski, Jr. (Eds.), *Interpersonal relationships and the self-concept* (pp. 69–88). Springer. ISBN: 978-3-030-43747-3. https://doi.org/10.1007/978-3-030-43747-3

Kumashiro, M., Rusbult, C. E., & Finkel, E. J. (2008). Navigating personal and relational concerns: The quest for equilibrium. *Journal of Personality and Social Psychology, 95,* 94–110.

Kural, A. I., & Kovacs, M. (2021). Attachment anxiety and resilience: The mediating role of coping. *Acta Psychologica, 221,* 103447. https://doi.org/10.1016/j.actpsy.2021.103447

Lades, L. K., Laffan, K., Daly, M., & Delaney, L. (2020). Daily emotional well-being during the COVID-19 pandemic. *British Journal of Health Psychology, 25,* 902–911.

Lemay, E. P., Jr., & Dudley, K. L. (2011). Caution: Fragile! Regulating the interpersonal security of chronically insecure partners. *Journal of Personality and Social Psychology, 100,* 681–702.

Lewandowski, G., & Murray, P. (2022). Relationship satisfaction returns to normal after pandemic bump. *Monmouth University Poll.* Relationship Satisfaction Returns to Normal After Pandemic Bump | Monmouth University Polling Institute.

Lewis, B., Dennes, M., Leach, C., Davison, A., & Vizard, T. (2022). Coronavirus and the social impacts on Great Britain: 18 March 2022. *Office of National Statistics.* https://www.ons.gov.uk/peoplepopulationandcommunity/healthandsocialcare/healthandwellbeing/bulletins/coronavirusandthesocialimpactsongreatbritain/latest

Li, T., & Chan, D. K. S. (2012). How anxious and avoidant attachment affect romantic relationships quality differently: A meta-analytic review. *European Journal of Social Psychology, 42,* 406–419.

Lillie, H. M., Chernichky-Karcher, S., & Venetis, M. K. (2021). Dyadic coping and discrete emotions during COVID-19: Connecting the communication theory of resilience with relational uncertainty. *Journal of Social and Personal Relationships, 38*(6), 1844–1868.

Lozano, E. B., & Fraley, R. C. (2021). Put your mask on first to help others: Attachment and sentinel behavior during the COVID-19 pandemic. *Personality and Individual Differences, 171,* 110487. https://doi.org/10.1016/j.paid.2020.110487

Luetke, M., Hensel, D., Herbenick, D., & Rosenberg, M. (2020). Romantic relationship conflict due to the COVID-19 pandemic and changes in intimate and sexual behaviors in a nationally representative sample of American adults. *Journal of Sex and Marital Therapy, 46,* 747–762.

Mazza, C., Colasanti, M., Ricci, E., Di Giandomenico, S., Marchetti, D., Fontanesi, L.,... & Roma, P. (2021). The COVID-19 outbreak and psychological distress in healthcare workers: The role of personality traits, attachment styles, and sociodemographic factors. *Sustainability, 13,* 4992. https://doi.org/10.3390/su13094992

Merolla, A. J., Otmar, C., & Hernandez, C. R. (2021). Day-to-day relational life during the COVID-19 pandemic: Linking mental health, daily relational experiences, and end-of-day outlook. *Journal of Social and Personal Relationships, 38,* 2350–2375.

Mikulincer, M., & Shaver, P. R. (2003). The attachment behavioral system in adulthood: Activation, psychodynamics, and interpersonal processes. *Advances in Experimental Social Psychology, 35,* 53–152.

Mikulincer, M., & Shaver, P. R. (2016). *Attachment in Adulthood: Structure, Dynamics, and Change* (2nd ed.). New York: Guildford Press.

Moccia, L., Janiri, D., Pepe, M., Dattoli, L., Molinaro, M., De Martin, V.,… & Di Nicola, M. (2020). Affective temperament, attachment style, and the psychological impact of the COVID-19 outbreak: an early report on the Italian general population. *Brain, Behavior, and Immunity, 87,* 75–79. https://doi.org/10.1016/j.bbi.2020.04.048

Overall, N. C., Chang, V. T., Pietromonaco, P. R., Low, R. S. T., & Henderson, A. M. E. (2022). Partners' attachment insecurity and stress predict poorer relationship functioning during COVID-19 quarantines. *Social Psychological and Personality Science, 13,* 285–298. https://doi.org/10.1177/1948550621992973

Overall, N. C., Girme, Y. U., Lemay, E. P. Jr., & Hammond, M. T. (2014). Attachment anxiety and reactions to relationship threat: The benefits and costs of inducing guilt in romantic partners. *Journal of Personality and Social Psychology, 106,* 235–256.

Overall, N. C., & Lemay, E. P. (2015). Attachment and dyadic regulation processes. In J. A. Simpson & W. S. Rholes (Eds.), *Attachment theory and research: New directions and emerging themes* (pp. 145–169). New York: Guildford Press.

Overall, N. C., Pietromonaco, P. R., Simpson, J. A. (2022). Buffering and spillover of adult attachment insecurity in couple and family relationships. *Nature Reviews Psychology, 1,* 101–111. https://doi.org/10.1038/s44159-021-00011-1

Park, Y., Johnson, M. D., MacDonald, G., & Impett, E. A. (2019). Perceiving gratitude from a romantic partner predicts decreases in attachment anxiety. *Developmental Psychology, 55,* 2692–2700.

Pieh, C., O' Rourke, T., Budimir, S., & Probst, T. (2020). Relationship quality and mental health during COVID-19 lockdown. *PloS One, 15*(9), e0238906. https://doi.org/10.1371/journal.pone.0238906

Randall, A. K., Leon, G., Basili, E., Martos, T., Boiger, M., Baldi, M., Hocker, L., Kline, K., Masturzi, A., Aryeetey, R., Bar-Kalifa, E., Boon, S. D., Botella, L., Burke, T., Carnelley, K. B., Carr, A., Dash, A., Fitriana, M., Gaines, S. O., … Chiarolanza, C. (2022). Coping with global uncertainty: Perceptions of COVID-19 psychological distress, relationship quality, and dyadic coping for romantic partners across 27 countries. *Journal of Social and Personal Relationships, 39,* 3–33. https://doi.org/10.1177/02654075211034236

Ray, C. D. (2021). The trajectory and determinants of loneliness during the early months of the COVID-19 pandemic in the United States. *Journal of Social and Personal Relationships, 38,* 1920–1938. https://doi.org/10.1177/02654075211016542

Ren, D., Arriaga, X. B., & Mahan, E. R. (2017). Attachment insecurity and perceived importance of relational features. *Journal of Social and Personal Relationships, 34,* 446–466. doi:10.1177/0265407516640604

Rholes, W. S., Eller, J., Simpson, J. A., & Arriaga, X. B. (2021). Support processes predict declines in attachment avoidance across the transition to parenthood. *Personality and Social Psychology Bulletin, 47,* 111–1134. doi.org/10.1177/0146167220960365

Rice, T. M., Kumashiro, M., & Arriaga, X. B. (2020). Mind the gap: Perceived partner responsiveness as a bridge between general and partner-specific attachment security. *International Journal of Environmental Research and Public Health, 17,* 7178. doi:10.3390/ijerph17197178

Righetti, F., Balliet, D., Molho, C., Columbus, S., Faure, R., Bahar, Y., Iqmal, M., Semenchenko, A., & Arriaga, X. B. (2020). Fostering attachment security: The role of interdependent situations. *International Journal of Environmental Research and Public Health, 17,* 7648. doi.org/10.3390/ijerph17207648

Ritchie, L., Cervone, D., & Sharpe, B. T. (2021). Goals and self-efficacy beliefs during the initial COVID-19 lockdown: A mixed methods analysis. *Frontiers in Psychology, 11*, 559114. https://doi.org/10.3389/fpsyg.2020.559114

Robinson, E., Sutin, A. R., Daly, M., & Jones, A. (2022). A systematic review and meta-analysis of longitudinal cohort studies comparing mental health before versus during the COVID-19 pandemic in 2020. *Journal of Affective Disorders, 296*, 567–576.

Romeo, A., Castelli, L., Benfante, A., & Tella, M. D. (2021). Love in the time of COVID-19: The negative effects of the pandemic on psychological well-being and dyadic adjustment. *Journal of Affective Disorders, 299*, 525–527.

Salvatore, J. E., Kuo, S. I., Steele, R. D., Simpson, J. A., & Collins, W. A. (2011). Recovering from conflict in romantic relationships: A developmental perspective. *Psychological Science, 22*, 376–383.

Sanders, L. (2021). COVID-19 pandemic has had a particularly negative impact on mothers' mental health. *YouGov*. https://today.yougov.com/topics/lifestyle/articles-reports/2021/03/04/coronavirus-impact-on-mothers-mental-health

Schmid, L., Wörn, J., Hank, K., Sawatzki, B., & Walper, S. (2021). Changes in employment and relationship satisfaction in times of the COVID-19 pandemic: Evidence from the German Family Panel. *European Societies, 23*, S743–S758. https://doi.org/10.1080/14616696.2020.1836385

Simpson, J. A., Beckes, L., & Weisberg, Y. J. (2008). Evolutionary accounts of individual differences in adult attachment orientations. In J. V. Wood, A. Tesser, & J. G. Holmes (Eds.), *The self and social relationships* (pp. 183–206). New York, NY: Psychology Press.

Simpson, J. A., & Overall, N. C. (2014). Partner buffering of attachment insecurity. *Current Directions in Psychological Science, 23*, 54–59.

Simpson, J. A., Rholes, W. S., & Nelligan, J. (1992). Support-seeking and support-giving within couples within an anxiety-provoking situation: The role of attachment styles. *Journal of Personality and Social Psychology, 62*, 434–446.

Stanton, S. C. E., Campbell, L., & Pink, J. C. (2017). Benefits of positive relationship experiences for avoidantly attached individuals. *Journal of Personality and Social Psychology, 113*, 568–588. https://doi.org/10.1037/pspi0000098

Tran, S., & Simpson, J. A. (2009). Prorelationship maintenance behaviors: The joint roles of attachment and commitment. *Journal of Personality and Social Psychology, 97*, 685–698.

Vigl, J., Strauss, H., Talamini, F., & Zentner, M. (2022). Relationship satisfaction in the early stages of the COVID-19 pandemic: A cross-national examination of situational, dispositional, and relationship factors. *PLoS ONE, 17*, e0264511. https://doi.org/10.1371/journal.pone.0264511

Von Mohr, M., Kirsch, L. P., & Fotopoulou, A. (2021). Social touch deprivation during COVID-19: Effects on psychological wellbeing and craving interpersonal touch. *Royal Society Open Science, 8*, 210287. https://doi.org/10.1098/rsos.210287

Vowels, L. M., & Carnelley, K. B. (2021). Attachment styles, negotiation of goal conflict, and perceived partner support during COVID-19. *Personality and Individual Differences, 171*, 110505. https://doi.org/10.1016/j.paid.2020.110505

Vowels, L. M., Carnelley, K. B., & Stanton, S. C. (2022). Attachment anxiety predicts worse mental health outcomes during COVID-19: Evidence from two studies. *Personality and Individual Differences, 185*, 111256. https://doi.org/10.1016/j.paid.2021.111256

Williamson, H. C. (2020). Early effects of the COVID-19 pandemic on relationship satisfaction and attributions. *Psychological Science, 31*(12), 1479–1487.

Wirkner, J., Christiansen, H., Knaevelsrud, C., Lüken, U., Wurm, S., Schneider, S., & Brakemeier, E. L. (2021). Mental health in times of the COVID-19 pandemic. *European Psychologist, 26,* 310–322.

Yang, S. W., Soltis, S. M., Ross, J. R., & Labianca, G. (J.). (2021). Dormant tie reactivation as an affiliative coping response to stressors during the COVID-19 crisis. *Journal of Applied Psychology, 106*(4), 489–500. https://doi.org/10.1037/apl0000909

14
SOCIAL IDENTITY DYNAMICS IN THE FACE OF OVERWHELMING UNCERTAINTY

Michael A. Hogg

CLAREMONT GRADUATE UNIVERSITY, UNIVERSITY OF KENT, USA

Amber M. Gaffney

CALIFORNIA POLYTECHNIC UNIVERSITY HUMBOLDT, USA

Abstract

Feelings of uncertainty are a part of everyday life that we generally seek to eradicate. How we reduce uncertainty is largely influenced by how we experience it. Feelings of uncertainty can differ in strength, persistence and focus, as well as the extent to which the uncertainty is experienced as an exhilarating challenge for which we possess the resources to resolve or an anxiety-ridden threat that we are ill-equipped to extinguish. It is the latter type of uncertainty that creates insecurity and has important ramifications for intragroup processes and intergroup relations. In this chapter, we focus on how people and society react to chronic and overwhelming uncertainty that is grounded in or reflects on who we feel we are in the world – our social identities. We build on *uncertainty-identity theory* and argue that feelings of self-uncertainty motivate group identification and that when uncertainty is experienced as unsolvable and manifests in deep insecurity, it generates social identity dynamics that have dark consequences for individuals and collectives. Insecure people often seek solace in distinctive groups that are homogeneous, ethnocentric, and have simple, unambiguous identities. Such groups suppress dissent and vilify outgroups and typically have autocratic leaders who promulgate unambiguous identity messages. Populism prevails, and factions polarize into distinctive identity-defined enclaves. Autocracy and ethno-nationalism triumph over democratic principles, and conspiracy theories and identity silos thrive. We end by suggesting circumstances that might help curtail people's seemingly extreme and unsolvable insecurities so that they do not turn to extreme group identities and world views.

Change and the prospect of change almost inevitably create a sense of uncertainty. Change often makes people question their well-established and often

DOI: 10.4324/9781003317623-17

habitual understanding of themselves and the social and physical world in which they live (see also Crano & Hohman, this volume). People feel they are no longer able to make reliable predictions, and therefore, plan adaptive actions. There is a loss of sense of mastery over one's ability to navigate one's world.

Change and uncertainty are ubiquitous and intrinsic features of the human condition. They cannot be completely avoided, and in certain circumstances, people can seek uncertainty. However, people can and do strive to reduce uncertainty – the process of reducing uncertainty is satisfying and adaptive. How they do this, and their success in doing so, depends on how strong and enduring the uncertainty is, what its primary focus and origin is, the extent to which it pervades many aspects of a person's life, and the resources and abilities that people believe they have to resolve the uncertainty.

Uncertainty can be generated by (perceived or actual) changes in one's close relationships with others (family, friends, and partners; see also Arriaga & Kumashiro; Mikulincer & Shaver; Murray & Lamarche, this volume), one's physical and mental health and abilities, and one's understanding of one's unique personality attributes. However, the focus of this chapter is dedicated primarily on wider societal-level change. In this respect, times of rapid and extensive change (e.g., mass migration/immigration, climate change, food and supply chain insecurity, economic recession and mass unemployment, identity instability, a changing world order, political and social chaos, pandemic, and wars and conflict) often provoke feelings of extreme and all-pervasive uncertainty and insecurity.

For example, on 24 June 2022, the US Supreme Court overturned its 1973 ruling Roe vs. Wade, which guaranteed women the federal right to legal and safe abortions. For women and Americans concerned with equal rights guaranteed to all citizens, the overturning of Roe calls into question the rights of American women to make decisions about their own health. What is the value placed on American women's ability to make autonomous decisions about their own bodies? Because 61% of Americans believe that abortions should be legal in most cases (Hartig, 2022), where does this place American trust in their political system, particularly in the Judicial Branch? How can the Supreme Court, whose power lies in the populace's willingness to abide by its rulings, be effective when its decision is discordant with the majority of Americans' opinions and raises profound uncertainty about national identity? Finally, with respect to the protests surrounding this decision and states and lawmakers seeking to defend women's rights (e.g., the California Senate ratifying the right to legal abortions in the state), how does the other side respond? Do they seek solace in their victory at the federal level, or do they tighten their convictions and seek harsher restrictions on a state-by-state basis?

Under such circumstances, people seeking greater certainty about themselves, their world, and their place within it can find it very difficult to satisfactorily reduce their uncertainty. They feel insecure and uncertain and look for ways to reduce it. They turn to leaders and cling to "people like us" who project consensus, unwavering conviction, and assuredness. This creates fractured societies in which intolerance, intergroup hostility, and extremist identities and groups can thrive. People inhabit polarized identity silos, are drawn to populist ideologies and leadership, and find conspiracy theories and narratives of victimhood attractive (see also Van Prooijen, this volume).

In this chapter, we provide a social-psychological account of how feelings of uncertainty about or reflecting on oneself can motivate people to identify with groups and make "extremist" groups, identities, worldviews, and behaviors attractive. We draw on uncertainty-identity theory (Hogg, 2021a; also see Hogg, 2007, 2012), specifically, its relevance to an explanation of societal extremism (Hogg, 2021b; also see Hogg, 2014, 2019), to argue that people who feel overwhelmed by irresolvable self-uncertainty can become so desperate that they seek solace from zealous identification with populist groups and identities.

Uncertainty

Uncertainty reduction has long been considered by social psychologists and behavioral scientists to play a fundamental role in motivating human behavior (e.g., Fromm, 1947); for example, in the context of stereotyping (e.g., Tajfel, 1969), decision-making (e.g., Kahneman, Slovic, & Tversky, 1982), and knowing one's attitudes and opinions (e.g., Festinger, 1954; see Križan & Gibbons, 2014). It is adaptive for people to feel they have true and valid knowledge about the physical and social world they live in – uncertainty makes it difficult for people to know what to expect and how to act. The process of confronting uncertainty, whether it be imposed or chosen, and successfully reducing it, is an adaptive learning process.

Typically, people cannot ever feel completely certain, only less uncertain. So, it is probably more accurate to talk about uncertainty reduction than the pursuit of certainty (Pollock, 2003). In the absence of complete objective certainty, particularly in a precarious and hazardous world, people can be prepared to go to great lengths to "feel" less uncertain (Dewey, 1929/2005) – for example, they can subscribe to preposterous conspiracy theories (Douglas & Sutton, 2018; Douglas, Sutton, & Cichoka, 2017) and isolate themselves from disconfirmation and objective truth (Frimer, Skitka, & Motyl, 2017; Wason, 1960).

The process of reducing uncertainty can require significant resources (for example, time, access to information, personal ability, and social connections) and therefore be cognitively demanding. People have relatively finite cognitive resources that they are strategic in allocating (e.g., Fiske & Taylor, 2017). They, therefore, expend cognitive energy resolving only those uncertainties that they consider most important to them, and do so only until they feel "sufficiently" certain. This provides "adequate" cognitive closure (e.g., Koffka,

1935; Kruglanski & Fishman, 2009; Kruglanski & Webster, 1996) and allows people to dedicate cognitive effort to other things, including addressing other uncertainties.

Taken together, cognitive and non-cognitive resources impose limits on people's abilities to reduce uncertainty. One consequence of this is that people's appraisals of their resources (cognitive, emotional, social, and material) to reduce uncertainty can impact their experience of uncertainty and how they approach uncertainty reduction. With adequate resources, people may view their uncertainty as easy to reduce and thus experience it as an exhilarating *challenge* to confront and resolve; without such resources, people may view uncertainty as difficult if not impossible to reduce and experience it as an anxiety-provoking and stressful *threat* to avoid and protect against – see Blascovich and associates' biopsychosocial model of challenge and threat (e.g., Blascovich, Mendes, Tomaka, Salomon, & Seery, 2003; Blascovich & Tomaka, 1996; Seery, 2013; for a related argument, see also Kruglanski & Ellenberg, this volume). One implication of this challenge versus threat distinction, which we explore later in the chapter in the context of Higgins's regulatory focus theory (e.g., Higgins, 1998), is that whether uncertainty is experienced as a challenge or a threat may impact how people approach the entire exercise of reducing uncertainty.

Self-uncertainty, Social Identity, and Group Identification

Self and Self-Uncertainty

Given that uncertainty can be aversive (e.g., Brown, Hohman, Niedbaka, & Stinnett, 2021; Jonas, McGregor, Klackl, et al., 2014; also see Sweeny, 2018; Sweeney & Falkenstein, 2015) and, as discussed above, uncertainty reduction is a core human motive that can require substantial cognitive resources to reduce, what influences which uncertainties are prioritized? Not all uncertainties are subjectively equal – some are more psychologically impactful and call out for more urgent and complete reduction than others.

A key premise of the uncertainty-identity theory (Hogg, 2007, 2012, 2021a) is that one of the most significant determinants of the psychological impact of uncertainty is the extent to which uncertainty reflects on or is directly about the self. This is because the self is a fundamental organizing and planning mechanism for how we represent and act within the world (e.g., Sedikides, Alicke, & Skowronski, 2021; Swann & Bosson, 2010).

In order to function adaptively, people need to know who they are, how to behave, and what to think, as well as who others are and how they might behave and what they might think (see also Cooper & Pearce, this volume). Uncertainties that matter most to us are those that involve our conception of who we are. People are particularly motivated to reduce uncertainty about themselves, their identity, how they relate to others, and how they are socially located, or about things that simply reflect on or are relevant to self. Reduced self-uncertainty allows us to feel we know ourselves, anticipate how others will perceive and treat us, and plan how to act.

Self-Uncertainty and Group Identification

Uncertainty-identity theory's other main premise is that group identification, via the process of self-categorization described by social identity theory (Turner, Hogg, Oakes, Reicher, & Wetherell, 1987; also see Abrams & Hogg, 2010; Hogg, 2018a; Hogg, Abrams, & Brewer, 2017), is one of the most effective ways to reduce self-uncertainty, particularly collective self-uncertainty. Feelings of uncertainty relating to who one is and how one should behave and motivate uncertainty reduction, and the process of self-categorization as a group member reduces self-uncertainty because it provides a coherent and consensual self-defining ingroup prototype that describes and prescribes who one is and how one should behave and also generates feelings of group identification, attachment, and belonging.

Group identification is so effective at reducing self-uncertainty because it provides us with a sense of who we are that prescribes what we should think, feel, and do, and it also reduces uncertainty about how others, both ingroup and outgroup members, will behave and about how social interactions will unfold. Identification also provides consensual validation of our worldview and sense of self, which further reduces uncertainty. Because people in a group tend to share the same prototype of "us" and the same prototype of "them", our expectations about the identity-based behavior of others are usually confirmed, and fellow group members agree with our perceptions, beliefs, attitudes, and values and approve of how we behave. The discovery that fellow ingroup members do not see the world as we do can create profound uncertainty about the group's identity and thus about self-conception (e.g., Wagoner, Belavadi, & Jung, 2017).

Because identification reduces and protects people from self-uncertainty, people who feel uncertain about themselves strive to successfully identify with a relevant group by "joining" new groups or identifying more strongly with a group they already belong to. Thus, uncertainty-identity theory's most basic prediction is that the more uncertain people are, the more likely they are to identify, and to identify more strongly, with a self-inclusive social category.

This prediction has been confirmed across many studies in which uncertainty is measured or manipulated in a variety of ways that indirectly or more directly focus on self-uncertainty, and identification is measured by widely-used and reliable group identification scales – see Choi and Hogg's (2020a) meta-analysis of 35 of these studies, involving 4,657 participants. Some studies experimentally manipulate self-uncertainty indirectly through perceptual uncertainty or information about identity clarity. Other studies prime self-uncertainty or directly prime collective self-uncertainty. Self-uncertainty is measured both directly and indirectly. Other research confirms that self-uncertainty is aversive (by measures of skin conductance and heart rate) and that group identification does indeed reduce the aversiveness associated with the uncertainty (using the same measures) (Brown, Hohman, Niedbaka, & Stinnett, 2021).

Uncertainty, in all of its forms, is a manifestation of insecurities: an insecure sense of self derived from feelings of security about one's group, one's place in a group, one's interpersonal relationships, or one's personality. The focus of uncertainty-identity theory and this chapter is on how self-uncertainty motivates and can be reduced by group identification. However, group identification and social identity dynamics are also motivated by self-enhancement (Hogg, 2018a). Self-enhancement, associated with the pursuit of self-esteem, explains why and how groups struggle over status and prestige (e.g., Abrams & Hogg, 1988; Ellemers, 1993; Tajfel & Turner, 1986). Uncertainty reduction explains why and how groups seek to establish an unambiguous, clearly defined, and distinct identity. Research examining the relationship between self-uncertainty and identification dis-confounded from self-enhancement-related group status considerations has shown that having a "certain" sense of self can sometimes take priority over having a favorable sense of self (Reid & Hogg, 2005) and in some cases create a need for self-handicapping (see Hirt, Eyink, & Heiman, this volume). People confronted by feelings of self-uncertainty will identify with a group that mediates undesirable status and lower self-esteem if such a group is their only social identity option.

The Many Faces of Self-uncertainty

Self-uncertainty can be experienced in different ways, and this can affect the manner in which people strive to reduce self-uncertainty and how difficult they may find it to satisfactorily reduce self-uncertainty. Here we discuss four factors that can influence the experience of self-uncertainty: (a) what aspect of self is primarily affected, (b) how much overlap is there among different aspects of self or different social identities, (c) what resources do people feel they have to deal with the uncertainty, and (d) how was the uncertainty caused.

Aspects of Self

The self has different aspects and manifestations. Brewer and Gardner (1996) have proposed three: (a) *individual self*, based on personal traits that differentiate "me" from all others; (b) *relational self*, based on connections and role relationships with significant others; and (c) *collective self*, based on group membership that differentiates "us" from "them" (also see Chen, Boucher, & Tapias, 2006; Sedikides & Brewer, 2001). Self-uncertainty can be associated with any of these aspects. You can primarily feel uncertain about your individual attributes, yourself in relation to specific other people, yourself as a group member, or the existence of your important group memberships (see Hirschberger, this volume).

Although these aspects are qualitatively different, uncertainty experienced in one domain may spread to other domains. For example, if you are primarily uncertain about your relational self, you may also become uncertain about your individual self. Research has found that self-uncertainty, irrespective of its primary focus, can motivate group identification to resolve the uncertainty. However, the strongest relationship between self-uncertainty and group identification exists when the focus of self-uncertainty is the collective self (Hogg & Mahajan, 2018).

It is likely easier to resolve self-uncertainty if it remains associated with only one aspect of self – you can invoke other aspects of self about which you feel more certain to feel generally less uncertain about your overall self. For example, if you feel uncertain about your relational self, you may be able to turn to an individual self-attribute that you are highly certain about. It is much more difficult to resolve the uncertainty that pervades the entire self-concept – you effectively have nowhere to turn to feel generally more certain about who you are. The fact that uncertainty primarily invoked in one self-domain has a strong tendency to spread to other domains (uncertainty surrounding a close relationship may make you also feel uncertain about yourself as a member of an organization) suggests that small, localized uncertainties can often become problematic. Thus, people who experience significant overlap between different selves should have a more insecure and uncertain global self than those who have a wider repertoire of more distinctive selves.

Overlap between Selves and Identities

One factor that influences, and may combat, the readiness and extent to which self-uncertainty in one domain pervades other domains of self is self- and social identity complexity – the degree to which attributes that define one aspect of self (or one social identity) overlap with or are the same as those that define other aspects of self (or other social identities) (Brewer & Pierce, 2005; Roccas & Brewer, 2002). A person has a complex self-concept and social identity if they have many discrete and diverse identities that do not overlap in terms of their identity-defining attributes; a person has a simple social identity if they have few identities and those they do have are largely isomorphic in terms of their identity-defining attributes.

A complex self-structure can quarantine identity-specific self-uncertainty and allow people to compensate by identifying more strongly with other identities (or aspects of self) that they believe are central to their overall sense of self. Providing some support for this idea, a pair of studies by Grant and

Hogg (2012) found that general self-uncertainty significantly strengthened identification with a central identity and did so most strongly when that identity was most self-conceptually prominent and also distinct from other identities.

Resources to Deal with Self-Uncertainty

As introduced earlier in this chapter, uncertainty can be experienced very differently depending on whether a person feels they have sufficient resources (cognitive, emotional, social, and material) to resolve the uncertainty (e.g., Blascovich, Mendes, Tomaka, Salomon, & Seery, 2003; Blascovich & Tomaka, 1996; Seery, 2013). This analysis also holds more specifically for self-uncertainty. Self-uncertainty in the presence of adequate resources is an exhilarating challenge to confront, even seek out, and resolve; self-uncertainty in the absence of such resources is an anxiety-provoking and stressful threat to protect oneself against and avoid.

How uncertainty is experienced may influence the behaviors people adopt to reduce the uncertainty – behaviors that can reflect a more promotive (e.g., self-promoting) or more preventative (e.g., self-protective) behavioral orientation (cf. Higgins's, 1998, regulatory focus theory). Uncertainty experienced as a challenge might encourage promotive behaviors (e.g., public assertion of one's identity and active attempts to directly address the uncertainty); uncertainty experienced as a threat might encourage more protective behaviors (e.g., retreat into identity echo chambers as a way to protect the self).

There is no research that directly explores the interactive effect of self-uncertainty and resource sufficiency (i.e., self-uncertainty experienced as a challenge or a threat) on regulatory focus (promotion versus prevention) and associated behavior. Existing research is only remotely relevant – studies of challenge and threat appraisals and intergroup relations (e.g., Scheepers, 2009; Scholl, Sassenrath, & Sassenberg, 2015); studies of expectancy-violating partners (e.g., Mendes, Blascovich, Hunter, Lickel, & Jost, 2007).

One study that is marginally relevant found that people in a prevention focus and challenge condition had an attentional bias toward negative words (for our purposes, this might suggest a more pessimistic orientation to life), whereas those in a promotion and threat condition did not (Sassenberg, Sassenrath, & Fetterman, 2015). In a similar vein, Sassenberg and Scholl (2019) refer to the literature that has focused on the interactive effect of regulatory focus (promotion/prevention) and challenge/threat on behavior. They conclude that the conjunction of threat and promotion sponsors processing of positive stimuli and a preference for social contexts requiring eager, rather than vigilant, goal striving, and the conjunction of challenge and prevention sponsors processing of negative stimuli and a preference for social contexts requiring vigilant rather than eager self-regulation. But without self-uncertainty and social identity information, it is difficult to infer the relevance of uncertainty-identity theory predictions. Sassenberg and Scholl, themselves, admit that a great deal more research is needed in this area. For example, is self-uncertainty, among those focused on prevention, experienced as a threat? And what happens when self-uncertainty is experienced as a challenge?

Causes of Self-uncertainty

Finally, there are many potential causes of self-uncertainty – some proximal, some distal; some transitory, some enduring. It is also important to recognize that self-uncertainty can be externally imposed on us (e.g., a recession and political upheaval) or sought out (e.g., a new job and going away to university) – in both cases, we can experience uncertainty as a challenge or a threat, as described above, and through reduction of the uncertainty, we learn and adapt.

Causes of uncertainty might include new social contexts, life crises, relationship changes, new work circumstances, technological and social change, immigration and emigration, socio-political and economic turmoil, war, pandemics, and natural disasters. These can all create uncertainty about one's collective self, one's social identity; though as discussed above self-uncertainty originating in one aspect of self can under some circumstances readily spread to other aspects and ultimately overwhelm a person's entire concept of self. Perhaps having the most direct impact on social identity are globalization, mass migration, climate crisis, automation and the reconfiguration of "work", postcolonialism and the new world order, national and intra-national conflict, and the realignment of super-national entities and alliances (e.g., the European Union).

Collective self-uncertainty can be especially aroused by uncertainty about (a) the defining attributes of a group that one identifies with (social identity clarity and distinctiveness is absent – Wagoner, Belavadi, & Jung, 2017), (b) how well one fits into and is accepted by a group that is central to one's sense of self (Choi & Hogg 2020b; Goldman, Giles, & Hogg, 2014; Goldman & Hogg, 2016; Hohman, Gaffney, & Hogg, 2017), and (c) how well one's group fits into a larger collective (for example a nation within the European Union – Wagoner, Antonini, Hogg, Barbieri, & Talamo, 2018; Wagoner & Hogg, 2016a). Most importantly, people are motivated to reduce self-uncertainty only when exogenous conditions create a sense of self-uncertainty.

In most of these cases, self-uncertainty is difficult to resolve (uncertainty saturates the self-concept, it takes time to resolve, there is social resistance to uncertainty reduction, there is a sense of desperation, and so forth) and therefore has the potential to be experienced as a threat. People are more likely to resort to relatively extreme and socially destructive measures (see below). For example, the pursuit of identity distinctiveness can lead to zealous identification with extremist groups and autocratic leaders, an individual's sense of social marginalization and exclusion can encourage radicalization, and (sub)groups that feel excluded and unheard can ultimately schism or engage in violent struggle that unravels the fabric of society, and thus create greater and more widespread uncertainty.

Groups, Identities, and Influence Processes

The uncertainty-identity relationship is moderated by general properties of groups, identities, and influence processes (Hogg, 2021a, 2021b). Because some qualities of groups, identities, and influence processes do a much better job of reducing self-uncertainty, people seeking self-uncertainty reduction are more strongly drawn to groups and group processes that possess these properties.

Distinctive Groups and Identities

One general group attribute that is well equipped to reduce self-uncertainty through group identification is *entitativity* – the property of a group that makes it a distinctive, coherent, and clearly structured unit with sharp intergroup boundaries, within which members share attributes and goals, have a shared fate, and interact with one another in a climate of interdependence (Campbell, 1958; Hamilton & Sherman, 1996). Highly entitative groups tend to have social identities that are simple, clear, unambiguous, prescriptive, focused, and consensual. Such identities, which can also be anchored in fixed underlying qualities of the group (e.g., Haslam, Bastian, Bain, & Kashima, 2006; Haslam, Rothschild, & Ernst, 1998), are more effective at reducing self-uncertainty than those that are vague, ambiguous, unfocused, and dissensual.

An unclearly structured, low-entitativity group with indistinct boundaries, ambiguous membership criteria, limited shared goals, and little agreement on group attributes will do a poor job of reducing or fending off self-uncertainty. In contrast, a clearly structured, high-entitativity group with sharp boundaries, unambiguous membership criteria, tightly shared goals, and consensus on group attributes will do an excellent job.

Uncertainty-identity theory predicts that self-uncertain people prefer to identify with, and identify more strongly with, highly entitative/distinctive groups than groups low in entitativity. This prediction is well supported across a large number of direct tests (e.g., Hogg, Sherman, Dierselhuis, Maitner, & Moffitt, 2007; Sherman, Hogg, & Maitner, 2009; for overviews, see Hogg, 2021a, 2021b). This research confirms that under conditions of uncertainty, people identify more strongly with high- than low-entitativity groups, and dis-identify from low-entitativity groups, or actively strive to make such groups appear more entitative. It also suggests that subgroups within a larger group that have an unclear social identity may, with the right leadership, pursue subgroup autonomy or separation from the larger group.

Although distinctive groups with clearly defined social identities are well equipped to reduce self-uncertainty, they can be quite exclusive. It can be difficult to feel welcomed and accepted and to have one's membership and identity fully validated. People who feel the group treats them as marginal members and who do not fit in with or embody the group's attributes will experience even greater self-uncertainty (e.g., Hohman, Gaffney, & Hogg, 2017; Wagoner & Hogg, 2016b). They may go to great extremes to try to win the group's trust (Goldman & Hogg, 2016), or dis-identify and seek identity confirmation elsewhere (see Choi & Hogg, 2020b). This latter option is readily available in the era of social media and widespread internet access. People can choose their own online "echo chamber" to validate their worldview and social identity (cf. Barberá, Jost, Nagler, Tucker, & Bonneau, 2015; Colleoni, Rozza, & Arvidsson, 2014; Peters, Morton, & Haslam, 2010).

Influence and Leadership

People who identify with a group to reduce the self-uncertainty need to know what the group's identity attributes are. They do this in many ways, most notably by directly or indirectly observing and communicating with fellow group members (Hogg & Giles, 2012), particularly the group's formal and informal leaders (Hogg, 2018b, 2020). Social identity reduces self-uncertainty by anchoring perceptions, attitudes, behaviors, and self-conceptions in a consensual, self-inclusive worldview that is validated by fellow ingroupers who see the world the same way as you do. However, all ingroup members are not equally reliable sources of identity information, so people show a preference for those they believe are the most prototypical members of the group and best embody the group's identity. People pay more attention to prototypical members as reliable sources of identity information.

Because prototypical members are also typically viewed as occupying leadership positions, leaders are often the most immediate and reliable sources of social identity information (Hogg, 2018b, 2020). They are highly influential, and, according to the social identity theory of leadership, prototypical leaders who embody the group's identity are more influential and well supported than less prototypical leaders (Hogg & Van Knippenberg, 2003; Hogg, Van Knippenberg & Rast, 2012; also see Haslam, Gaffney, Hogg, Rast, & Steffens, in press). This prediction is well supported by a meta-analysis of 35 studies with 6,678 participants (Barreto & Hogg, 2017) and a more widely cast meta-analysis of 32,834 participants (Steffens, Munt, Van Knippenberg, Platow, & Haslam, 2021).

Under the conditions of uncertainty, people need and expect to be able to trust fellow group members (e.g., Van Vugt & Hart, 2004; for a related argument, see also Van den Bos, this volume), particularly those who are highly prototypical and thus viewed as "one of us". This allows prototypical leaders to play a prominent communicative role in defining the group's identity (Hogg, 2018b, 2020), and because they are trusted, the group allows them to be innovative in defining the group's identity (e.g., Abrams, Randsley de Moura, Marques, & Hutchison, 2008). They can act as entrepreneurs of identity who frame and construct perceptions of the group's identity (e.g., Bos, Schemer, Corbu, Hameleers, Andreadis, Schulz, et al., 2020; Seyranian, 2014). They can also use uncertainty strategically as an identity threat (Marris, 1996) – raising uncertainty and then resolving it in order to secure their position of power and influence in the group. However, the loss of a prototypical leader predicts a reduction in group members' own prototypicality, which in turn can predict increased levels of uncertainty (Kuljian, Hohman, & Gaffney, in press). Because leaders who embody the group's prototype can become the physical and metaphorical face of the group, groups with strong figureheads with populist tendencies may dissipate after the loss of the leader (see Crano & Gaffney, 2021).

What Happens When Certainty Is Hard to Find

Self-uncertainty can motivate group identification, and group identification and associated group phenomena have extensive positive consequences for individuals, groups, and society (see Gaffney & Hogg, 2022; Griffin & Grote, 2020; Hogg, in press). However, self-uncertainty can also have toxic consequences (Forgas, this volume; Hogg, 2014, 2021b; Hogg & Gøtzsche-Astrup, 2021). These are largely extreme expressions and manifestations of people's preferences under conditions of uncertainty for distinctive groups and identities and directive leadership (see above).

This dark side surfaces when self-uncertainty is extreme and chronic and experienced as a threat because people feel they do not have the resources to resolve it; and when people have a simple identity structure with few discrete (and positive) identities (e.g., Roccas & Brewer, 2002), and their sense of self is grounded in a single social identity that saturates the self-concept (e.g., Swann, Jetten, Gomez, Whitehouse, & Bastian, 2012). Under these circumstances, people are desperate to identify and belong and yearn for leadership to help resolve their uncertainty and make them feel included and validated.

Not only are distinctive groups with unambiguously defined identities and directive leadership particularly attractive, but people develop a social identity and group-membership preference for partisan, xenophobic groups that are polarized, intolerant of internal dissent and have demagogic leaders (e.g., Gaffney, Rast, Hackett, & Hogg, 2014; Hogg, Meehan, & Farquharson, 2010; Rast, Gaffney, Hogg, & Crisp, 2012; Rast, Hogg, & Giessner, 2013). Groupcentrism becomes entrenched (e.g., Kruglanski, Pierro, Mannetti, & De Grada, 2006). People selectively expose themselves only to the opinions of other group members; endorse a central authority that dictates social identity; suppress dissent, shun diversity, and promote ingroup favoritism; and venerate and fiercely adhere to their group's norms and traditions. They embrace ethnocentrism (Brewer & Campbell, 1976), mistrust and fear outsiders (Stephan, 2014), view group attributes as fixed essences (Haslam, Bastian, Bain, & Kashima, 2006), and harbor the potential to dehumanize outgroups (Haslam, 2006; Haslam, Loughnan, & Kashima, 2008).

Populism

Populism prevails when populist ideologies and leaders strengthen the group's perceived ability to resolve uncertainty. From a social-psychological perspective, populism has a number of components (Hogg, 2021b; Hogg & Gøtzsche-Astrup, 2021). People believe that the *will and sovereignty* of the people (the group's autonomy) are supreme but are subverted by the deliberate actions of outsiders who represent an *antagonistic system or elite* that is determined to destroy "us" (Bakker, Rooduijin, & Schumacher, 2016). They also believe in *conspiracy theories* that identify outgroup actions that victimize the ingroup (e.g., Douglas & Sutton, 2018; Douglas, Sutton, & Cichoka, 2017).

There is a sense of *collective narcissism* and underappreciated superiority and a narrative of *collective victimhood* that unites the group, recruits third-party sympathy and support (e.g., Belavadi & Hogg, 2018), and raises the specter of an existential threat to the ingroup that invites and justifies violence against the outgroup (e.g., Belavadi, Rinella, & Hogg, 2020; see also Forgas, this volume). There is also support for *hierarchy, social dominance, and authoritarianism,* and a preference for leaders who *fuel zealotry and embody and promote populist attributes* in an *autocratic and authoritarian* manner that projects strength and conviction and a simple and unambiguous identity message (e.g., Gaffney, Hackett, Rast, Hohman, & Jaurique, 2018; Jetten, Ryan, & Mols, 2017).

Leadership and Group Fragmentation

These extreme properties of groups and social identity are highly effective at reducing uncertainty when people experience uncertainty as an irresolvable threat. Leadership plays an important role (e.g., Hogg, 2018b, 2021b; Rast & Hogg, 2017). People just need leadership (Rast, Gaffney, Hogg, & Crisp, 2012), but they prefer leaders who are dominant (Kakkar & Sivanathan, 2017) and autocratic (Rast, Hogg, & Giessner, 2013) and who exemplify and promote populist ideology (Bos et al., 2020; Jetten et al., 2017; see Ernst, Engesser, Büchel, Blassnig, & Esser, 2017). Uncertainty also creates an environment in which leaders who exhibit the Dark Triad attributes of Machiavellianism, narcissism and psychopathy (a personality mix associated with autocratic, toxic and dysfunctional leadership) secure and thrive in leadership positions (e.g., Guillén, Jacquart, & Hogg, in press; Kakkar & Sivanathan, 2017; Lipman-Blumen, 2005; Nevicka, De Hoogh, Van Vianen, & Ten Velden, 2013).

Self-uncertainty that is grounded in irresolvable identity uncertainty (due to a lack of clarity and consensus about the group's identity-defining attributes) can lead people to dissociate from the group, identify more strongly with more distinctive and clearly defined subgroups they belong to, and pursue subgroup autonomy or separation from the larger group. This is most likely when the subgroup is self-conceptually important and is viewed as being relatively more entitative and having a less ambiguous and dissensual identity than the superordinate group. This reasoning has been investigated and largely supported by research on Sardinia within Italy (Wagoner, Antonini, Hogg, Barbieri, & Talamo, 2018), Texas within the US (Wagoner & Hogg, 2016a), South Korea within superordinate Korean identity (Jung, Hogg, & Choi, 2016), and Scotland within the UK (Jung, Hogg, & Lewis, 2018).

Concluding Comments

In this chapter, we have explored how people may try to reduce feelings of overwhelming insecurity about their identities, which stem from self-uncertainty. We focus specifically on uncertainty that reflects on or is directly about one's self and identity – about who one is and how one fits into social relationships and society. Our analysis is framed by and grounded in uncertainty-identity theory (e.g., Hogg, 2021a, 2021b).

People need to reduce feelings of self-uncertainty to function adaptively, and one very powerful way to do this is through group identification. Group identification provides people with an identity that prescribes behavior and allows them to reliably anticipate how others will behave and treat them. All groups and identities are not the same. Distinctive groups with clearly and consensually defined identity attributes do a much better job at reducing self-uncertainty through identification. They also provide an environment in which "people like you" validate your identity and who you are. Under uncertainty people prefer to identify with and identify more strongly with these types of groups.

However, self-uncertainty can sometimes be experienced as overwhelming and almost impossible to resolve. This is most likely to be the case when (a) people have a simple self-concept in which they have few distinct identities and those identities they do have overlap so substantially that they are in effect one identity; (b) due to identity overlap, uncertainty in one aspect of self rapidly metastasizes to affect one's entire sense of self; and (c) people feel they do not have the cognitive, emotional, social, and material resources to resolve the uncertainty – uncertainty is experienced as an irresolvable and anxiety-ridden threat rather than an easily resolved and exciting challenge.

Under these circumstances, identification to resolve uncertainty generates social identity dynamics that have dark consequences for individuals, groups, and society. People seek solace in distinctive groups that are homogeneous and ethnocentric and have simple, unambiguous identities. Such groups suppress dissent and vilify outgroups and typically have autocratic leaders who construct and promulgate unambiguous identity messages. Populism prevails, and factions polarize into distinctive identity-defined enclaves. Autocracy and ethno-nationalism triumph over democratic principles, and conspiracy theories and identity silos thrive. In these uncertain times, without the proper tools to regain a sense of security, we run the risk of retreating and, to borrow from Bob Dylan, waiting for some bleachers to be set up in the sun from which we can witness the next world war, pandemic, or decision limiting our rights.

This is a dystopic prognosis for a world in which we are assailed by large-scale uncertainties that, to varying degrees, impact who we are. However, knowing a little about what can make self-uncertainty unbearable and how this steers people toward identification-contingent behaviors and outcomes that are personally and socially destructive allows us to know what to be vigilant for and what sorts of provisions might help. A focus on providing people with resources that frame uncertainties and self-insecurities as challenges that can be overcome rather than as threats that cannot be resolved may be important ingredients in reducing polarization and support for autocratic regimes and leaders. These might include facilitating the development and maintenance of complex self-concepts, providing readily available prosocial resources to resolve uncertainty, contesting or breaking down identity silos and (social) media platforms that make false identity-confirming information easily accessible, and avoiding behaviors that make people feel marginalized, irrelevant, invisible, and disrespected in society. Actually implementing these provisions is, of course, a whole different story.

References

Abrams, D., & Hogg, M. A. (1988). Comments on the motivational status of selfesteem in social identity and intergroup discrimination. *European Journal of Social Psychology, 18*, 317334.

Abrams, D., & Hogg, M. A. (2010). Social identity and self-categorization. In J. F. Dovidio, M. Hewstone, P. Glick & V. M. Esses (Eds.), *Handbook of prejudice, stereotyping, and discrimination* (pp. 179–193). London: Sage.

Abrams, D., Randsley de Moura, G., Marques, J. M., & Hutchison, P. (2008). Innovation credit: When can leaders oppose their group's norms? *Journal of Personality and Social Psychology, 95*, 662–678.

Bakker, B. N., Rooduijn, M., & Schumacher, G. (2016). The psychological roots of populist voting: Evidence from the United States, the Netherlands and Germany. *European Journal of Political Research, 55*(2), 302–320.

Barberá, P., Jost, J. T., Nagler, J., Tucker, J. A., & Bonneau, R. (2015). Tweeting from left to right: Is online political communication more than an echo chamber? *Psychological Science, 26*, 1531–1542.

Belavadi, S., & Hogg, M. A. (2018). We are victims! How observers evaluate a group's claim of collective victimhood. *Journal of Applied Social Psychology, 48*, 651–660. doi: 10.1002/jts5.37

Belavadi, S., Rinella, M., & Hogg, M. A. (2020). When social identity-defining groups become violent: Collective responses to identity uncertainty, status erosion, and resource threat. In C. A. Ireland, M. Lewis, A. C. Lopez, & J. L. Ireland (Eds.), *The handbook of collective violence: Current developments and understanding* (pp. 17–30). New York: Routledge.

Barreto, N. B., & Hogg, M. A. (2017). Evaluation of and support for group prototypical leaders: A meta-analysis of twenty years of empirical research. *Social Influence, 12*, 41–55. doi: 10.1080/15534510.2017.1316771

Bos, L., Schemer, C., Corbu, N., Hameleers, M., Andreadis, I., Schulz, A., et al. (2020). The effects of populism as a social identity frame on persuasion and mobilization: Evidence from a 15-country experiment. *European Journal of Political Research, 59*, 3–24.

Brewer, M. B., & Campbell, D. T. (1976). *Ethnocentrism and intergroup attitudes: East African evidence*. New York: Sage.

Brewer, M. B., & Gardner, W. (1996). Who is this "we"? Levels of collective identity and self representations. *Journal of Personality and Social Psychology, 71*, 83–93.

Brewer, M. B., & Pierce, K. P. (2005). Social identity complexity and outgroup tolerance. *Personality and Social Psychology Bulletin, 31*, 428–437.

Blascovich, J., Mendes, W. B., Tomaka, J., Salomon, K., & Seery, M. (2003). The robust nature of the biopsychosocial model of challenge and threat: A reply to Wright and Kirby. *Personality and Social Psychology Review, 7*, 234–243.

Blascovich, J., & Tomaka, J. (1996). The biopsychosocial model of arousal regulation. *Advances in Experimental Social Psychology, 28*, 1–51. Doi: 10.1016/S0065-2601(08)60235-X

Brown, J. K., Hohman, Z. P., Niedbaka, E. M., & Stinnett, A. J (2021). Sweating the big stuff: Arousal and stress as functions of self-uncertainty and identification. *Psychophysiology, 00*:e13836.

Campbell, D. T. (1958). Common fate, similarity, and other indices of the status of aggregates of persons as social entities. *Behavioral Science, 3*, 14–25.

Chen, S., Boucher, H. C., & Tapias, M. P. (2006). The relational self revealed: Integrative conceptualization and implications for interpersonal life. *Psychological Bulletin, 132*, 151–179.

Choi, E. U., & Hogg, M. A. (2020a). Self-uncertainty and group identification: A meta-analysis. *Group Processes and Intergroup Relations, 23*, 483–501.

Choi, E. U., & Hogg, M. A. (2020b). Who do you think you are? Ingroup and outgroup sources of identity validation. *Journal of Theoretical Social Psychology, 4*(3), 125–134. https://doi.org/10.1002/jts5.66

Colleoni, E., Rozza, A., & Arvidsson, A. (2014). Echo chamber or public sphere? Predicting political orientation and measuring political homophily in Twitter using big data. *Journal of Communication, 64*, 317–332.

Crano, W. D., & Gaffney, A. M. (2021). Populism in the west as a form of influence. In J. P. Forgas, W. D. Crano, & K. Fiedler (Eds.), *The psychology of populism: The tribal challenge to liberal democracy* (pp. 297–318). London: Psychology Press.

Dewey, J. (1929/2005). *The quest for certainty: A study of the relation of knowledge and action*. Whitefish, MT: Kessinger Publishing.

Douglas, K. M., & Sutton, R. M. (2018). Why conspiracy theories matter: A social psychological analysis. *European Review of Social Psychology, 29*, 256–298.

Douglas, K. M., Sutton, R. M., & Cichoka, A. (2017). The psychology of conspiracy theories. *Current Directions in Psychological Science, 26*, 538–542.

Ellemers, N. (1993). The influence of socio-structural variables on identity management strategies. *European Review of Social Psychology, 4*, 27–57.

Festinger, L. (1954). A theory of social comparison processes. *Human Relations, 7*, 117–140.

Ernst, N., Engesser, S., Büchel, F., Blassnig, S., & Esser, F. (2017). Extreme parties and populism: An analysis of Facebook and Twitter across six countries. *Information, Communication and Society, 20*, 1347–1364.

Fiske, S. T., & Taylor, S. E. (2017). *Social cognition: From brains to culture* (3rd ed.). Thousand Oaks, CA: Sage.

Frimer, J. A., Skitka, L. J., & Motyl, M. (2017). Liberals and conservatives are similarly motivated to avoid exposure to one another's opinions. *Journal of Experimental Social Psychology, 72*, 1–12.

Fromm, E. (1947). *Man for himself: An inquiry into the psychology of ethics*. New York: Rinehart.

Gaffney, A. M., Hackett, J. D., Rast, D. E., Hohman, Z. P., & Jaurique, A. (2018). The state of American protest: Shared anger and populism. *Analyses of Social Issues and Public Policy, 18*, 11–33.

Gaffney, A. M. & Hogg, M. A. (2022). A social identity analysis of sociability: Making, breaking, and shaping groups and societies. In J. P. Forgas, W. D. Crano, & K. Fiedler (Eds.), *The psychology of sociability: Understanding human attachment* (pp. 140–161). New York: Routledge.

Gaffney, A. M., Rast, D. E. III, Hackett, J. D., & Hogg, M. A. (2014). Further to the right: Uncertainty, political polarization and the American "Tea Party" movement. *Social Influence, 9*(4), 272–288. doi: 10.1080/15534510.2013.842495

Goldman, L., Giles, H., & Hogg, M. A. (2014). Going to extremes: Social identity and communication processes associated with gang membership. *Group Processes and Intergroup Relations, 17*, 813–832.

Goldman, L., & Hogg, M. A. (2016). Going to extremes for one's group: The role of prototypicality and group acceptance. *Journal of Applied Social Psychology, 46*, 544–553.

Grant, F., & Hogg, M. A. (2012). Self-uncertainty, social identity prominence and group identification. *Journal of Experimental Social Psychology, 48*, 538–542.

Griffin, M. A., & Grote, G. (2020). When is uncertainty better? A model of uncertainty regulation and effectiveness. *Academy of Management Review, 45*, 745–765.

Guillén, L., Jacquart, P., & Hogg, M. A. (in press). To lead, or to follow? How self-uncertainty and the dark triad of personality influence leadership motivation. *Personality and Social Psychology Bulletin*, in press. https://doi.org/1177/01461672221086771

Hamilton, D. L., & Sherman, S. J. (1996). Perceiving persons and groups. *Psychological Review, 103*, 336–355.

Hartig, H. (13 June, 2022). *About six-in-ten Americans say abortion should be legal in all or most cases*. Pew Research Center. https://www.pewresearch.org/fact-tank/2022/06/13/about-six-in-ten-americans-say-abortion-should-be-legal-in-all-or-most-cases-2/

Haslam, N. (2006). Dehumanization: An integrative review. *Personality and Social Psychology Review, 10*, 252–264.

Haslam, N., Bastian, B., Bain, P., & Kashima, Y. (2006). Psychological essentialism, implicit theories, and intergroup relations. *Group Processes and Intergroup Relations, 9*, 63–76.

Haslam, N., Loughnan, S., & Kashima, Y. (2008). Attributing and denying humanness to others. *European Review of Social Psychology, 19*, 55–85.

Haslam, N., Rothschild, L., & Ernst, D. (1998). Essentialist beliefs about social categories. *British Journal of Social Psychology, 39*, 113–127.

Haslam, S. A., Gaffney, A. M., Hogg, M. A., Rast, D. E. III, & Steffens, N. K. (in press). Reconciling identity leadership and leader identity: A dual-identity framework. *The Leadership Quarterly*, in press. https:/doi.org/10.1016/j.leaqua.2022.101620

Higgins, E. T. (1998). Promotion and prevention: Regulatory focus as a motivational principle. *Advances in Experimental Social Psychology, 30*, 1–46. doi: 10.1016/S0065-2601(08)60381-0

Hogg, M. A. (2007). Uncertainty-identity theory. *Advances in Experimental Social Psychology, 39*, 69–126. doi: 10.1016/s0065-2601(06)39002-8

Hogg, M. A. (2012). Uncertainty-identity theory. In P. A. M. Van Lange, A. W. Kruglanski, & E. T. Higgins (Eds.), *Handbook of theories of social psychology* (Vol. 2, pp. 62–80). Thousand Oaks, CA: Sage.

Hogg, M. A. (2014). From uncertainty to extremism: Social categorization and identity processes. *Current Directions in Psychological Science, 23*, 338–342. https://doi.org/10.1177/0963721414540168

Hogg, M. A. (2018a). Social identity theory. In P. J. Burke (Ed.), *Contemporary social psychological theories* (2nd ed., pp. 112–138). Stanford, CA: Stanford University Press.

Hogg, M. A. (2018b). Self-uncertainty, leadership preference, and communication of social identity. *Atlantic Journal of Communication, 26*, 111–121.

Hogg, M. A. (2019). Radical change. Uncertainty in the world threatens our sense of self: To cope, people embrace populism. *Scientific American, 321*(3), 85–87.

Hogg, M. A. (2020). Learning who we are from our leaders: How leaders shape group and organizational norms and identities. In L. Argote & J. M. Levine (Eds.), *The Oxford handbook of group and organizational learning* (pp. 587–602). New York: Oxford University Press.

Hogg, M. A. (2021a). Self-uncertainty and group identification: Consequences for social identity, group behavior, intergroup relations, and society. *Advances in Experimental Social Psychology, 64*, 263–316. https://doi.org/10.1016/bs.aesp.2021.04.004

Hogg, M. A. (2021b). Uncertain self in a changing world: A foundation for radicalization, populism, and autocratic leadership. *European Review of Social Psychology, 32*(2), 235–268. https://doi.org/10.1080/10463283.2020.1827628

Hogg, M. A. (in press). Self-uncertainty and social identity processes in organizations: An uncertainty-identity theory perspective. In M. Griffin & G. Grote (Eds.), *The Oxford handbook of uncertainty management in work organizations*. New York: Oxford University Press.

Hogg, M. A., Abrams, D., & Brewer, M. B. (2017). Social identity: The role of self in group processes and intergroup relations. *Group Processes and Intergroup Relations, 20*, 570–581.

Hogg, M. A., & Giles, H. (2012). Norm talk and identity in intergroup communication. In H. Giles (Ed.), *The handbook of intergroup communication* (pp. 373–387). New York: Routledge.

Hogg, M. A., & Gøtzsche-Astrup, O. (2021). Self-uncertainty and populism: Why we endorse populist ideologies, identify with populist groups, and support populist leaders. In J. P. Forgas, W. D. Crano, & K. Fiedler (Eds.), *The psychology of populism: The tribal challenge to liberal democracy* (pp. 197–218). New York: Routledge.

Hogg, M. A., & Mahajan, N. (2018). Domains of self-uncertainty and their relationship to group identification. *Journal of Theoretical Social Psychology, 2*, 67–75.

Hogg, M. A., Meehan, C., & Farquharson, J. (2010). The solace of radicalism: Self-uncertainty and group identification in the face of threat. *Journal of Experimental Social Psychology, 46*, 1061–1066.

Hogg, M. A., Sherman, D. K., Dierselhuis, J., Maitner, A. T., & Moffitt, G. (2007). Uncertainty, entitativity, and group identification. *Journal of Experimental Social Psychology, 43*, 135–142.

Hogg, M. A., & Van Knippenberg, D. (2003). Social identity and leadership processes in groups. *Advances in Experimental Social Psychology, 35*, 1–52. doi: 10.1016/S0065-2601(03)01001-3

Hogg, M. A., Van Knippenberg, D., & Rast, D. E. III. (2012). The social identity theory of leadership: Theoretical origins, research findings, and conceptual developments. *European Review of Social Psychology, 23*, 258–304.

Hohman, Z. P., Gaffney, A. M., & Hogg, M. A. (2017). Who am I if I am not like my group? Self-uncertainty and feeling peripheral in a group. *Journal of Experimental Social Psychology, 72*, 125–132. doi: 10.1016/j.jesp.2017.05.002

Jetten, J., Ryan, R., & Mols, F. (2017). Stepping in the shoes of leaders of populist right-wing parties: Promoting anti-immigrant views in times of economic prosperity. *Social Psychology, 48*, 40–46.

Jonas, E., McGregor, I., Klackl, J., Agroskin, D., Fritsche, I., Holbrook, C., Nash, K., Proulx, T., & Quirin, M. (2014). Threat and defense: From anxiety to approach. *Advances in Experimental Social Psychology, 49*, 219–286. doi: 10.1016/B978-0-12-800052-6.00004-4

Jung, J., Hogg, M. A., & Choi, H.-S. (2016). Reaching across the DMZ: Identity uncertainty and reunification on the Korean peninsula. *Political Psychology, 37*, 341–350.

Jung, J., Hogg, M. A., & Lewis, G. J. (2018). Identity uncertainty and UK-Scottish relations: Different dynamics depending on relative identity centrality. *Group Processes and Intergroup Relations, 21*, 861–873.

Kahneman, D., Slovic, P., & Tversky, A. (Eds.) (1982). *Judgment under uncertainty: Heuristics and biases.* New York: Cambridge University Press.

Kakkar, H., & Sivanathan, N. (2017). When the appeal of a dominant leader is greater than a prestige leader. *Proceedings of the National Academy of Sciences, 114*(26), 6734–6739. doi: 10.1073/pnas.1617711114

Koffka, K. (1935). *Principles of Gestalt psychology.* New York: Harcourt, Brace & Co.

Križan, Z., & Gibbons, F. X. (Eds.) (2014). *Communal functions of social comparison.* New York: Cambridge University Press.

Kruglanski, A. W., & Fishman, S. (2009). The need for cognitive closure. In M. R. Leary & R. H. Hoyle (Eds.), *Handbook of individual differences in social behavior* (pp. 343–353). New York: Guilford Press.

Kruglanski, A. W., Pierro, A., Mannetti, L., & De Grada, E. (2006). Groups as epistemic providers: Need for closure and the unfolding of group-centrism. *Psychological Review, 113*(1), 84–100. doi: 101037/0033-295X.113.1.84

Kruglanski, A. W., & Webster, D. M. (1996). Motivated closing of the mind: 'Seizing' and 'freezing'. *Psychological Review, 103*, 263–283.

Kuljian, O., Hohman, Z. P., & Gaffney, A. M. (in press). Who are we if we do not know who our leader is? Perceptions of leaders' prototypicality affects followers' self-prototypicality and uncertainty. *Social Psychological and Personality Science.*

Lipman-Blumen, J. (2005). Toxic leadership: When grand illusions masquerade as noble visions. *Leader to Leader, 36*, 29–36.

Marris, P. (1996). *The politics of uncertainty: Attachment in private and public life*. London: Routledge.

Mendes, W. B., Blascovich, J., Hunter, S. B., Lickel, B., & Jost, J. T. (2007). Threatened by the unexpected: Physiological responses during social interactions with expectancy-violating partners. *Journal of Personality and Social Psychology, 92*(4), 698–716. https://doi.org/10.1037/0022-3514.92.4.698

Nevicka, B., De Hoogh, A. H. B., Van Vianen, A. E. M., & Ten Velden, F. S. (2013). Uncertainty enhances the preference for narcissistic leaders. *European Journal of Social Psychology, 43*, 370–380.

Peters, K., Morton, T. A., & Haslam, S. A. (2010). Communication silos and social identity complexity in organizations. In H. Giles, S. A. Reid, & J. Harwood (Eds.), *Dynamics of intergroup communication* (pp. 221–234). New York: Peter Lang.

Pollock, H. N. (2003). *Uncertain science … uncertain world*. Cambridge, UK: Cambridge University Press.

Rast, D. E. III, Gaffney, A. M., Hogg, M. A., & Crisp, R. J. (2012). Leadership under uncertainty: When Leaders who are non-prototypical group members can gain support. *Journal of Experimental Social Psychology, 48*, 646–653.

Rast, D. E. III, & Hogg, M. A. (2017). Leadership in the face of crisis and uncertainty. In J. Storey, J. Hartley, J.-L. Denis, P. 't Hart, & D. Ulrich (Eds.), *The Routledge companion to leadership* (pp. 52–64). New York: Routledge.

Rast, D. E. III, Hogg, M. A., & Giessner, S. R. (2013). Self-uncertainty and support for autocratic leadership. *Self and Identity, 12*, 635–649.

Reid, S. A., & Hogg, M. A. (2005). Uncertainty reduction, self-enhancement, and ingroup identification. *Personality and Social Psychology Bulletin, 31*, 804–817.

Roccas, S., & Brewer, M. B. (2002). Social identity complexity. *Personality and Social Psychology Review, 6*, 88–106.

Sassenberg, K., & Scholl, A. (2019). Linking regulatory focus and threat–challenge: Transitions between and outcomes of four motivational states. *European Review of Social Psychology, 30*(1), 174–215. https://doi.org/10.1080/10463283.2019.1647507

Sassenberg, K., Sassenrath, C., & Fetterman, A. K. (2015). Threat ≠ prevention, challenge ≠ promotion: The impact of threat, challenge and regulatory focus on attention to negative stimuli. *Cognition and Emotion, 29*(1), 188–195. https://doi.org/10.1080/02699931.2014.898612

Scheepers, D. (2009). Turning social identity threat into challenge: Status stability and cardiovascular reactivity during inter-group competition. *Journal of Experimental Social Psychology, 45*(1), 228–233. https://doi.org/10.1016/j.jesp.2008.09.011

Scholl, A., Sassenrath, C., & Sassenberg, K. (2015). Attracted to power: Challenge/Threat and Promotion/prevention focus differentially predict the attractiveness of group power. *Frontiers in Psychology, 6*. https://doi.org/10.3389/fpsyg.2015.00397

Sedikides, C., Alicke, M. D., & Skowronski, J. J. (2021). On the utility of the self in social perception: An egocentric tactician model. *Advances in Experimental Social Psychology, 63*, 247–298.

Sedikides, C., & Brewer, M. B. (Eds.) (2001). *Individual self, relational self, and collective self*. Philadelphia, PA: Psychology Press.

Seery, M. D. (2013). The biopsychosocial model of challenge and threat: Using the heart to measure the mind. *Social and Personality Psychology Compass, 7*(9), 637–653. https://doi.org/10.1111/spc3.12052

Seyranian, V. (2014). Social identity framing communication strategies for mobilizing social change. *The Leadership Quarterly, 25*, 468–486.

Sherman, D. K., Hogg, M. A., & Maitner, A. T. (2009). Perceived polarization: Reconciling ingroup and intergroup perceptions under uncertainty. *Group Processes and Intergroup Relations, 12*, 95–109.

Steffens, N. K., Munt, K. A., Van Knippenberg, D., Platow, M. J., & Haslam, S. A. (2021). Advancing the social identity theory of leadership: A meta-analytic review of leader group prototypicality. *Organizational Psychology Review, 11*(1), 35–72.

Stephan, W. G. (2014). Intergroup anxiety: Theory, research, and practice. *Personality and Social Psychology Review, 18*, 239–255.

Swann, W. B., & Bosson, J. K. (2010). Self and identity. In S. T. Fiske, D. T. Gilbert, & G. Lindzey (Eds.), *Handbook of social psychology* (5th ed., Vol. 1, pp. 589–628). New York: Wiley.

Swann, W. B., Jetten, J., Gomez, A., Whitehouse, H., & Bastian, B. (2012). When group membership gets personal: A theory of identity fusion. *Psychological Review, 119*, 441–456.

Sweeny, K. (2018). On the experience of awaiting uncertain news. *Current Directions in Psychological Science, 27*(4), 281–285. https://doi.org/10.1177/0963721417754197

Sweeny, K., & Falkenstein, A. (2015). Is waiting really the hardest part? Comparing the emotional experiences of awaiting and receiving bad news. *Personality and Social Psychology Bulletin, 41*(11), 1551–1559. https://doi.org/10.1177/0146167215601407

Tajfel, H. (1969). Cognitive aspects of prejudice. *Journal of Social Issues, 25*, 79–97.

Tajfel, H., & Turner, J. C. (1986). The social identity theory of intergroup behaviour. In S. Worchel & W. G. Austin (Eds.), *Psychology of intergroup relations* (2nd ed., pp. 7–24). Chicago: Nelson-Hall.

Turner, J. C., Hogg, M. A., Oakes, P. J., Reicher, S. D., & Wetherell, M. S. (1987). *Rediscovering the social group: A self-categorization theory.* Oxford, UK: Blackwell.

Van Vugt, M., & Hart, C. M. (2004). Social identity as social glue: The origins of group loyalty. *Journal of Personality and Social Psychology, 86*, 585–598.

Wagoner, J. A., Antonini, M., Hogg, M. A., Barbieri, B., & Talamo, A. (2018). Identity-centrality, dimensions of uncertainty, and pursuit of subgroup autonomy: The case of Sardinia within Italy. *Journal of Applied Social Psychology, 48*, 582–589. doi: 10.1111/jasp.12549

Wagoner, J. A., Belavadi, S., & Jung, J. (2017). Social identity uncertainty: Conceptualization, measurement, and construct validity. *Self and Identity, 16*, 505–530.

Wagoner, J. A., & Hogg, M. A. (2016a). Normative dissensus, identity-uncertainty, and subgroup autonomy. *Group Dynamics: Theory, Research, and Practice, 20*, 310–322.

Wagoner, J. A., & Hogg, M. A. (2016b). Uncertainty and group identification: Moderation by warmth and competence as cues to inclusion and identity validation. *Self and Identity, 15*, 525–535.

Wason, P. C. (1960). On the failure to eliminate hypotheses in a conceptual task. *Quarterly Journal of Experimental Psychology, 12*(3), 129–140.

15
FROM INDIVIDUAL INSECURITY TO COLLECTIVE SECURITY

The Group Survival Motivation

Gilad Hirschberger

REICHMAN UNIVERSITY, ISRAEL

Abstract

Individuals are predestined to die, but human groups may persevere for millennia. Because individual humans are inferior to other animals in many physical aspects such as size, strength, speed, and senses, the human ability to coalesce in large and efficient groups has provided humans with an evolutionary advantage. Consequently, humans have been motivated to perpetuate the existence of groups that safeguard their own existence. From this *group survival motivation*, I draw three hypotheses pertaining to inter- and intra-group relations: First, group survival, not intergroup relations, is the main force driving group behavior. Intergroup conflict resolution interventions that fail to consider implications for group survival may inadvertently compromise group safety by reducing vigilance to collective threats. Second, the memory of collective victimization, often depicted as an obstacle to intergroup conflict resolution, may have group survival advantages in the sense of "once burned, twice cautious." Third, ideological polemics within a group that often seem negative and disruptive may function as a system of checks and balances between different strategies that serve group survival needs. Because the perception of threats and opportunities is contingent on political affiliation, groups that are ideologically diverse may have an advantage in perceiving threats and seizing opportunities. In this chapter, I present research supporting these hypotheses and show how the group survival perspective may help understand burning social issues, from the obstinate nature of intergroup conflict to the strengths and vulnerabilities of liberal democracies.

From the moment of its inception, social psychology has devoted a great deal of effort to understanding intergroup relations and intergroup conflict resolution (see also Hogg & Gaffney; Kreko, this volume). However, in spite of our best efforts, intergroup relations remain strenuous, and protracted intergroup conflicts are as tenacious as ever. The Russian invasion of Ukraine in 2022 has cast even further doubt on some of the basic assumptions about what propels groups into conflict and what repels them from fighting their neighbors. In this chapter, I propose that social psychology's focus on intergroup relations and conflict resolution may have obscured the possibility that human groups are first and foremost motivated by survive and thrive motivations. Relations with other groups may not be a primary group motivation but a possible (albeit not always necessary) means toward an existential end – to promote security and the continued survival of the group. If we assume such a group survival perspective, we may gain new insights into the intransient nature of intergroup conflicts that seem impervious to our best efforts at conflict resolution.

In this chapter, I briefly reexamine the basic tenets of the social identity approach and propose that adding a group survival element to it may increase its relevance to intergroup conflict resolution. I then suggest that some of the research on groups and intergroup relations that ostensibly contributes to the resolution of intergroup conflict may inadvertently increase the potential for conflict. Specifically, I argue that the emotion-in-conflict approach (suggesting that certain negative emotions stand at the heart of intergroup conflict and therefore should be regulated) may have overlooked the possible negative implications of emotion regulation on group survival and long-term security. Similarly, the collective victimization literature indicates that the shared memory of historical trauma and victimhood has long-term negative effects on the resolution of contemporary intergroup conflict. But the memory of collective trauma may also convey lessons for group survival. Finally, I argue that the ideological polemics within a society may be more adaptive to the group than they seem at first glance and that the balance between the political right and left may enable groups to protect themselves and at the same time seize opportunities for conflict resolution with other groups, ultimately increasing rather than reducing security. The integration of these disparate literatures on intergroup relations supports a group survival explanation of why intergroup conflict persists and when conflict resolution efforts may succeed or fail.

The Social and Existential Function of Groups

Social identity theory developed against the backdrop of the Holocaust at a time when biological and evolutionary models of human groups were too reminiscent of Social Darwinism to even be considered (Brewer & Caporeal, 2006; see also Hogg & Gaffney, this volume). Psychology was eager to find a social explanation that could shed light on why people organize in groups and why members of these groups dislike each other so often. It is probably no coincidence that two of the main figures responsible for shaping our understanding of social groups were stateless people forced to live in exile by harsh political realities. Muzafer Sherif, one of the founders of social psychology and of the famous Robbers' Cave Experiment, was a Turkish intellectual and social activist who opposed the rise of fascism and pro-Nazi sentiments in Turkey in the 1940s. After being dismissed from his university in Ankara, he chose to leave for the US (Dost-Gozkan, 2017), where he rose to prominence in the social psychology community. Henry Tajfel, father of social identity theory, was a Polish Jew who studied chemistry at the Sorbonne in Paris because numerus clausus (quota) policies restricted the number of Jews in Polish universities. He soon became a French citizen, fought in the French army in WWII, was captured by the Germans, and survived in spite of his Jewish origins. Later, he moved to Britain, once again changed his nationality, and conducted his important work on social identity (Hogg, 2016). These two refugees, scholars in exile, had learned an important lesson from their traumatic personal experiences. Identity is arbitrary, fluid, and often random; people can shift from one social identity to another; and the adoption of a certain social identity, as arbitrary as it may be, has far-reaching consequences.

The minimal group paradigm (Tajfel, 1970; Tajfel, Billig, Bundy, & Flament, 1971) empirically demonstrating these ideas was not just an elegant experimental procedure but a natural extension of Tajfel's personal experiences and struggles with social identity (Hogg, 2016). The minimal group studies demonstrate that even the most artificially contrived groups display the dynamics observed among ostensibly "real" groups. The research was particularly efficient in identifying the conditions that are sufficient to produce hostility between groups that have no past, no future, and nothing in common aside from a random group that they were arbitrarily assigned to. Tajfel thought that the reason allocation to meaningless groups influences behavior so dramatically is because people have a spontaneous preference to define themselves and obtain feelings of worth based on their group membership, and once assigned to groups, any group, these processes are set in motion.

Uncertainty identity theory that stems from the social identity approach (Hogg & Gaffney, this volume) effectively demonstrates the contingencies between individual certainty (or lack thereof) and feelings of attachment and pride to the social group that provides a coherent explanation to a plethora of phenomena from intergroup relations to collective action. Although the social identity approach has unquestionably contributed greatly to our understanding of groups and intergroup processes, there seems to be something lacking in the notion that people create social groups just to obtain feelings of worth and value. Could that alone explain why people throughout history and in every known culture readily sacrifice themselves for God and country?

To the notion that group identity provides a sense of identification and self-worth, terror management theory (TMT; Pyszczynski & Sundby, this volume) added that extending the self from the individual level to the collective level also helps assuage existential concerns. Whereas the individual is inevitably mortal and destined to die, the group is potentially everlasting and may persist indefinitely. Individuals are thus motivated to extend their self to the collective level and to take comfort in symbolic immortality – the notion that some aspects of the self will persevere after physical death (Lifton, 1973). TMT provides an important extension to the social identity approach by showing that group membership is not merely about feelings of self-worth but touches the core aspects of human existence.

Although TMT comes close to understanding the fundamentally existential protective function of groups, it limits this function to the symbolic level alone by empirically demonstrating how adherence to social groups regulates death anxiety. By focusing exclusively on the symbolic level, TMT does not cover the full existential meaning of group membership. Whereas regulating death anxiety is undoubtedly important, groups have a much more fundamental role in human existence – they increase the odds of actual individual survival through attachment to close others (see also Mikulincer & Shaver, this volume).

The Group Survival Perspective

Humans are puny and weak, and at the individual level the human ability to deal with threat is severely limited (Chapman & Chapman, 2000). Individual humans suffer from significant disadvantages in almost every physical aspect compared to other species: they have a small and fragile body; they are slow to move; they have poor eyesight, poor hearing, and a poor sense of smell. They cannot fly, have no tail to swing from, and have no claws or sharp teeth (e.g., Bergman, 2004; Blanchard, 2009). Humans, therefore, should be no match for nature's many predators. Yet, not only have humans survived and prevailed, but they have also ascended to the top of the animal kingdom and dominated the planet. How did such an ostensibly puny creature attain such remarkable evolutionary success? Humans survived due to their superior cognitive faculties, their unique ability to imagine and prepare for the future (von Hippel & Merakovsky, this volume), and by facing danger as a group (Ein-Dor & Hirschberger, 2016). The evolution of foresight and the ability to simulate the future gave rise to anxiety that has prompted humans to overcome their disadvantages through elaborate and effective cooperation (Krueger & Gruening, this volume).

The group survival perspective suggests that because groups have played such an important role in the actual survival of humans, people are motivated to defend the groups that have served over time to protect them. Emile Durkheim (1893) provided a useful metaphor to explain the dialectical relationships between individual and collective motivations and how they contribute to survival. He suggested that culture functions as a superorganism, with each cell (individual) being part of a larger system that works in synchrony and is guided by a self-preservation motivation. Accordingly, individuals who were motivated to perpetuate the existence of their groups increased the likelihood that these groups would survive over time, and group survival in turn contributed to individual survival and evolutionary success. The notion that groups are motivated to survive and that individuals are propelled to ensure the existence of groups that keep them safe may have been overlooked in the social psychological literature, perhaps because the predominant theories of intergroup processes, realistic conflict theory, and social identity theory, stem from a social constructivist perspective. The view that groups are inherently fluid and often arbitrary symbolic constructions leaves little room to consider them as unique historical entities.

Unlike social identity theory that tends to view all groups as essentially the same, the group survival perspective makes a critical distinction between the historical-cultural group and transient groups. The historical-cultural group is unique in having a long history, providing a sense of historical continuity (Sani et al., 2007), in having an expectation for surviving into the future, in providing a sense of entitativity (Yzerbyt et al., 1998), and particularly in the willingness to sacrifice current group members for the sake of the eternal group (Kahn, Klar, & Roccas, 2017). Research indicates that brief reminders of existential threat prompt people from different cultures to declare a willingness to sacrifice their own life for the historical-cultural group (Caspi-Berkowicz et al., 2019; Pyszczynski et al., 2006). It is unlikely that such intentions would be expressed for the benefit of groups that provide only social identity (e.g., one's alma mater). The evolutionary logic of such behavior may be somewhat confusing:

> At first sight, self-sacrificing for a stranger may seem foolish and obtuse, and from an evolutionary perspective highly maladaptive. When examining this intention ... at the group level, however, it may be a highly adaptive response because a group that shows a high level of altruistic behavior to the point of self-sacrificing for another will be a highly adaptive and safe group.
>
> *(Sober & Wilson, 1998)*

The distinction between transient or arbitrary groups and historical-cultural groups is important to understanding the function of groups in managing insecurity because, although the experience of exile and statelessness that sparked the development of social identity theory is truly remarkable, it is not representative of the experiences of most humans. The majority of the people in the world today (e.g., China, India, and Indonesia) still live in local communities, confined to a specific geographic location, with a long history of their ancestral group members living a similar life in the same place. Definitions of groups as fluid, amorphous social constructions do not fully describe these tight and enduring historical and cultural groups very well (Byford & Tileaga, 2014).

Intergroup relations have been at the heart of social identity theory from the very beginning (Tajfel & Turner, 1979), but the focus on the resolution of intergroup conflict has led to a narrow view of collective threats and their implications. Threat perception is often viewed as a bias, as a problem that needs to be mitigated, and not as a legitimate concern (Lüders, Jonas, Fritsche, & Agroskin, 2016). Anxiety, insecurity, and uncertainty, according to this view, lead to lack of control, which in turn breeds xenophobia and support for authoritarian leaders (see also Forgas; van den Bos; van Prooijen, this volume). While there is little doubt that the sense of threat could promote a tribalism that has many hazardous implications, including prejudice, hatred, and violence, it is surprising that there is little consideration in the literature of legitimate feelings of threat. When groups face real, formidable threats, their sense of threat is not only reasonable but also highly adaptive to group survival. The Ukrainians, who are currently fending off a brutal Russian invasion, may have prejudiced feelings toward the Russians; they may even hate them, but there is more to the story than that. They are mainly taking the necessary and perhaps inevitable psychological steps to secure the existence of their nation and people.

Social identity theory has contributed greatly to understanding some, but not all, aspects of the function of groups. While there are many processes that can be elegantly explained with this theory, the explanatory power for other phenomena is limited. Intergroup conflict may be one such case wherein a social identity explanation alone cannot account for why people throughout the ages seem keen on sacrificing their young and killing their neighbors. A better understanding of the role of the historical-cultural group in safeguarding individual existence and security, and the individual motivation to defend the groups that protect them by all means necessary, may be a good starting point.

Intergroup Relations and Conflict Resolution

The intergroup relations and conflict resolution literatures are heavily influenced by social identity theory (Hogg, Abrams, & Brewer, 2017; Reicher et al., 2016) and view the peaceful resolution of intergroup conflict as a primary goal of social psychological research (Bar-Tal, 2000). If groups are arbitrary and fluid social constructions, and conflicts among them reflect psychological dynamics similar to those found in the minimal group paradigm, then it makes sense to focus on conflict resolution while placing less emphasis on the integrity of the groups involved in these conflicts.

The reality of intergroup conflicts, however, is that conflicts are often difficult to resolve, and many attempts at conflict resolution fail. Signing a peace accord does not guarantee the termination of threats and hostilities between groups (see Harbom et al., 2006), and between one-third and one-half of all conflicts revert to warfare within five years (Lomborg, 2004; Walter, 2004; Doyle & Sambanis, 2006). This is not to say that conflict resolution efforts are pointless or that warfare is an acceptable state; it means that the resolution of conflict is often temporary and does not constitute what Fukuyama (1992) proclaimed as the "end of history." Therefore, the need to remain vigilant and to consider the implications of any agreement on group survival is paramount.

Currently, many of the solutions offered in the conflict resolution literature consist of manipulations and interventions that increase participants' support for conflict resolution on self-reported measures. There may be an important distinction, however, between the individual desire for peace in these decontextualized experimental settings and in actually obtaining peace in the complicated and somewhat chaotic real world (Li, Rovenpor, & Leidner, 2016). Even if we assume that these studies have some external validity, increasing people's support for peace without accounting for other conceivable effects may possibly backfire and increase the likelihood that conflict will escalate.

Over the past years, an impressive literature has shown that in search for security, emotion regulation strategies, such as cognitive reappraisal, increase peaceful cognitions, hope, and optimism about intergroup reconciliation (Halperin et al., 2013; Porat et al., 2016). This literature has gone as far as to claim that intergroup conflict resolution is, in its very essence, an emotion regulation process involving positive affective change (Čehajić-Clancy et al., 2016). The notion that intergroup conflict is the outcome of internal individual psychological processes unencumbered by context or history, and that the resolution of intergroup conflict requires internal individual change, implies that conflict is inherently irrational. Moreover, if conflict-eliciting processes occur in individual minds, efforts to change the external realities of conflict would seem futile (Klar & Branscombe, 2016).

The emotion-in-conflict literature suggests that negative emotions such as anger and fear that reach elevated levels at times of conflict are the main culprits in strenuous and violent intergroup relations: "these emotions constitute a barrier to peace because they motivate uncompromising positions that block repeated peace efforts" (Gross et al., 2013, p. 423). They are myopically seen as perpetuating intergroup conflict (Halperin, Sharvit, & Gross, 2011), increasing support for aggressive actions (Lerner et al., 2003; Skitka et al., 2006), and standing in the way of progress toward conflict resolution. The solution to these undesirable affective states is that people should be "active regulators not passive victims of their emotions" (Goldenberg, Halperin, van Zomeren, & Gross, 2015, p. 118).

This portrayal of the emotion system as a problem that requires constant external intervention to keep negative emotions in check, however, does not bide well with an evolutionary perspective on emotion. From an evolutionary perspective, the emotion system is a highly adaptive system that was crafted over millennia of natural selection to address a fundamental human problem: survival (e.g., Porges, 1997). Negative emotions, therefore, are not a problem but an important part of the solution. Emotions, positive or negative, may also have undesirable effects at times (e.g., Tangney et al., 1996; Unkelbach, Forgas, & Denson, 2008), and may be falsely triggered or activated by bias and manipulation (Gross, 2013). But more often than not, emotions provide important information that increases adaptivity and human success, especially in highly adverse environments (Lerner, Yi, Valdesolo, & Kassam, 2015). Intergroup conflict creates an insecure, threatening, and hostile environment for people living in it. The importance of the emotion system is particularly poignant in such an environment wherein safety and even survival depend on the ability to detect threats on time and respond to them swiftly and effectively. Negative affect influences information processing and recruits vigilant attention to details in the environment (Forgas, 1998). Specific emotions, such as fear, increase the detection of threats in complex environments (Ohman, Lundqvist, & Esteves, 2001), and groups with a higher rate of anxious members do not merely detect more threats but are more accurate in detecting threats and responding to them effectively (Ein-Dor, Perry, & Hirschberger, 2017). Anger is the expression of a neurocognitive system that developed for bargaining for better treatment (Sell et al., 2009) – a vital component of any peace negotiation. Anger also focuses attention to threatening stimuli in the environment (Novaco, 2016) and reduces the threshold for detecting threatening stimuli (Baumann & DeSteno, 2010). Regulating these emotions without considering the consequences may be hazardous to the safety and survival of individuals and groups.

In the political realm, reappraisal may have short-term benefits in restoring individual well-being, but at the expense of productive political action (Ford et al., 2019). In recent research conducted in my lab, we show that reappraisal of fear and anger in the context of the Israeli-Palestinian conflict may have even more severe consequences than the political complacency found by Ford and colleagues (2019). It decreases threat detection and effective responses to life-threatening situations (Schuster, Meyers, Leidner, & Hirschberger, 2022), supporting the claim that regulating emotion may increase support for peace, at the expense of detecting threats on time and responding to them effectively.

When British Prime Minister Neville Chamberlain hopefully proclaimed "peace in our time" following the signing of the 1938 Munich Agreement between the United Kingdom, France, Italy, and Nazi Germany, he could not have imagined that millions would perish due to his complacency in the face of the looming Nazi threat. Sometimes an overzealous desire to resolve intergroup conflict may blind us to the possibility that seemingly undesirable emotions are reacting to signals that we cannot consciously detect or prefer to ignore, and that these emotions serve an adaptive function by turning our attention to threats that may slow down our progress toward the desired goal of reconciliation, but for valid reasons.

If the emotion system in general, and negative emotions in particular, were designed to protect humans from threats, it follows that the regulation of emotion may have a paradoxical effect by reducing vigilance and effective responses to threats. At times, this regulatory response is highly functional. In the context of close relationships, for example, regulating negative emotions to better manage relationship conflict and achieve reconciliation is generally beneficial, because in this context, lowering one's guard and vigilance toward threats is instrumental in restoring intimacy and closeness in a relationship that is normally not life-threatening (see Murray & Lamarche; Arriaga & Kumashiro, this volume). In abusive close relationships, however, regulating negative emotions may render a victim of abuse even more susceptible to harm because they will be impaired in their ability to detect subtle cues in the behavior of the abuser that indicate an imminent threat (Marx, Heidt, & Gold, 2005). The violent and existentially threatening context of intergroup conflict is probably more similar to an abusive relationship than to a functional one, as the desire to subdue, defeat, and even annihilate an adversary is explicit. In this environment, therefore, negative emotions play an important signaling role. They keep group members vigilant and attentive to any slight signal of threat or danger and activate response systems that enable individuals to react effectively to threats when they do emerge (Ein-Dor & Hirschberger, 2016). Thus, emotion regulation in intergroup conflict may have the somewhat paradoxical outcome of increasing peaceful cognitions while lowering a group's guard against potential threats, thus increasing insecurity and rendering the group unprepared to deal with potentially disingenuous adversaries.

The emotion-in-conflict literature views peaceful intergroup relations rather than group survival as the main goal and then confounds the adaptivity of negative emotions with their adverse effects on conflict resolution (for a critique, see Ein-Dor & Hirschberger, 2018). Even research that does recognize the nuances of negative emotion and considers constructive and not just maladaptive functions of negative emotions such as anger erroneously equates constructiveness with conflict resolution and not with security and survival (Shuman et al., 2018). Conflict resolution may, in some specific cases, be an important means to increase the chances of individual and group survival, but the naïve desire to end conflict may backfire if an adversary is disingenuous and concealing belligerent intent.

Further, from a group survival perspective, conflict may sometimes be an unavoidable aspect of defending the group's existence. As tragic as conflict may be, it may be the only possible mode of operation in the face of a relentless and aggressive adversary. From Operation Overlord (the D-Day invasion) to present-day Ukrainians fighting with all their might to forestall a Russian invasion of their country, there are times wherein fighting for existence remains the only option. Winston Churchill clearly understood this logic when he faced Nazism and proclaimed: "We shall defend our island whatever the cost may be … we shall never surrender!" (Speech before Commons June 4, 1940). Throughout human history, group members who felt rage because of an attack against them and retaliated were more likely to survive, due to the deterrent effects of retaliatory violence (McCullough, Kurzban, & Tabak, 2013). This is not to say that conflict is preferable to conflict resolution, but that sometimes the peaceful resolution of conflict is not possible, and the desire for peace may ironically bring about greater insecurity and more, not less, conflict.

Collective Victimization

If emotion regulation is a solution to intergroup conflict, collective victimization is often presented as part of the problem (Bar-Tal et al., 2009). Collective victimization is in essence an intergroup phenomenon wherein one group harms another group. Much of the research, therefore, on collective victimization has naturally taken an intergroup relations and conflict resolution perspective. This perspective, however, may be somewhat misleading because an intergroup relations perspective is likely to view the long-term effects of collective victimization as incompatible with the goal of ameliorating relations between groups and resolving conflict. When we shift focus from intergroup relations to group survival and understand that collective existence is a fundamental human goal that takes precedence over many other motivations, the long-term effects of collective victimization may start to make more sense (Bilewicz, 2020; Hirschberger & Ein-Dor, 2020).

The literature clearly shows that collective victimization has a transformative effect on the way group members understand insecurity and the relationship between their group and other groups (Vollhardt, 2012). This is true for both victim and perpetrator groups (Hirschberger et al., 2022; Li et al., 2022). At the individual cognitive level, the memory of collective trauma serves as a prism through which people perceive and analyze new information (Bar-Tal et al., 2009), such that victims of collective trauma tend to be more vigilant and cautious in situations that are perceived as threatening (Shnabel & Noor, 2012). These seemingly adverse effects of the sense of victimhood are sometimes depicted as a psychological barrier and a "distorted lens" (Schori-Eyal et al., 2017) because they are detrimental to the primary collective goal that is considered in this literature – intergroup relations.

The historical victimization literature indicates that the memory of victimization constitutes a defensive stance that has negative implications on relations with other groups, especially adversarial groups (Adelman, Leidner, Unal, Nahhas, & Shnabel, 2016; Bilali & Vollhardt, 2019), and on relations between descendants of historical adversaries (Rimé, Bouchat, Klein, & Licata, 2015; see also Forgas, this volume). This collective memory strengthens in-group identity, in-group favoritism, and a sense of affiliation among victims and their descendants (Canetti et al., 2018, Wohl & Branscombe, 2008). It reduces intergroup trust and the willingness to forgive past offenses; increases support for violent and uncompromising solutions to intergroup conflict (Hirschberger et al., 2017; Schori-Eyal et al., 2017); increases a desire for revenge against the perpetrator group (Hirschberger et al., 2021); and reduces a sense of moral obligation to other victim groups (Warner et al., 2014).

It is easy to see why this grim depiction of the long-term effects of collective victimization would lead to the conclusion that the memory of past collective trauma is, for the most part, detrimental. From a group survival perspective, however, the memory of trauma also has clear group-protective benefits. It leads to greater vigilance, more suspicion of the intents of other groups, and a reduced willingness to blindly trust others and simply let bygones be bygones. It promotes a hostile and sometimes violent stance toward current adversaries, which may serve as an effective deterrent that could prevent conflict escalation. For instance, Lebanese Maronite Christians who highly identified with their group (and were therefore probably more committed to the protection of the group) were more likely to perceive continuity between past and current members of their main adversaries, Lebanese Muslims (Licata, Klein, Saade, Azzi, & Branscombe, 2012). Keeping the memory of past conflict alive and remaining skeptical about the leopard changing its spots is frustrating from a conflict resolution perspective but may be highly adaptive when considering group survival. Dwelling on past traumas, however, may also overshoot the goal of group survival and lead the group to miss opportunities for peace. Finding the right balance between caution and calculated risk-taking for peace is a challenge for groups in conflict.

Group survival mechanisms are activated primarily when reminded of the unique victimization of the group and less so when victimization is discussed in general terms or is expanded to include other groups. The literature on collective victimization makes a distinction between *inclusive framings* that describe the trauma as a crime against all of humanity and *exclusive framings* that portray the victimization as pertaining exclusively to the victimized group (Vollhardt, 2009). Research has indicated that exclusive framings of historical victimization have a particularly noxious effect on current conflict resolution (Hirschberger et al., 2017). From a group survival perspective, however, it is precisely exclusive victimization that increases vigilance and alertness, and may, thus, be highly adaptive to groups as they serve as a warning system against complacency in the face of looming existential threats. Future research should take a more nuanced look at exclusive victimization, considering both the costs and benefits to the continued survival of victimized groups.

The Ideological Debate

Political polarization is on the rise around the world, and it seems that it is reaching an all-time high (Finkel et al., 2020). The current state of polarization leaves many pessimistic and concerned about the future of relations between political partisans (Pew Research Center, 2019); it is described by some as "toxic" (Moore-Berg, Hameiri, & Bruneau, 2020); it is seen as a threat to the future and stability of democracy (McCoy et al., 2018); and it is as a threat not only to *intra*-group relations but to *inter*group relations as well (Harel, Maoz, & Halperin, 2020). Some have gone as far as to suggest that political polarization today has reached levels where loathing of the opponent exceeds feelings of affinity toward co-partisans (Iyengar & Krupenkin, 2018).

Although there is little doubt that the rift between the political left and right threatens the stability of democracies at present, the ideological debate in and of itself may not be entirely destructive and is perhaps even beneficial for group survival. The value of ideological diversity in social psychology was recently debated between Yoel Inbar and John Jost over SPSP's e-dialogue. In this debate, Jost maintained that "both sidesism" is both morally and scientifically flawed: "It is often asserted – without any evidence whatsoever – that ideological diversity is necessarily beneficial to the advancement of science, but this is plainly false" (Jost, 2018). There is now ample evidence that the opposite is the case – tolerance for viewpoint diversity and heterodoxy are essential prerequisites for any group, including scientific groups, to maintain a broad and inclusive interest in a variety of problems (see also Jussim, Finkelstein & Stevens, this volume).

If conflict resolution research is a part of science, then there is reason to believe that Jost may have overstated his claim, and that ideological diversity is instrumental at least in this regard. In order to manage insecurity, the violent, threatening nature of intractable conflict requires the courage to take a chance for peace along with constant vigilance against threats. Recent research conducted by my colleagues and I found ideological differences in threat perception between rightists and leftists (or conservatives and liberals) in more than 20 different countries. In these studies, conservatives were more attuned to threats of commission (i.e., threats with a clear perpetrator) that were local, and liberals were more sensitive to threats of omission (threats that may emerge without malintent) that were global (Kahn et al., 2022). Liberals also view threats to democracy and tolerance as more psychologically close and concrete than security threats, whereas conservatives show the opposite trend (Kahn et al., 2021). In the context of the Israeli-Palestinian conflict, right-wing Israelis showed concern that withdrawal from the West Bank and the establishment of a Palestinian state would constitute a severe security threat to Israel, whereas left-wing Israelis are concerned that the continued occupation of the Palestinians endangers Israel's identity as a Jewish and democratic state (Hirschberger et al., 2016). Because all of these threats are essentially valid and real (at least to some extent), groups would benefit from having a diverse ideological spectrum of individuals who are concerned about different types of threat.

We recently utilized a behavioral threat and opportunity detection paradigm to examine whether Jewish-Israeli participants would pick up insinuations of threat and opportunity for peace when exposed to a Palestinian's monologue on Arab-Jewish relations. Participants were asked to press response buttons when they detected any insinuation of threat or opportunity for peace.

Results indicated that right-wingers (i.e., participants who voted for right-wing parties in the 2019 general elections in Israel) detected more threats with a high level of accuracy, whereas left-wingers (those who voted for left and center parties) detected more opportunities for peace (Meyers, Meshulam, & Hirschberger, 2022). These findings underscore the value of diversity for optimal group functioning. Groups embroiled in conflict need to stay vigilant and safe. But if all group members were concerned about security threats, the group would never be able to resolve its conflicts with other groups. Some people in the group need to be able to detect opportunities and seize the chance for conflict resolution. The delicate balance between political partisans may determine whether the group is incautious and needlessly exposing itself to threat, defensive and missing opportunities to stop the cycle of violence, or able to take measured risks for peace while staying alert for unexpected surprises. The benefit of ideological diversity during times of conflict was eloquently described by Winston Churchill:

> *I have always urged fighting wars and other contentions with might and main till overwhelming victory, and then offering the hand of friendship to the vanquished. Thus, I have always been against the Pacifists during the quarrel, and against the Jingoes at its close....*
>
> *(Churchill, 1930, p. 346)*

Conclusions

The group survival perspective sheds new light on intergroup conflict and indicates that violent, intractable intergroup conflict is not just a social identity issue but a struggle to be secure, to survive, and to thrive. Security in this sense should be broadly conceptualized. It includes physical security – the confidence that a peace partner will not take advantage of the calm to launch an attack against the group. It also includes symbolic-identity security – the confidence that one's culture, language, customs, and traditions will be preserved. The group survival perspective may help navigate the tension between the need to maintain the group's character and contend with processes of change that require an ongoing reassessment of that character. It may also provide a framework for understanding the needs of majority and minority groups and for negotiating security in asymmetric relations between groups. In this sense, the needs-based model of reconciliation (Shnabel & Nadler, 2008) that focuses on victims' needs for status and power and perpetrator' needs for moral legitimacy can be expanded to include group survival needs. In the aftermath of intergroup conflict or collective trauma, groups need to find new modes of survival that satisfy both physical needs and the need for identity and meaning (Hirschberger, 2018).

The group survival perspective may provide a new understanding of insecurity and conflict resolution by suggesting that survival supersedes intergroup reconciliation and that efforts to increase peaceful attitudes may sometimes backfire by reducing vigilance and attention to threat, thereby placing the group in jeopardy (see also Crano & Hohman, this volume). Establishing the security needs of groups as a prerequisite to conflict resolution may not only increase the relevance of social psychological research to real-world conflict resolution, it may also attract the attention of members of groups in conflict that have become wary of peace attempts that often turn sour. Viewpoint diversity and heterodoxy within groups may also be intrinsically beneficial to survival by promoting attention to a wide variety of different problems, a principle that also applies to our very own field of psychology (see also Jussim et al., this volume).

The group survival perspective treats a group's history as a potential asset, not just a liability, and contends that the lessons of the past serve a group well to protect itself in the present. Conflict resolution does not entail denying or obfuscating a history of trauma, but providing the necessary assurances that compromises for peace will safeguard the group's future. At the intra-group level, group survival delineates how ideological polemics within groups that seem negative and disruptive may, in fact, serve group survival goals by augmenting the group's ability to identify various threats and opportunities.

As the third decade of the 21st century unfolds, the world is becoming less stable and more violent and volatile than it has been since WWII. From the Ukraine to the Middle East, from the treatment of the Uyghurs in China to intergroup tensions in North America, feelings of fear and insecurity abound (Forgas; Kreko, this volume). A group survival perspective that recognizes the fundamental security needs of all people may stimulate research that seeks not only to end conflict but to recognize collective survival needs on the turbulent and often dangerous route to peace.

References

Adelman, L., Leidner, B., Ünal, H., Nahhas, E., & Shnabel, N. (2016). A whole other story: Inclusive victimhood narratives reduce competitive victimhood and intergroup hostility. *Personality and Social Psychology Bulletin, 42*(10), 1416–1430.

Bar-Tal, D. (2000). From intractable conflict through conflict resolution to reconciliation: Psychological analysis. *Political Psychology, 21*, 351–365.

Bar-Tal, D., Chernyak-Hai, L., Schori, N., & Gundar, A. (2009). A sense of self-perceived collective victimhood in intractable conflicts. *International Review of the Red Cross, 91*(874), 229–258.

Baumann, J., & DeSteno, D. (2010). Emotion guided threat detection: Expecting guns where there are none. *Journal of Personality and Social Psychology, 99*(4), 595–602.

Bergman, J. (2004). Why mammal body hair is an evolutionary enigma. *Creation Research Society Quarterly, 40*(3), 240–243.

Bilali, R., & Vollhardt, J. R. (2019). Victim and perpetrator groups' divergent perspectives on collective violence: Implications for intergroup relations. *Political Psychology, 40*, 75–108.

Bilewicz, M. (2016). The dark side of emotion regulation: Historical defensiveness as an obstacle in reconciliation. *Psychological Inquiry, 27*(2), 89–95.

Blanchard, D. C. (2009). Of lion manes and human beards: some unusual effects of the interaction between aggression and sociality. *Frontiers in Behavioral Neuroscience, 3*.

Brewer, M. B., & Caporael, L. R. (2006). An evolutionary perspective on social identity: Revisiting groups. *Evolution and Social Psychology, 143*, 161.

Byford, J., & Tileagă, C. (2014). Social psychology, history, and the study of the Holocaust: The perils of interdisciplinary "borrowing". *Peace and Conflict: Journal of Peace Psychology, 20*(4), 349–364.

Canetti, D., Hirschberger, G., Rapaport, C., Elad-Strenger, J., Ein-Dor, T., Rosenzveig, S., & Hobfoll, S. E. (2018). Collective trauma from the lab to the real world: The effects of the holocaust on contemporary Israeli political cognitions. *Political Psychology, 39*, 3–21.

Caspi-Berkowitz, N., Mikulincer, M., Hirschberger, G., Ein-Dor, T., & Shaver, P. R. (2019). To die for a cause but not for a companion: Attachment-related variations in the terror management function of self-sacrifice. *Journal of Personality and Social Psychology, 117*(6), 1105–1126.

Čehajić-Clancy, S., Goldenberg, A., Gross, J. J., & Halperin, E. (2016). Socialpsychological interventions for intergroup reconciliation: An emotion regulation perspective. *Psychological Inquiry, 27*, 73–88.

Chapman, C. A., & Chapman, L. J. (2000). Determinants of group size in primates: The importance of travel costs. In S. Boinski & P. A. Garber (Eds.), *On the move: How and why animals travel in groups* (pp. 24–42). The University of Chicago Press.

Churchill, W. S. (1930). *My early life*. T. Butterworth.

Dost-Gozkan, A. (2017). *Norms, groups, conflict, and social change: rediscovering Muzafer Sherif's psychology*. Routledge.

Doyle, M. W., & Sambanis, N. (2006). *Making war and building peace: United Nations peace operations*. Princeton University Press.

Durkheim, E. (1893). *The division of labor in society*. Free Press.

Ein-Dor, T., & Hirschberger, G. (2016). Rethinking attachment theory: From a theory of relationships to a theory of individual and group survival. *Current Directions in Psychological Science, 25*, 223–227.

Ein-Dor, T., & Hirschberger, G. (2018). On sentinels and rapid responders: The adaptive functions of emotion dysregulation. In H. Lench (Ed.). *The Function of Emotion* (pp. 25–43): New York: Springer.

Ein-Dor, T., Perry-Paldi, A., & Hirschberger, G. (2017). Friend or foe? Evidence that groups of anxious people make accurate shooting decisions. *European Journal of Social Psychology, 47*, 783–788.

Finkel, E. J., Bail, C. A., Cikara, M., Ditto, P. H., Iyengar, S., Klar, S.,... & Druckman, J. N. (2020). Political sectarianism in America. *Science, 370*(6516), 533–536.

Ford, B. Q., Feinberg, M., Lam, P., Mauss, I. B., & John, O. P. (2019). Using reappraisal to regulate negative emotion after the 2016 US Presidential election: Does emotion regulation trump political action? *Journal of Personality and Social Psychology, 117*(5), 998–1015.

Forgas, J. P. (1998). On being happy and mistaken: Mood effects on the fundamental attribution error. *Journal of Personality and Social Psychology, 75*(2), 318–331.

Fukuyama, F. (1989). The end of history? *The National Interest, 16*, 3–18.

Goldenberg, A., Halperin, E., van Zomeren, M., & Gross, J. J. (2016). The process model of group-based emotion: Integrating intergroup emotion and emotion regulation perspectives. *Personality and Social Psychology Review, 20*(2), 118–141.

Gross, J. J. (Ed.). (2013). *Handbook of emotion regulation*. New York: Guilford Publications.

Gross, J. J., Halperin, E., & Porat, R. (2013). Emotion regulation in intractable conflicts. *Current Directions in Psychological Science, 22*(6), 423–429.

Halperin, E., Cohen-Chen, S., & Goldenberg, A. (2014). Indirect emotion regulation in intractable conflicts: A new approach to conflict resolution. *European Review of Social Psychology, 25*, 1–31.

Halperin, E., Sharvit, K., & Gross, J. J. (2011). Emotion and emotion regulation in intergroup conflict: An appraisal-based framework. In D. Bar-Tal (Ed.), *Intergroup conflicts and their resolution: A social psychological perspective* (pp. 83–103). Psychology Press.

Harel, T. O., Maoz, I., & Halperin, E. (2020). A conflict within a conflict: Intragroup ideological polarization and intergroup intractable conflict. *Current Opinion in Behavioral Sciences, 34*, 52–57.

Harbom, L., Högbladh, S., & Wallensteen, P. (2006). Armed conflict and peace agreements. *Journal of Peace Research, 43*(5), 617–631.

Hirschberger, G. (2018). Collective trauma and the social construction of meaning. *Frontiers in Psychology*, 1441.

Hirschberger, G., & Ein-Dor, T. (2020). A temporal account of collective victimization as existential threat: Reconsidering adaptive and maladaptive responses. In J. R. Vollhardt (Ed.), *The social psychology of collective victimhood*. New York: Oxford University Press.

Hirschberger, G., Ein-Dor, T., Leidner, B., & Saguy, T. (2016). How is existential threat related to intergroup conflict? Introducing the multidimensional existential threat (MET) model. *Frontiers in Psychology, 7*, 1877.

Hirschberger, G., Imhoff, R., Kahn, D., & Hanke, K. (2022). Making sense of the past to understand the present: Attributions for historical trauma predict contemporary social and political attitudes. *Group Processes & Intergroup Relations, 25*(2), 509–526.

Hirschberger, G., Lifshin, U., Seeman, S., Ein-Dor, T., & Pyszczynski, T. (2017). When criticism is ineffective: The case of historical trauma and unsupportive allies. *European Journal of Social Psychology, 47*, 304–319.

Hirschberger, G., Lifshin, U., Dellus, V., Shuster, B., & Kretzschmar, M. (2021). German desire for historical closure indirectly affects Israelis' intergroup attitudes. *European Journal of Social Psychology, 51*(4–5), 784–799.

Hogg, M. A. (2016). Social identity theory. In S. McKeown, R. Haji and N. Ferguson (Eds.), *Understanding peace and conflict through social identity theory* (pp. 3–17). Springer, Cham.

Hogg, M. A., Abrams, D., & Brewer, M. B. (2017). Social identity: The role of self in group processes and intergroup relations. *Group Processes & Intergroup Relations, 20*(5), 570–581.

Iyengar, S., & Krupenkin, M. (2018). The strengthening of partisan affect. *Political Psychology, 39*, 201–218.

Jost, J. (2018). The ideology of social psychologists and why in matters. E-dialogue. Retrieved from https://spsp.org/ideology-social-psychologists-and-why-it-matters.

Kahn, D., Björklund, F., & Hirschberger, G. (2021). Why are our political rivals so blind to the problems facing society? Evidence that political leftists and rightists in

Israel mentally construe collective threats differently. *Peace and Conflict: Journal of Peace Psychology, 27*(3), 426–435.

Kahn, D. T., Björklund, F., & Hirschberger, G. (2022). The intent and extent of collective threats: A data-driven conceptualization of collective threats and their relation to political preferences. *Journal of Experimental Psychology: General, 151*(5), 1178–1198.

Kahn, D. T., Klar, Y., & Roccas, S. (2017). For the sake of the eternal group: perceiving the group as trans-generational and endurance of ingroup suffering. *Personality and Social Psychology Bulletin, 43*, 272–283.

Klar, Y., & Branscombe, N. R. (2016). Intergroup reconciliation: Emotions are not enough. *Psychological Inquiry, 27*(2), 106–112.

Lerner, J. S., Gonzalez, R. M., Small, D. A., & Fischhoff, B. (2003). Effects of fear and anger on perceived risks of terrorism: A national field experiment. *Psychological Science, 14*(2), 144–150.

Lerner, J. S., Li, Y., Valdesolo, P., & Kassam, K. S. (2015). Emotion and decision making. *Annual Review of Psychology, 66*, 799–823.

Li, M., Rovenpor, D. R., & Leidner, B. (2016). Regulating the scope of an emotion regulation perspective on intergroup reconciliation. *Psychological Inquiry, 27*(2), 117–123.

Licata, L., Klein, O., Saade, W., Azzi, A. E., & Branscombe, N. R. (2012). Perceived out-group (dis) continuity and attribution of responsibility for the Lebanese civil war mediate effects of national and religious subgroup identification on intergroup attitudes. *Group Processes & Intergroup Relations, 15*(2), 179–192.

Lifton, R. J. (1973). The sense of immortality: On death and the continuity of life. *American Journal of Psychoanalysis, 33*(1), 3–15.

Lomborg, B. (2004). *Global crises, global solutions.* Cambridge University Press.

Lüders, A., Jonas, E., Fritsche, I., & Agroskin, D. (2016). Between the lines of us and them: Identity threat, anxious uncertainty, and reactive in-group affirmation: How can antisocial outcomes be prevented? In S. McKeown, R. Haji, & N. Ferguson (Eds.), *Understanding peace and conflict through social identity theory: Contemporary global perspectives* (pp. 33–53). Springer International Publishing.

Marx, B. P., Heidt, J. M., & Gold, S. D. (2005). Perceived uncontrollability and unpredictability self-regulation, and sexual revictimization. *Review of General Psychology, 9*(1), 67–90.

McCoy, J., Rahman, T., & Somer, M. (2018). Polarization and the global crisis of democracy: Common patterns, dynamics, and pernicious consequences for democratic polities. *American Behavioral Scientist, 62*(1), 16–42.

McCullough, M. E., Kurzban, R., & Tabak, B. A. (2013). Putting revenge and forgiveness in an evolutionary context. *Behavioral and Brain Sciences, 36*(1), 41–58.

Moore-Berg, S. L., Hameiri, B., & Bruneau, E. (2020). The prime psychological suspects of toxic political polarization. *Current Opinion in Behavioral Sciences, 34*, 199–204.

Meyers, S., Meshulam, A., & Hirschberger, G. (2022). Ideological differences in threat and opportunity detection in the context of intergroup conflict. Unpublished manuscript. Reichman University.

Novaco, R. W. (2016). Anger. In G. Fink (Ed.), *Stress: Concepts, cognition, emotion, and behavior* (pp. 285–292). Elsevier Academic Press.

Öhman, A., Lundqvist, D., & Esteves, F. (2001). The face in the crowd revisited: a threat advantage with schematic stimuli. *Journal of Personality and Social Psychology, 80*(3), 381–396.

Pew Research Center: Public Highly Critical of State of Political Discourse in the U.S. Pew Research Center; 2019. Available at https://www.people-press.org/2019/06/19/ public-highly-critical-of-state-of-political-discourse-in-the-u-s/ (Accessed October 2019).

Porat, R., Halperin, E., & Tamir, M. (2016). What we want is what we get: Group-based emotional preferences and conflict resolution. *Journal of Personality and Social Psychology, 110*(2), 167–190.

Porges, S. W. (1997). Emotion: an evolutionary by-product of the neural regulation of the autonomic nervous systema. *Annals of the New York Academy of Sciences, 807*(1), 62–77.

Pyszczynski, T., Abdollahi, A., Solomon, S., Greenberg, J., Cohen, F., & Weise, D. (2006). Mortality salience, martyrdom, and military might: The great Satan versus the axis of evil. *Personality and Social Psychology Bulletin, 32*(4), 525–537.

Reicher, S. D., Haslam, S. A., Platow, M., & Steffens, N. (2016). Tyranny and leadership. In S. McKeown, R. Haji, & N. Ferguson (Eds.), *Understanding peace and conflict through social identity theory: Theoretical, contemporary and worldwide perspectives* (pp 71–87). New York: Springer.

Rimé, B., Bouchat, P., Klein, O., & Licata, L. (2015). When collective memories of victimhood fade: Generational evolution of intergroup attitudes and political aspirations in Belgium. *European Journal of Social Psychology, 45*(4), 515–532.

Sani, F., Bowe, M., Herrera, M., Manna, C., Cossa, T., Miao, X., & Zhou, Y. (2007). Perceived collective continuity: Seeing groups as entities that move through time. *European Journal of Social Psychology, 37*, 1118–1134.

Schori-Eyal, N., Klar, Y., & Ben-Ami, Y. (2017). Perpetual ingroup victimhood as a distorted lens: Effects on attribution and categorization. *European Journal of Social Psychology, 47*(2), 180–194.

Sell, A., Tooby, J., & Cosmides, L. (2009). Formidability and the logic of human anger. *Proceedings of the National Academy of Sciences, 106*(35), 15073–15078.

Shnabel, N., & Nadler, A. (2008). A needs-based model of reconciliation: Satisfying the differential emotional needs of victim and perpetrator as a key to promoting reconciliation. *Journal of Personality and Social Psychology, 94*(1), 116–132.

Shnabel, N., & Noor, M. (2012). Competitive victimhood among Jewish and Palestinian Israelis reflects differential threats to their identities: The perspective of the needs-based model. In K. J. Jonas & T. A. Morton (Eds.), *Restoring civil societies: The psychology of intervention and engagement following crisis* (pp. 192–207). Wiley Blackwell.

Shuman, E., Halperin, E., & Reifen Tagar, M. (2018). Anger as a catalyst for change? Incremental beliefs and anger's constructive effects in conflict. *Group Processes & Intergroup Relations, 21*(7), 1092–1106.

Shuster, B., Meyers, S., & Hirschberger, G. (2022). Reappraisal may increase support for peace, but also vulnerability to threat: The dead hippie hypothesis. Unpublished manuscript. Reichman University.

Sober, E., & Wilson, D. S. (1998). *Unto others.* Cambridge University Press.

Skitka, L. J., Bauman, C. W., Aramovich, N. P., & Morgan, G. S. (2006). Confrontational and preventative policy responses to terrorism: Anger wants a fight and fear wants "them" to go away. *Basic and Applied Social Psychology, 28*(4), 375–384.

Tangney, J. P., Hill-Barlow, D., Wagner, P. E., Marschall, D. E., Borenstein, J. K., Sanftner, J.,.... & Gramzow, R. (1996). Assessing individual differences in constructive versus destructive responses to anger across the lifespan. *Journal of Personality and Social Psychology, 70*(4), 780–796.

Tajfel, H. (1970). Experiments in intergroup discrimination. *Scientific American, 223*, 96–102.

Tajfel, H., Billig, M. G., Bundy, R. P., & Flament, C. (1971). Social categorization and intergroup behaviour. *European Journal of Social Psychology, 1*(2), 149–178.

Tajfel, H., & Turner, J. C. (1979). An integrative theory of intergroup conflict. In W. G. Austin & S. Worchel (Eds.), *The social psychology of intergroup relations* (pp. 33–47). Monterey, CA: Brooks/Cole.

Unkelbach, C., Forgas, J. P., & Denson, T. F. (2008). The turban effect: The influence of Muslim headgear and induced affect on aggressive responses in the shooter bias paradigm. *Journal of Experimental Social Psychology, 44*(5), 1409–1413.

Vollhardt, J. R. (2009). The role of victim beliefs in the Israeli–Palestinian conflict: Risk or potential for peace? *Peace and Conflict: Journal of Peace Psychology, 15*(2), 135–159.

Vollhardt, J. R. (2012). Collective victimization. In L. Tropp (Ed.), *Oxford handbook of intergroup conflict* (pp. 136–157). New York, NY: Oxford University Press.

Walter, B. F. (2004). Does conflict beget conflict? Explaining recurring civil war. *Journal of Peace Research, 41*(3), 371–388.

Warner, R. H., Wohl, M. J., & Branscombe, N. R. (2014). When do victim group members feel a moral obligation to help suffering others? *European Journal of Social Psychology, 44*(3), 231–241.

Wohl, M. J., & Branscombe, N. R. (2008). Remembering historical victimization: collective guilt for current ingroup transgressions. *Journal of Personality and Social Psychology, 94*(6), 988–1006.

Yzerbyt, V. Y., Rogier, A., & Fiske, S. T. (1998). Group entitativity and social attribution: On translating situational constraints into stereotypes. *Personality and Social Psychology Bulletin, 24*, 1089–1103.

PART IV
The role of Insecurity and Uncertainty in Politics and Public Life

16
TRUST IN SOCIAL INSTITUTIONS

The Role of Informational and Personal Uncertainty

Kees van den Bos

UTRECHT UNIVERSITY, NETHERLANDS

Abstract

In this chapter, I focus on trust in social institutions such as government, law, and science. I propose that one of the reasons trust in these institutions is decreasing is that many people experience several personal uncertainties. Personal uncertainty can be an alarming experience, making people start responding in more distrusting ways toward those who have power over them and can exclude them from important goods or relationships. Providing good, reliable, and accessible information about how the institutions actually work can help mitigate this process. However, judgments about the working of social institutions are often formed under conditions of high levels of informational uncertainty. This analysis has implications for the science and practice of trust in institutions and the associated constructs of personal and informational uncertainty.

It can be good to critically monitor those who hold positions of power in society. In fact, adopting a somewhat skeptical view on powerholders is underlying important assumptions of the proper functioning of the rule of law and often may be quite appropriate and indeed warranted (Hobbes, 1651). Furthermore, some social institutions do not work that well and thus should be viewed even more critically, with a keen eye toward necessary improvements. This being said, there are several reasons why we should worry about waning trust in institutions that are intended to give social structure and to help our societies to function in open manner and fulfill important human needs (see also Forgas; Kreko; Van Prooijen, this volume). After all, trust in certain norms and values is also needed when we want to maintain social order and stability and keep our societies as open as possible (Popper, 1945).

In the present paper, I examine trust in institutions such as government, law, and science. I study these issues following the observation that trust in these institutions may be decreasing (Albright, 2018; see also Forgas; Jussim et al.; and Van Prooijen, this volume). Another reason why studying these issues is important has to do with the assumption that many surveys and trust barometers tend to overestimate the level of trust in these institutions and sometimes tend to miss outright, unwarranted distrust in these important domains of human life (Van den Bos, Hulst, Robijn, Romijn, & Wever, in press). Obviously, the subject of trust in social institutions involves many issues. In this chapter, I focus on the role of informational and personal uncertainty.

I propose that one of the reasons why low levels of trust and increasing levels of distrust exist is because many people experience several personal uncertainties. Experiencing personal uncertainty can be quite alarming, making people start responding in more distrustful ways toward those who have power over them and can exclude them from important goods or relationships (Van den Bos, 2018; see also Arriaga & Kumashiro; Murray & Lamarche, this volume). Information about how the institutions work can sometimes help mitigate this process, especially when the information is reliable and easily accessible. However, often people need to form judgments about the functioning of social institutions during high levels of informational uncertainty (Van den Bos, 2011).

In what follows, I define the concept of trust in social institutions and then examine the role that informational and personal uncertainty have in the process by which people form judgments of trust in social institutions. I close this chapter by formulating some warnings and encouraging notes for the science and practice of trust in institutions and the role of informational and personal uncertainty.

Trust

Trust is a complex issue (see, e.g., Alesina & La Ferrera, 2002; Das, Echambadi, McCardle & Luckett, 2003; Evans & Krueger, 2009; Fukuyama, 1995; Nummela, Sulander, Rahkonen & Uutela, 2009; Warren, 1999; Zaheer, McEvily & Perrone, 1998). It has been defined in many ways, building on various conceptual perspectives (see, e.g., Castaldo, Premazzi & Zerbini, 2010; Deutsch, 1958; Ely, 1980; Evans & Krueger, 2009; Gambetta, 1987; Goold, 2002; Johnson, 1996; Kramer, 1999; Kramer & Cook, 2004; Kramer & Isen, 1994; Maddox, 1995; Messick, Wilke, Brewer, Kramer, Zemke, & Lui, 1983; Rotter, 1980; Stanghellini, 2000). In this chapter, I rely on an earlier, Dutch, and more extensive treatment of this issue (Van den Bos, 2011) and define trust as the conviction that others are well-intentioned toward us, will consider our interests if possible, and will not harm us intentionally if they can avoid doing so (Sztompka, 1999; see also Colquitt, Scott, & LePine, 2007).

To a certain extent, people's willingness to rely on others reflects a personal disposition. Their trust propensity is also affected by the situations in which they find themselves (Van den Bos, 2011). Furthermore, a distinction is often drawn in the psychological literature between trust and trustworthiness. For example, Colquitt et al. (2007) regard trust as the intention to accept vulnerability toward a trustee based on positive expectations of his or her actions. Trustworthiness, on the other hand, depends on the ability, benevolence, and integrity of the trustee and, in particular, on the extent to which these characteristics are ascribed to the trustee (Van den Bos, 2011). According to the Concise Oxford Dictionary, "trustworthy" means "worthy of trust," while "trust" is defined as "a firm belief in the reliability, honesty, veracity, justice, strength, etc., of a person or thing." This suggests that trust and trustworthiness are closely related in English—with the important distinction that trust is an action performed by the person concerned, while trustworthiness is a characteristic ascribed by that person to the trustee.

Trustworthiness is regarded by Brugman, Oskam, and Oosterlaken (2010) as the most important moral trait for the assessment of others. Trust propensity is a personal characteristic that affects not only the extent of trust itself but also all three perceived pillars of trustworthiness (ability, benevolence, and integrity; Brower, Lester, Korsgaard, & Dineen, 2009). Furthermore, I want to emphasize that it is important to distinguish between trust in social institutions and trust by these institutions. It is striking that while relatively much is known about citizens' trust in institutions, the influence of trust by institutions, such as the government, law, and science, in citizens has not yet been widely investigated. I will come back to this point.

I draw a distinction between trust in institutions and trust in other people. The former is often referred to as "political trust" and the latter as "social trust" (Hetherington, 1998; Newton, 2007; Schyns & Koop, 2010). Political and social trust typically operate in different directions: Political trust is generally vertically oriented, toward people or organizations at a higher hierarchical level (such as politicians or government agencies), while social trust often acts horizontally, toward people at the same social level in one's living environment (such as spouses, partners, or neighbors). I therefore refer to political trust as vertical trust and social trust as horizontal trust. The main focus in this chapter will be on understanding vertical trust, and I note that there has been much more research on the psychological processes underlying horizontal trust (see, e.g., Richell et al., 2005; Said, Baron, & Todorov, 2009; Spezio et al., 2008). I will further argue that with the necessary caveats (see, e.g., Brehm & Rahn, 1997; Hetherington, 1999), insights gained from the study of horizontal trust can be used to understand vertical trust.

Here I assume that the basic psychological mechanisms underlying vertical and horizontal trust overlap to a certain extent. I also point out that there are important differences between vertical and horizontal trust. In particular, vertical trust exists in hierarchical settings in which important power differences exist (Lind, 1995; see also Murray & Lamarche, this volume). Furthermore, it involves trust in abstract entities and organizations (Van den Bos, 2011). Nevertheless, I will argue here that because direct information about trust in institutions is often missing (Van den Bos, Van Schie, & Colenberg, 2002; Van den Bos, Wilke, & Lind, 1998), political or vertical trust is often personalized: When forming judgments of political or vertical trust, people frequently zoom in on trust in persons representing social institutions. In particular, how fairly individuals such as civil servants, politicians, judges, or scientists act serves as an important indication whether the institution the person represents can be trusted or not (Van den Bos, 2011, 2018; see also Blumer, 1969; Mead, 1934).

Social Institutions

Social institutions, their genesis, and their functioning are principal objects of study in the social and behavioral sciences (Durkheim, 1895). As with trust, there are many definitions of social institutions. Different definitions of institutions emphasize varying levels of formality and organizational complexity (Calvert, 1995; Streeck & Thelen, 2005). In this chapter on the social psychology of social institutions, I focus on institutions as mechanisms that govern the behavior of people within a given community or society, with the purpose of giving direction to important rules that direct or are supposed to direct people's behaviors. I note that institutions often tend to involve integrated systems of rules that structure social interactions (Hodgson, 2015). Social institutions can also consist of stable, valued, recurring patterns of behavior (Huntington, 1996). Thus, how I use the term "institutions" most of the time applies to formal institutions created by law as well as custom and that have a distinctive permanence in ordering social behaviors. When talking about "institutions," I also refer to informal institutions such as customs or behavior patterns important to a society.[1]

One type of trust in institutions concerns trust in government. Government as an institution can be defined as the machinery that is set up by the state to administer its functions and duties. The function of the government as an institution, thus defined, is to keep the state-organized, run its affairs, and administer its various functions and duties. Viewed in this manner, a government is an institution through which leaders exercise power to make and enforce laws. A government's basic functions are to provide leadership, maintain order, provide public services, provide national security, provide economic security, and provide economic assistance.[2] As we shall see, both personal and informational uncertainty play an important role in people's trust in government (Van den Bos, 2011).

Another important concern has to do with trust in the law (Tyler & Huo, 2002; Van den Bos, 2021). The law as a system can be defined as a codified set of rules developed to regulate interactions and exchanges among people (Tyler & Jost, 2007). As such, the law constitutes an arrangement of rules and guidelines that are created and enforced through social and governmental institutions to regulate behavior. This regulation of behavior includes conflict resolution and sentencing decisions, and ideally takes place in such a way that a community shows respect to its members (Robertson, 2013). Personal uncertainty certainly plays an important role in how people experience court hearings, but the role of informational uncertainty is especially important in the evaluation of many legal issues and people's trust in law, so I argue. After all, many lay citizens do not have access to formal jurisprudence or have a hard time interpreting earlier legal rulings and verdicts (Van den Bos, 2021). As a consequence, so I propose, people's judgments of trust in law are often formed under conditions of informational uncertainty.

A final issue that I would like to examine here is trust in science. Science has important characteristics of an institution, as it can be "regarded as a body of rules and related objects which exist prior to and independently of a given person and which exercise a constraining influence upon the person's behavior" (Hartung, 1951, p. 35). Science constitutes an important domain of human life, in part because it involves reliability of insight on which we want to build our lives. Science also involves the trustworthiness of scientists and the integrity of research findings (see also Jussim et al., this volume). Thus, I argue that when trust in science is shaken, this increases levels of personal uncertainty. Furthermore, when scientific findings are difficult to understand or not accessible because they are put behind paywalls, people form their judgments of trust in science under important conditions of informational uncertainty. This also includes trust in scientific organizations and persons representing those organizations, such as organizations and scientists that try to manage certain crises (such as the COVID-19 crisis) while they are still learning about the causes of the crises under consideration. In what follows, I examine some implications of these introductory notes on trust and social institutions.

Informational and Personal Uncertainty

It is important to examine briefly what the concept of uncertainty entails. In doing so, I rely on earlier conceptual discussions of this issue, in particular Forgas (this volume) and Van den Bos (2009) and Van den Bos and Lind (2002; see also Van den Bos, 2001, 2004; Van den Bos & Lind, 2009; Van den Bos & Loseman, 2011, and Van den Bos, McGregor, & Martin, 2015).

There are many different types of uncertainties that people can encounter, and it is important not to confuse them (Van den Bos & Lind, 2002, 2009). In our work, my colleagues and I have focused on two important varieties (Van den Bos, 2009). One noteworthy type of uncertainty that people often face when forming social judgments is informational uncertainty, which involves having less information available than one ideally would like to have in order to be able to confidently form a given social judgment. For example, work on human decision-making reveals that human judgments are often formed under conditions of incomplete information and that these conditions can lead to predictable effects on human decision and social judgment processes (e.g., Kahneman, Slovic, & Tversky, 1982). Thus, when studying how people make social judgments, a pivotal issue is what information people have available.

Informational uncertainty is important and may be what psychologists come up most frequently when they think of the concept of uncertainty, partly because of the success of the decision-making literature and the well-known work of Nobel laureates such as Kahneman and Phelps (e.g., Kahneman et al., 1982; Phelps, 1970). However, I argue that while informational uncertainty is important, we should not confuse the concept with personal uncertainty. Personal uncertainty is another type of uncertainty and is important to understand self-regulation, existential sense-making, and worldview defense. I define personal uncertainty as a subjective sense of doubt or instability in self-views, worldviews, or the interrelation between the two (Arkin, Oleson, & Carroll, 2009). Furthermore, personal uncertainty, as I conceive of it, involves the implicit and explicit feelings and other subjective reactions people experience as a result of being uncertain about themselves (Van den Bos, 2001, 2007; Van den Bos, Poortvliet, Maas, Miedema, & Van den Ham, 2005). In short, personal uncertainty is the feeling that you experience when you feel uncertain about yourself, and I argue that typically experiencing personal uncertainty constitutes an aversive or at least an uncomfortable feeling (Hogg, 2007; Van den Bos & Lind, 2002; see also Hogg & Gaffney, this volume).

The difference between informational and personal uncertainty is related to the distinction that has been drawn between epistemic and affective dimensions of uncertainty. In other words, *knowing* that you are uncertain about something is different from *feeling* uncertain (Hogg, 2007). Personal uncertainty entails both stable individual differences, such as differences in emotional uncertainty (Greco & Roger, 2001; Sedikides, De Cremer, Hart, & Brebels, 2009), and situational fluctuations, such as conditions in which people's personal uncertainties have (versus have not) been made salient (Van den Bos, 2001). After all, personal uncertainty can be produced by contextual factors that challenge people's certainty about their cognitions, perceptions, feelings, behaviors, and ultimately, their certainty about and confidence in their sense of self (Hogg, 2001). This self-certainty is very important because the self-concept is the critical organizing principle, referent point, or integrative framework for diverse perceptions, feelings, and behaviors (Sedikides & Strube, 1997; see also Loseman, Miedema, Van den Bos, & Vermunt, 2009). The locus of uncertainty can be found in many aspects of the social context, and therefore we are all susceptible to personal uncertainty. However, biographical factors also create stable individual differences in levels of uncertainty, and they can impact people's approaches to how they manage uncertainty (Sorrentino, Hodson, & Huber, 2001; Sorrentino, Short, & Raynor, 1984). Furthermore, people strive more strongly, of course, for certainty about those aspects of life that are important to them (Hogg & Mullin, 1999; see also Hogg & Gaffney, this volume).

Informational Uncertainty and Trust in Institutions

A key question that people often struggle with concerns the issue of whether other people or institutions with which they are involved are to be trusted (Van den Bos, 2011). Lind (1995) has characterized this as the fundamental social dilemma: Can I trust others, and especially societal authorities and institutions (Tyler & Lind, 1992), not to exploit or exclude me from important relationships or social connections (see also Hirschberger; Mikulincer & Shaver; and Murray & Lamarche, this volume)? Furthermore, people's trust in social institutions, including government, law, and science, has an important bearing on the legitimacy of these institutions. But quite often, direct information about this issue is lacking as well (Van den Bos, 2011).

Trust is thus related to an important building block of our society, a foundation on which our society rests. And being able to trust others and institutions is very important for people. However, contrary to what is assumed in the literature on trust and trustworthiness (Brewer, 2008; Damasio, 2005; Giffin, 1967; Güth, Ockenfels, & Wendel, 1997; Kramer, 2001), people often lack the information they need to decide whether others (including abstract entities such as institutions) can be trusted and regarded as reliable interaction partners (Van den Bos, 2000; Van den Bos et al., 1998). It is indeed often difficult to determine whether you really can trust another party. You need, for example, a lot of experience with the other party before you can reach such a decision with certainty, and in general, we do not have such information. Under such circumstances, when you have less information than you would like to have about the other party's trustworthiness, you will have to make do with the information you do have at your disposition. This often concerns fairness: people can often form a good impression of how fairly they are being treated based on relatively little information (Lind, 1995). The impressions gained from some encounters are often enough to allow people to decide whether they are being treated fairly and in a just manner by the other party.

People thus often use information about how fairly or unfairly they are treated by persons representing social institutions as a proxy for the lacking information on the institution's trustworthiness (Van den Bos, 2011; Van den Bos et al., 1998). If the representative behaves fairly, this is viewed as an important indication that the institution is legitimate and can be trusted. And if a person representing the social institution acts in clearly unfair ways, then important doubts about the legitimacy and trustworthiness of the institution will remain or arise. Perceived fairness is thus important as a substitute for institutional legitimacy (Van den Bos, 2011) and because it gives people information about the extent to which they can trust other people, the government, law, science, and other institutions (Van den Bos et al., 2002).

Personal Uncertainty and Trust in Institutions

Apart from informational uncertainty, personal uncertainty is one of the other main reasons why fairness is important to people (Van den Bos & Lind, 2002, 2009, 2011). People often feel uncertain and insecure in their dealings with authorities and institutions, for example, because these agencies can exert power or influence over them (Tyler & Huo, 2002) and may even exploit them or cut off social links that are important to them, excluding them from society as a whole or important groups within society (Tyler & DeGoey, 1996; see also Mikulincer & Shaver, this volume). Furthermore, people in modern society often have experiences that make them feel unsure of themselves. This feeling of personal uncertainty is experienced by people as unpleasant (Hogg, 2007), often as very unpleasant (Van den Bos & Lind, 2002, 2009). In fact, personal uncertainty is often seen as an alarm signal ("What's going on here?"; "I'll have to watch out in this situation: it doesn't feel good"; Van den Bos, Ham, Lind, Simonis, Van Essen & Rijpkema, 2008; see also Mikulincer & Shaver, and Von Hippel & Merakovsky, this volume).

Stable individual differences exist in the extent to which people experience personal insecurity as emotionally threatening (Greco & Roger, 2001; Van den Bos, Euwema, Poortvliet & Maas, 2007). In fact, some people regard uncertainty as an enjoyable challenge rather than a threat (Sorrentino, Bobocel, Gitta, Olson & Hewitt, 1988; see also Fiedler & McCaughey; and Kruglanski & Ellenberg, this volume), although I view this as an exception that generally involves informational uncertainty (Weary & Jacobson, 1997; Wilson, Centerbar, Kermer & Gilbert, 2005) rather than personal uncertainty (Van den Bos, 2001, 2011), and applies especially when people can trust other people in their environments and institutions in their society (see also Mikulincer & Shaver, this volume).

Most people regard personal uncertainty as unpleasant and try to cope with it in some way. A possible coping mechanism is to explore the extent to which one forms part of one's social environment and the society in which one lives, or, in other words, to explore the extent of one's social integration (Hogg & Gaffney, 2022; Hogg, Sherman, Dierselhuis, Maitner & Moffitt, 2007; see also Arriaga & Kumashiro, this volume). This makes it important for people to feel that they are accepted and respected by important people or groupings in their environment or in wider society. A key indicator of this acceptance and respect is being fairly and decently treated by important people in society or important individuals in the group to which one belongs (Lind & Tyler, 1988; Lind & Van den Bos, 2002; see also See, 2009; Thau, Aquino & Wittek, 2007;

Thau, Bennett, Mitchell & Marrs, 2009). In this way, perceived fairness can help people cope with personal uncertainty. It may even turn uncertainty from a threat into an agreeable challenge (Van den Bos & Lind, 2009), perhaps because people associate fair treatment with positive affect (Van den Bos, 2007, 2009, 2011).

The requirement of fairness is an important norm in practically every society and subculture (Van den Bos, Brockner et al., 2010). The precise form of fairness required varies from one culture (Van den Bos, Brockner et al., 2010) or subculture (Doosje, Loseman, & Van den Bos, 2013) to another. For example, some cultures attach greater importance to the fair treatment of all members of the group, while others focus more on the fair treatment of individuals (Brockner, De Cremer, Van den Bos & Chen, 2005; see also Hofstede, 2001; Markus & Kitayama, 1991; Triandis, Bontempo, Villareal, Asai & Lucca, 1988). With this proviso, fair, decent treatment appears to be an important norm and cultural value in practically any culture or subculture (Lind, Tyler, & Huo, 1997). A main reason why this is the case is because being treated fairly and justly by important members of your group or society, such as representatives of your society's institutions, indicates that you are viewed as an important member of your group and society (Lind & Tyler, 1988). In short, group values are important in perceiving treatment fairness and how you respond to social institutions.

I further assume that situations in which people are interacting and coordinating their behaviors with others play a major role in processes that people go through when forming judgments of trust in social institutions (Van den Bos, 2018). People's social values are important in this respect. Findings suggest that most (but certainly not all) humans tend to be oriented toward cooperation. Indeed, in many studies, a small majority of 60–70% of participants tend to adhere to cooperative value orientations and as such can be characterized as prosocial beings (Van Lange, Otten, De Bruin, & Joireman, 1997; Van den Bos et al., 2011). Ironically, the social quality of people may inhibit them from showing their prosociality, especially when they are busy trying to sort out what is going on, how to behave in the situation at hand, and how others will view their behaviors. Having made sense of how to interpret the situation at hand and what constitutes appropriate behavior in the situation may help people to free themselves and engage in prosocial behaviors, including putting trust in other people, such as persons representing social institutions that have power over them and play an important role in the societies in which people live. However, overcoming inhibitory constraints can be difficult, which constitutes an important reason why the prosocial or trusting qualities of people may not always show in public circumstances (Van den Bos & Lind, 2013; Van den Bos et al., 2011; Van den Bos, Müller, & Van Bussel, 2009). Furthermore, when people are very uncertain about themselves, their cooperative intentions can easily come under pressure (Van den Bos, 2018). It is often the combination of personal and informational uncertainty that will have the strongest impact on people's reactions (Van den Bos, 2011, 2018).

A Warning Note on Distrust, Weird Studies, and the Internet

It is important to emphasize that low levels of trust or the absence of trust are not the same as outright distrust in institutions such as government, law, and science. Furthermore, judgments of trust and distrust in institutions are not made in a vacuum. Instead, they are formed under conditions that change in dynamic ways over time (Jansma, Van den Bos, & De Graaf, 2022). An important issue that I want to note here is how high distrust in our social institutions has grown over the years, how social psychology and the behavioral sciences may miss this development, and how easily distrust is exacerbated on the Internet.

I indeed think we should not be naive about the growing and sometimes hidden levels of unwarranted distrust in institutions that aim, or should aim, to hold our society together. An important reason why distrust in institutions may occur is that many modern institutions, parts of these institutions, or people affiliated with these institutions do not function as well as they should. For example, in many countries, government agencies are now run with much attention given to issues of process management but with decreasing expertise in the areas of content, they are supposed to govern. Furthermore, judges sometimes have a hard time dealing with modern citizens, who demand and expect to be involved much more actively and intensively during the handling of their cases in court. Moreover, some individual scientists clearly failed to live up to the high levels of scientific quality and research integrity that society expected them to adhere to.[3] These observations can be good and valid reasons why trust in important institutions that aim, or should aim, to hold societies together is waning or may even turn into judgments of distrust in these institutions. It is important, indeed crucial, to remain critical about the current state of social institutions such as government, law, and science. It would be wrong to take any form of distrust in these and other institutions to be inaccurate and misguided.

Informational uncertainty about the workings of institutions, when combined with high levels of personal uncertainty regarding one's role in society, can lead to growing levels of distrust (Van den Bos, 2011). It seems clear that distrust leads to resentment, anger, complaints (from citizens who are capable of looking after themselves; Van den Bos, 2007), aggressive behavior (from citizens who need help looking after themselves; Van den Bos, 2007), and individual or collective protest (Klandermans, 1997). Furthermore, distrust leads to activation of the amygdala (Van den Bos, 2011), which is probably related to feelings of fear elicited by distrust (for example, triggered by the sight of faces that one distrusts). Oxytocin deactivates the amygdala, thus reducing distrust (Kirsch et al., 2005; see also Richell et al., 2005).

De Gruijter, Smits van Waesberghe, and Boutellier (2010) studied the dissatisfaction of citizens of Dutch extraction with new immigrants and with government policy on this point. The idea of "active citizenship," as propagated by the Dutch government, implies that citizens are actively involved in society and that they can cope with social differences. The results achieved in practice are different, however, as the study by De Gruiter and colleagues shows. People who live in mixed neighborhoods see the government as mainly to blame for all their problems. An important finding of the study was the perception of local residents that the government was very distant from their concerns. The respondents regarded government officials, figures of authority, and politicians as privileged people who had no idea how the common man or woman lives and no feeling for the real economic and social problems of citizens. This can easily lead to misunderstandings and poor communication, especially when doubt exists as to how things are arranged in modern society (Boutellier, 2010).

The differences between citizens with high and low educational levels appear to play an important role here (Bovens & Wille, 2017). Educational degrees divide many societies nowadays. There are marked differences between the extents to which people with high and low educational levels trust politics and the constitutional state. Dissatisfaction and cynicism about profiteers and social climbers are found mainly among white people with low educational levels, who feel neglected by the upper classes (Bovens & Wille, 2017; De Gruijter et al., 2010). Furthermore, Trzesniewski and Donnellan (2010) comment that young people trust society less than older people.

Some scholars propagate the idea that distrust can fulfill a constructive function and that reasonable, well-organized distrust of those elites is to be applauded (Hobbes, 1651). Inquiring whether matters are properly arranged and whether the government, law, and science are to be completely trusted at all times is indeed part of the democratic scrutiny that may be expected of citizens. Nevertheless, too much distrust in government, law, or science is often undesirable, both at a social level (see, e.g., Ely, 1980; Warren, 1999) and at a psychological level (see, e.g., Kramer, 1994; Kramer & Cook, 2004).

I believe that we should not enthusiastically embrace simplified notions about the constructive value of distrust. I am particularly skeptical about the extent to which such conflict models (see also Dahrendorf, 1959) actually describe the real behavior of citizens, and I suspect that they may naively overestimate the positive role conflict can play in society and interpersonal relationships (see also Etzioni, 2004).

Another issue that deserves attention is that social psychology needs to broaden its scope of attention in order not to miss the possible growth of distrust in institutions. For example, many studies in social psychology, and indeed in the behavioral sciences more generally, rely too strongly on Western, Educated, Industrialized, Rich, and Democratic (WEIRD) participants (Henrich, 2020; Henrich, Heine, & Norenzayan, 2010a, 2010b). In fact, social psychological findings may be missing crucial patterns because WEIRD participants are tested by WEIRD interviewers. For example, Van den Bos et al. (in press) show that when answering questionnaires on trust in judges that were given to them by interviewers from law schools or psychology departments, lower-educated people indicated that they hold high levels of trust. That pattern replicates many known findings. Yet, when the same interviewers presented themselves as coming from lower-educated backgrounds, participants' responses changed such that they reported much lower levels of trust. These findings suggest that experimentally varying the "WEIRD-ness" of interviewers may help to detect deeply held but rarely expressed feelings about mainstream institutions.

Finally, I would like to argue that we should not be naive about the Internet as an important moderator of unwarranted distrust in institutions. Already in 1999, this issue was discussed in a now-famous interview of David Bowie by Jeremy Paxman on BBC Newsnight. Bowie, an Internet pioneer, talked about the fragmentation of society that he saw beginning in the 1970s and correctly predicted that the Internet would further fragment things away from a world where there were "known truths and known lies" toward a world where there are "two, three, four sides to every question," something that would be simultaneously "exhilarating and terrifying" and would "crush our ideas of what mediums are all about."[4] Indeed, the Internet and so-called "social media" can easily lead people to start adopting exaggerated levels of distrust in social institutions, letting go of self-control, inflaming emotional responses, and starting to sympathize with attempts to break the law in order to reach their goals (Van den Bos, 2018; Van den Bos et al., 2021).

An Encouraging Note on Legitimacy and Perceived Fairness

I want to close with some encouraging words. In this chapter, I proposed that trust in these social institutions may be decreasing because many people experience personal uncertainties, which constitute an alarming experience to most people, leading to lower levels of trust in institutions that have power over them. The provision of good, reliable, and accessible information about how institutions actually work can lead to calmer responses and higher levels of trust in institutions. This is not an easy process that always works, for one thing, because there tends to be a lot of informational uncertainty about how social institutions operate and function. Furthermore, whether institutions have legitimacy is often difficult to ascertain with certainty. From the literature on perceived treatment fairness follows that in circumstances in which personal and informational uncertainty are high, people tend to rely on the perceived fairness of persons representing social institutions (Van den Bos, 2005, 2011,

2015; Van den Bos & Lind, 2002, 2009; Van den Bos et al., 1998, 2002). This means that the individual civil servant, politician, judge, lawyer, and scientific researcher and teacher have important responsibilities: When they act in ways that are truly fair and honest, giving people opportunities to voice their opinions at appropriate times, carefully listening to these opinions, and thus treating people with respect as full-fledged citizens of their society, this can increase trust in institutions and prevent unwarranted levels of distrust (Van den Bos, Van der Velden, & Lind, 2014). I hope that the social psychology of informational and personal uncertainty, combined with the associated literature on perceived fairness, may help to firmly build or rebuild warranted trust in social institutions.

Notes

1 For more information, see, for example, https://en.wikipedia.org/wiki/Institution.
2 See, for instance, https://philosophy-question.com/library/lecture/read/352409-what-is-government-as-institution.
3 See, for instance, https://www.tilburguniversity.edu/nl/over/gedrag-integriteit/commissie-levelt.
4 See, for example, https://tidbits.com/2020/11/01/david-bowies-1999-insights-into-the-internet/, https://www.bbc.com/news/av/entertainment-arts-35286749, https://www.theguardian.com/technology/2016/jan/11/david-bowie-bowienet-isp-internet.

References

Albright, M. (2018). *Fascism: A warning*. London, UK: William Collins.
Alesina, A., & La Ferrera, E. (2002). Who trusts others? *Journal of Public Economics, 85*, 207–234.
Arkin, R. M., Oleson, K. C., & Carroll, P. J. (Eds.). (2009). *Handbook of the uncertain self*. New York: Psychology Press.
Blumer, H. (1969). *Symbolic interactionism: Perspective and method*. Englewood Cliffs, NJ: Prentice-Hall.
Boutellier, H. (2010). *De improvisatiemaatschappij [The improvisation society]*. Amsterdam: Boom Lemma.
Bovens, M., & Wille, A. (2017). *Diploma democracy: The rise of political meritocracy*. New York: Oxford University Press.
Brehm, J., & Rahn, W. (1997). Individual-level evidence for the causes and consequences of social capital. *American Journal of Political Science, 41*, 999–1024.
Brewer, M. B. (2008). Depersonalized trust and intergroup cooperation. In J. I. Krueger (Ed.), *Rationality and social responsibility* (pp. 215–232). New York: Psychology Press.
Brockner, J., De Cremer, D., Van den Bos, K., & Chen, Y. (2005). The influence of interdependent self-construal on procedural fairness effects. *Organizational Behavior and Human Decision Processes, 96*, 155–167.
Brower, H. H., Lester, S. W., Korsgaard, M. A., & Dineen, B. R. (2009). A closer look at trust between managers and subordinates: Understanding the effects of both trusting and being trusted on subordinate outcomes. *Journal of Management, 35*, 327–347.

Brugman, D., Oskam, H., & Oosterlaken, A. (2010, May). *Moral identity and the relationship between moral judgment and antisocial behavior in adolescents*. Paper presented at the Second Conference of the Netherlands Society of Development Psychologists, Wageningen, The Netherlands.

Calvert, R. (1995). Rational actors, equilibrium and social institutions. In J. Knight & I. Sened (Eds.), *Explaining social institutions* (pp. 58–60). Ann Arbor: University of Michigan Press.

Castaldo, S., Premazzi, K., & Zerbini, F. (2010). The meaning(s) of trust: A content analysis on the diverse conceptualizations of trust in scholarly research on business relationships. *Journal of Business Ethics, 96*, 657–668.

Colquitt, J. A., Scott, B. A., & LePine, J. A. (2007). Trust, trustworthiness, and trust propensity: A meta-analytic test of their unique relationships with risk taking and job performance. *Journal of Applied Psychology, 92*, 909–927.

Dahrendorf, R. (1959). *Class and class conflict in industrial society*. London: Routledge & Paul.

Damasio, A. (2005). Brain trust. *Nature, 435*, 571–572.

Das, S., Echambadi, R., McCardle, M., & Luckett, M. (2003). The effect of interpersonal trust, need for cognition, and social loneliness on shopping, information seeking and surfing on the web. *Marketing Letters, 14*, 185–202.

De Gruijter, M., Smits van Waesberghe, E., & Boutellier, H. (2010). *"Een vreemde in eigen land": Ontevreden autochtone burgers over nieuwe Nederlanders en de overheid [A stranger in your own country: Dissatisfied autochtonous citizens on newcomers and the government]*. Amsterdam: Aksant.

Deutsch, M. (1958). Trust and suspicion. *Journal of Conflict Resolution, 2*, 265–279.

Doosje, B., Loseman, A., & Van den Bos, K. (2013). Determinants of radicalization of Islamic youth in the Netherlands: Personal uncertainty, perceived injustice, and perceived group threat. *Journal of Social Issues, 69*, 586–604.

Durkheim, E. (1982). *The rules of sociological method*. New York: Simon and Schuster. (Original work published in 1895 as *Les Règles de la méthode sociologique*).

Ely, J. H. (1980). *Democracy and distrust: A theory of judicial review*. Cambridge, MA: Harvard University Press.

Etzioni, A. (2004). *From empire to community*. New York: Pallgrave Macmillan.

Evans, A. M., & Krueger, J. I. (2009). The psychology (and economics) of trust. *Social and Personality Psychology Compass, 3*, 1003–1017.

Fukuyama, F. (1995). *Trust: The social virtues and the creation of prosperity*. New York: Free Press.

Gambetta, D. (1987). *Trust: Making and breaking cooperative relations*. Cambridge, MA: Oxford.

Giffin, K. (1967). The contribution of studies of source credibility to a theory of interpersonal trust in the communication process. *Psychological Bulletin, 68*, 104–120.

Goold, S. D. (2002). Trust, distrust and trustworthiness: Lessons from the field. *Journal of General Internal Medicine, 17*, 79–81.

Greco, V., & Roger, D. (2001). Coping with uncertainty: The construction and validation of a new measure. *Personality and Individual Differences, 31*, 519–534.

Güth, W., Ockenfels, P., & Wendel, M. (1997). Cooperation based on trust: An experimental investigation. *Journal of Economic Psychology, 18*, 15–43.

Hartung, F. E. (1951). Science as an institution. *Philosophy of Science, 18*, 35–54.

Henrich, J. (2020). *The WEIRDest people in the world: How the West became psychologically peculiar and particularly prosperous*. New York: Macmillan.
Henrich, J., Heine, S. J., & Norenzayan, A. (2010a). Most people are not WEIRD. *Nature, 466*, 29.
Henrich, J., Heine, S. J., & Norenzayan, A. (2010b). The weirdest people in the world? *Behavioral and Brain Sciences, 33*, 61–83.
Hetherington, M. J. (1998). The political relevance of political trust. *American Political Science Review, 92*, 791–808.
Hetherington, M. J. (1999). The effect of political trust on the presidential vote, 1968–96 *American Political Science Review, 93*, 311–326.
Hobbes, T. (1985). *Leviathan*. London: Penguin. (Original work published in 1651.)
Hodgson, G. M. (2015). On defining institutions: Rules *versus* equilibria. *Journal of Institutional Economics, 11*, 497–505.
Hofstede, G. (2001). *Culture's consequences: Comparing values, behaviors, institutions, and organizations across nations* (2nd ed.). Thousand Oaks, CA: Sage.
Hogg, M. A. (2001). Self-categorization and subjective uncertainty resolution: Cognitive and motivational facets of social identity and group membership. In J. P. Forgas, K. D. Williams, & L. Wheeler (Eds.), *The social mind: Cognitive and motivational aspects of interpersonal behavior* (pp. 323–349). New York: Cambridge University Press.
Hogg, M. A. (2007). Uncertainty-identity theory. In M. P. Zanna (Ed.), *Advances in experimental social psychology* (Vol. 39, pp. 70–126). San Diego, CA: Academic Press.
Hogg, M. A., & Mullin, B.-A. (1999). Joining groups to reduce uncertainty: Subjective uncertainty reduction and group identification. In D. Abrams & M. A. Hogg (Eds.), *Social identity and social cognition* (pp. 249–279). Oxford: Blackwell.
Hogg, M. A., Sherman, D. K., Dierselhuis, J., Maitner, A. T., & Moffitt, G. (2007). Uncertainty, entitativity, and group identification. *Journal of Experimental Social Psychology, 43*, 135–142.
Huntington, S. P. (1996). *Political order in changing societies*. New Haven, CT: Yale University Press.
Jansma, A., Van den Bos, K., & De Graaf, B. A. (2021). *Unfairness in society and over time: How insights from social psychology and history can help to understand possible radicalization for climate change*. Manuscript submitted for publication.
Johnson, C. B. (1996). Distrust of science. *Nature, 380*, 18.
Kahneman, D., Slovic, P., & Tversky, A. (Eds.). (1982). *Judgment under uncertainty: Heuristics and biases*. New York: Cambridge University Press.
Kirsch, P., Esslinger, C., Chen, Q., Mier, D., Lis, D., Siddhanti, S., Gruppe, H., Mattay, V. S., Gallhofer, B., & Meyer-Lindenberg, A. (2005). Oxytocin modulates neural circuitry for social cognition and fear in humans. *The Journal of Neuroscience, 25*, 11489–11493.
Klandermans, B. (1997). *The social psychology of protest*. Oxford, UK: Blackwell.
Kramer, R. M. (1994). The sinister attribution error: Paranoid cognition and collective distrust in organizations. *Motivation and Emotion, 18*, 199–230.
Kramer, R. M. (1999). Trust and distrust in organizations: Emerging perspectives, enduring questions. *Annual Review of Psychology, 50*, 569–598.
Kramer, R. M. (2001). Identity and trust in organizations: One anatomy of a productive but problematic relationship. In M. A. Hogg & D. J. Terry (Eds.), *Social identity processes in organizational contexts* (pp. 167–180). Philadelphia, PA: Psychology Press.

Kramer, R. M., & Cook, K. S. (2004). *Trust and distrust in organizations: Dilemmas and approaches.* New York: Russell Sage Foundation.

Kramer, R. M., & Isen, A. M. (1994). Trust and distrust: Its psychological and social dimensions. *Motivation and Emotion, 18,* 105–107.

Lind, E. A. (1995). *Social conflict and social justice: Lessons from the social psychology of justice judgments.* Inaugural oration, Leiden University, Leiden, The Netherlands.

Lind, E. A., & Tyler, T. R. (1988). *The social psychology of procedural justice.* New York: Plenum.

Lind, E. A., & Van den Bos, K. (2002). When fairness works: Toward a general theory of uncertainty management. In B. M. Staw & R. M. Kramer (Eds.), *Research in organizational behavior* (Vol. 24, pp. 181–223). Greenwich, CT: JAI Press.

Lind, E. A., Tyler, T. R., & Huo, Y. J. (1997). Procedural context and culture: Variation in the antecedents of procedural justice judgments. *Journal of Personality and Social Psychology, 73,* 767–780.

Loseman, A., Miedema, J., Van den Bos, K., & Vermunt, R. (2009). Exploring how people respond to conflicts between self-interest and fairness: The influence of threats to the self on affective reactions to advantageous inequity. *Australian Journal of Psychology, 61,* 13–21.

Maddox, J. (1995). The prevalent distrust of science. *Nature, 378,* 435–437.

Markus, H. R., & Kitayama, S. (1991). Culture and the self: Implications for cognition, emotion, and motivation. *Psychological Review, 98,* 224–253.

Mead, G. H. (1934). *Mind, self, and society: From the standpoint of a social behaviorist.* Chicago: University of Chicago Press.

Messick, D. M., Wilke, H., Brewer, M. B., Kramer, R. M., Zemke, P., & Lui, L. (1983). Individual adaptations and structural changes as solutions to social dilemmas. *Journal of Personality and Social Psychology, 44,* 294–309.

Newton, K. (2007). Social and political trust. In R. J. Dalton & H. Klingemann (Eds.), *Oxford handbook of political behaviour* (pp. 342–361). Oxford: Oxford University Press.

Nummela, O., Sulander, T., Rahkonen, O., & Uutela, A. (2009). The effect of trust and change in trust on self-rated health: A longitudinal study among aging people. *Archives of Gerontology and Geriatrics, 49,* 339–342.

Phelps, E. S. (1970). *Microeconomic foundations of employment and inflation theory.* New York: Norton.

Popper, K. R. (1945). *The open society and its enemies.* London: Routledge and Kegan Paul.

Richell, R. A., Mitchell, D. G. V., Peschardt, K. S., Winston, J. S., Leonard, A., Dolan, R. J., & Blair, R. J. R. (2005). Trust and distrust: The perception of trustworthiness of faces in psychopathic and non-psychopathic offenders. *Personality and Individual Differences, 38,* 1735–1744.

Robertson, G. (2013). *Crimes against humanity: The struggle for global justice* (4th ed.). New York: The New Press.

Rotter, J. B. (1980). Interpersonal trust, trustworthiness, and gullibility. *American Psychologist, 35,* 1–7.

Said, C. P., Baron, S., & Todorov, A. (2009). Nonlinear amygdala response to face trustworthiness: Contributions of high and low spatial frequency information. *Journal of Cognitive Neuroscience, 21,* 519–528.

Schyns, P., & Koop, C. (2010). Political distrust and social capital in Europe and the USA, *Social Indicators Research, 96,* 145–167.

Sedikides, C., & Strube, M. J. (1997). Self-evaluation: To thine own self be good, to thine own self be sure, to thine own self be true, and to thine own self be better. In M. P. Zanna (Ed.), *Advances in experimental social psychology* (Vol. 29, pp. 209–269). San Diego, CA: Academic Press.

Sedikides, C., De Cremer, D., Hart, C. M., & Brebels, L. (2009). Procedural fairness responses in the context of self-uncertainty. In R. M. Arkin, K. C. Oleson, & P. J. Carroll (Eds.), *Handbook of the uncertain self* (pp. 142–159). New York: Psychology Press.

See, K. E. (2009). Reactions to decisions with uncertain consequences: Reliance on perceived fairness versus predicted outcomes depends on knowledge. *Journal of Personality and Social Psychology, 96,* 104–117.

Sorrentino, R. M., Bobocel, D. R., Gitta, M. Z., Olson, J. M., & Hewitt, E. C. (1988). Uncertainty orientation and persuasion: Individual differences in the effects of personal relevance on social judgments. *Journal of Personality and Social Psychology, 55,* 357–371.

Sorrentino, R. M., Hodson, G., & Huber, G. L. (2001). Uncertainty orientation and the social mind: Individual differences in the interpersonal context. In J. P. Forgas, K. D. Williams, & L. Wheeler (Eds.), *The social mind: Cognitive and motivational aspects of interpersonal behavior* (pp. 199–227). New York: Cambridge University Press.

Sorrentino, R. M., Short, J.-A. C., & Raynor, J. O. (1984). Uncertainty orientation: Implications for affective and cognitive views of achievement behavior. *Journal of Personality and Social Psychology, 46,* 189–201.

Spezio, M., Rangel, A., Alvarez, M., O'Doherty, J., Mattes, K., Todorov, A., Kim, H., & Adolphs, R. (2008). A neural basis for the effect of candidate appearance on election outcomes. *Social, Cognitive, and Affective Neuroscience, 3,* 344–352.

Stanghellini, G. (2000). Vulnerability to schizophrenia and lack of common sense. *Schizophrenia Bulletin, 26,* 775–787.

Streeck, W., & Thelen, K. A. (Eds.). (2005). *Beyond continuity: Institutional change in advanced political economies.* New York: Oxford University Press.

Sztompka, P. (1999). *Trust: A sociological theory.* Cambridge: Cambridge University Press.

Thau, S., Aquino, K., & Wittek, R. (2007). An extension of uncertainty management theory to the self: The relationship between justice, social comparison orientation, and antisocial work behaviors. *Journal of Applied Psychology, 92,* 286–295.

Thau, S., Bennett, R. J., Mitchell, M. S., & Marrs, M. B. (2009). How management style moderates the relationship between abusive supervision and workplace deviance: An uncertainty management theory perspective. *Organizational Behavior and Human Decision Processes, 108,* 79–92.

Triandis, H., Bontempo, R., Villareal, M., Asai, M., & Lucca, N. (1988). Individualism and collectivism: Cross-cultural perspectives on self-ingroup relationships. *Journal of Personality and Social Psychology, 54,* 323–338.

Trzesniewski, K. H., & Donnellan, M. B. (2010). Rethinking "Generation Me": A study of cohort effects from 1976–2006. *Perspectives on Psychological Science, 5,* 58–75.

Tyler, T. R., & DeGoey, P. (1996). Trust in organizational authorities: The influence of motive attributions on willingness to accept decisions. In R. Kramer & T. R. Tyler (Eds.), *Trust in organizations: Frontiers of theory and research* (pp. 331–356). Thousand Oaks, CA: Sage.

Tyler, T. R., & Huo, Y. J. (2002). *Trust in the law: Encouraging public cooperation with the police and courts.* New York: Russell Sage Foundation.

Tyler, T. R., & Jost, J. T. (2007). Psychology and the law: Reconciling normative and descriptive accounts of social justice and system legitimacy. In A. W. Kruglanski & E. T. Higgins (Eds.), *Social psychology: Handbook of basic principles* (2nd ed., pp. 807–825). New York: Guilford Press.

Tyler, T. R., & Lind, E. A. (1992). A relational model of authority in groups. In M. P. Zanna (Ed.), *Advances in experimental social psychology* (Vol. 25, pp. 115–191). San Diego, CA: Academic Press.

Van den Bos, K. (2001). Uncertainty management: The influence of uncertainty salience on reactions to perceived procedural fairness. *Journal of Personality and Social Psychology, 80*, 931–941.

Van den Bos, K. (2004). An existentialist approach to the social psychology of fairness: The influence of mortality and uncertainty salience on reactions to fair and unfair events. In J. Greenberg, S. L. Koole, & T. Pyszczynski (Eds.), *Handbook of experimental existential psychology* (pp. 167–181). New York: Guilford.

Van den Bos, K. (2005). What is responsible for the fair process effect? In J. Greenberg & J. A. Colquitt (Eds.), *Handbook of organizational justice: Fundamental questions about fairness in the workplace* (pp. 273–300). Mahwah, NJ: Erlbaum.

Van den Bos, K. (2007). Hot cognition and social justice judgments: The combined influence of cognitive and affective factors on the justice judgment process. In D. de Cremer (Ed.), *Advances in the psychology of justice and affect* (pp. 59–82). Greenwich, CT: Information Age Publishing.

Van den Bos, K. (2009). Making sense of life: The existential self-trying to deal with personal uncertainty. *Psychological Inquiry, 20*, 197–217.

Van den Bos, K. (2011). *Vertrouwen in de overheid: Wanneer hebben burgers het, wanneer hebben ze het niet, en wanneer weten ze niet of de overheid te vertrouwen is?* [Trust in government: When do people have it, when don't they, and when don't they know whether they trust the government or not?]. The Hague: Netherlands Ministry of the Interior and Kingdom Relations.

Van den Bos, K. (2015). Humans making sense of alarming conditions: Psychological insight into the fair process effect. In R. S. Cropanzano & M. L. Ambrose (Eds.), *Oxford handbook of justice in work organizations* (pp. 403–417). New York: Oxford University Press.

Van den Bos, K. (2018). *Why people radicalize: How unfairness judgments are used to fuel radical beliefs, extremist behaviors, and terrorism.* New York: Oxford University Press.

Van den Bos, K. (2021). Social psychology and law: Basic social psychological principles in legal contexts. In P. A. M. van Lange, E. T. Higgins, & A. W. Kruglanski (Eds.), *Social psychology: Handbook of basic principles* (3rd ed., pp. 513–531). New York: Guilford Press.

Van den Bos, K., Brockner, J., Stein, J. H., Steiner, D. D., Van Yperen, N. W., & Dekker, D. M. (2010). The psychology of voice and performance capabilities in masculine and feminine cultures and contexts. *Journal of Personality and Social Psychology, 99*, 638–648.

Van den Bos, K., Euwema, M. C., Poortvliet, P. M., & Maas, M. (2007). Uncertainty management and social issues: Uncertainty as important determinant of reactions to socially deviating people. *Journal of Applied Social Psychology, 37*, 1726–1756.

Van den Bos, K., Ham, J., Lind, E. A., Simonis, M., Van Essen, W. J., & Rijpkema, M. (2008). Justice and the human alarm system: The impact of exclamation points and flashing lights on the justice judgment process. *Journal of Experimental Social Psychology, 44*, 201–219.

Van den Bos, K., Hulst, L., Robijn, M., Romijn, S., & Wever, T. (in press). Field experiments examining trust in law: Interviewer effects on participants with lower educational backgrounds. *Utrecht Law Review*.

Van den Bos, K., Hulst, L., Van Sintemaartensdijk, I., Schuurman, B., Stel, M., & Noppers, M. (2021). *Copycatgedrag bij terroristische aanslagen: Een verkenning* [Copycatting behavior following terrorist attacks: An exploration]. The Hague: Research and Documentation Centre of the Netherlands Ministry of Justice and Security.

Van den Bos, K., & Lind, E. A. (2002). Uncertainty management by means of fairness judgments. In M. P. Zanna (Ed.), *Advances in experimental social psychology* (Vol. 34, pp. 1–60). San Diego, CA: Academic Press.

Van den Bos, K., & Lind, E. A. (2009). The social psychology of fairness and the regulation of personal uncertainty. In R. M. Arkin, K. C. Oleson, & P. J. Carroll (Eds.), *Handbook of the uncertain self* (pp. 122–141). New York: Psychology Press.

Van den Bos, K., & Lind, E. A. (2013). On sense-making reactions and public inhibition of benign social motives: An appraisal model of prosocial behavior. In J. M. Olson & M. P. Zanna (Eds.), *Advances in experimental social psychology* (Vol. 48, pp. 1–58). San Diego, CA: Academic Press.

Van den Bos, K., & Loseman, A. (2011). Radical worldview defense in reaction to personal uncertainty. In M. A. Hogg & D. L. Blaylock (Eds.), *Extremism and the psychology of uncertainty* (pp. 71–89). Oxford, UK: Wiley-Blackwell.

Van den Bos, K., McGregor, I., & Martin, L. L. (2015). Security and uncertainty in contemporary delayed-return cultures: Coping with the blockage of personal goals. In P. J. Carroll, R. M. Arkin, & A. L. Wichman (Eds.), *Handbook of personal security* (pp. 21–35). New York: Psychology Press.

Van den Bos, K., Müller, P. A., & Van Bussel, A. A. L. (2009). Helping to overcome intervention inertia in bystander's dilemmas: Behavioral disinhibition can improve the greater good. *Journal of Experimental Social Psychology, 45*, 873–878.

Van den Bos, K., Poortvliet, P. M., Maas, M., Miedema, J., & Van den Ham, E.-J. (2005). An enquiry concerning the principles of cultural norms and values: The impact of uncertainty and mortality salience on reactions to violations and bolstering of cultural worldviews. *Journal of Experimental Social Psychology, 41*, 91–113.

Van den Bos, K., Van der Velden, L., & Lind, E. A. (2014). On the role of perceived procedural justice in citizens' reactions to government decisions and the handling of conflicts. *Utrecht Law Review, 10*(4), 1–26.

Van den Bos, K., Van Lange, P. A. M., Lind, E. A., Venhoeven, L. A., Beudeker, D. A., Cramwinckel, F. M., Smulders, L., & Van der Laan, J. (2011). On the benign qualities of behavioral disinhibition: Because of the prosocial nature of people, behavioral disinhibition can weaken pleasure with getting more than you deserve. *Journal of Personality and Social Psychology, 101*, 791–811.

Van den Bos, K., Van Schie, E. C. M., & Colenberg, S. E. (2002). Parents' reactions to child day care organizations: The influence of perceptions of procedures and the role of organizations' trustworthiness. *Social Justice Research, 15*, 53–62.

Van den Bos, K., Wilke, H. A. M., & Lind, E. A. (1998). When do we need procedural fairness? The role of trust in authority. *Journal of Personality and Social Psychology, 75*, 1449–1458.

Van Lange, P. A. M., Otten, W., De Bruin, E. M. N., & Joireman, J. A. (1997). Development of prosocial, individualistic, and competitive orientations: Theory and preliminary evidence. *Journal of Personality and Social Psychology, 73*, 733–746.

Warren, M. E. (1999). Democratic theory and trust. In M. E. Warren (Ed.), *Democracy and trust* (pp. 310–346). Cambridge: Cambridge University Press.

Weary, G., & Jacobson, J. A. (1997). Causal uncertainty beliefs and diagnostic information seeking. *Journal of Personality and Social Psychology, 73*, 839–848.

Wilson, T. D., Centerbar, D. B., Kermer, D. A., & Gilbert, D. T. (2005). The pleasures of uncertainty: Prolonging positive moods in ways people do not anticipate. *Journal of Personality and Social Psychology, 88*, 5–21.

Zaheer, A., McEvily, B., & Perrone, V. (1998). Does trust matter? Exploring the effects of interorganizational and interpersonal trust on performance. *Organization Science, 9*, 141–159.

17
THE POLITICS OF INSECURITY
How Uncertainty Promotes Populism and Tribalism

Joseph P. Forgas

UNIVERSITY OF NEW SOUTH WALES, SYDNEY, AUSTRALIA

Abstract

The experience of insecurity plays an important role in political affairs, an issue already recognized in Plato's Republic. The rise of social, economic, or existential insecurity often fuels populist movements, as long as effective propaganda can provide voters with a suitable psychological narrative to channel fear and uncertainty. The chapter reviews recent evidence for the rise of populist politics both on the left and on the right of the political spectrum. It is suggested that insecurity potentiates an evolutionary need for tribal belonging (see also Hirschberger; Hogg & Gaffney, this volume). Insecurities can be manipulated and channeled by populist leaders by promoting tribal ideologies such as ethno-nationalism, xenophobia, Marxism, woke-ism, critical race theory, and others (see also Kreko; van Prooijen, this volume). Using an evolutionary psychological framework, the chapter suggests that tribalism helps alleviate insecurity and uncertainty by offering epistemic certainty and simplicity, tribal belonging and identification, moral superiority, the comforts of autocracy, and charismatic leadership (see also Kruglanski & Ellenberg; Pyszczynsky & Sundby, this volume). The psychological principles leading from insecurity to political populism are illustrated using an empirical case study of Hungary, an European Union (EU) member country that descended from democracy to populist autocracy in the last decade. The chapter argues that an evolutionary understanding of the paleolithic characteristics of tribalism as a fundamental feature of human nature offers a constructive way to understand the links between uncertainty, tribalism, and the rise of populist movements.

Throughout human history, insecurity has been our lot. Human groups have always faced dangers and threats such as violence, warfare, poverty, and famine. As Hobbes argued (1968, p. 186), life is "continual fear, ... solitary, poor, nasty, brutish, and short". Indeed, it is arguable that fear has shaped our politics and culture since time immemorial (Robin, 2004). There is considerable historical and anthropological evidence suggesting that in the face of such threats, humans typically respond by affirming group norms and seeking stronger sanctions against deviants (McCann, 1997; Doty, Peterson, & Winter, 1991; see also Hirschberger; Hogg & Gaffney; Kreko; Pyszczynski & Sundby, this volume).

The role of insecurity in driving support for authoritarian regimes has received renewed interest in recent years (Forgas et al., 2021; Marcus, 2021; Kruglanski, Molinari & Sensales, 2021). Analysis of archival US data confirms that authoritarianism significantly increases in high-insecurity periods (Doty et al., 1991). Early studies found, for example, that economic insecurity increases tribal violence: "Lynchings were more frequent in years when the ... price of cotton was declining and inflationary pressure was increasing" (Figure 17.1; Beck & Tolnay, 1990, p. 526).

The European and World Value Surveys (total N = 134,516; Onraet, Alain and Cornelis, 2013) also found a link between insecurity and right-wing attitudes across 91 nations, and the same also holds true for left-wing politics (Jussim et al., this volume). It seems that psychological insecurity often drives

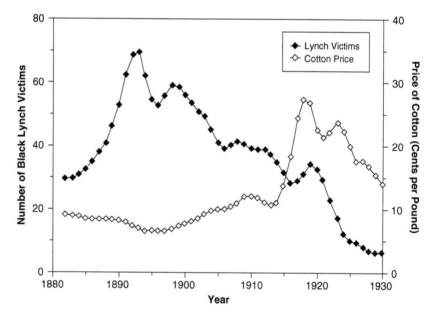

FIGURE 17.1 The relationship between economic insecurity and tribal violence: the link between the price of cotton and the number of Black lynching victims in the period 1882–1930 (three year smoothed averages; after Beck & Tolnay, 1990).

political demands for safety, protection, and authoritarianism (Feldman & Stenner, 1997; Feldman, 2021). Consistent with this idea, Burke, Kosloff and Landau (2013) found that insecurity induced by mortality salience had a huge effect on political attitudes ($r = .50$), a view also supported by other research (Castano et al., 2011; see Pyszczynski & Sundby; and Hirschberger, this volume).

Why this matters: The rise of populism. In the last few decades, we have seen a marked deterioration of democracy in many countries and the rise of populist, authoritarian movements (Fukuyama, 2022). The values of liberalism as identified by John Stuart Mill – freedom of speech, open exchange, universal humanism, individualism, and tolerance – are under aggressive challenge by shrill populist narratives. Populist activism is thriving both on the political left (woke ideology, critical race theory, Black Lives Matter (BLM), Antifa; see also Jussim et al., this volume) and on the political right (Trump, Orban, Edogan, AfD, Le Pen, etc.). These movements represent the latest historical challenge to the values of the Enlightenment – the atavistic re-emergence of the stone-age psychology of tribalism as an alternative to individualism. What role does the experience of psychological insecurity and uncertainty play in these developments?

Insecurity and fear are perennial problems for running effective political systems. Plato (1974) warned about the dangers to democracy when fears and passions produce irrational decisions. Echoing this view, Brennan (2017) suggested that democracy has become the rule of the ignorant and the irrational, a view also supported by Caplan (2007). Davies (2019) explicitly argues that insecurity – 'nervous states' – is part of the reason why contemporary politics has become so fractious and polarized, as feelings rather than facts dominate discussion and decisions. This chapter will present empirical evidence linking insecurity to political populism in one country, Hungary.

Narratives of insecurity. Uncertainty and insecurity undermine the established consensual worldview in democratic societies and reduce trust in existing systems and institutions (see also Cooper & Pearce, this volume). To be sure, there are objective reasons for the recent upsurge in feelings of insecurity. The past two decades saw a marked increase in threatening economic and social developments (inequality, economic crises, pandemics, terrorism, uncontrolled migration, threats to individual, and collective identity) that challenged many people's fundamental needs and values (Bar-Tal & Magal, 2021; see also Hogg & Gaffney; Kreko; van den Bos; van Prooijen, this volume).

Policies imposed by out-of-touch elites further fuel populist resentment (Norris & Inglehart, 2018; Oesch, 2008). But any explanation of populism also requires a psychological understanding of how people mentally represent their political reality. Understanding the psychology of these movements is critical because economic, social, or racial grievances are not in themselves sufficient for radical populism to flourish. Humans mostly lived in abysmal conditions throughout history, yet populist revolts were rare. What is also essential is understanding the psychological *narratives* that turn dissatisfaction into potent political forces.

The idea that political systems must be based on an understanding of human psychology originated with Plato, and populism indeed benefits from a basic human vulnerability to emotionalism and tribalism. In the 17th and 18th centuries, philosophers like Hobbes, Locke, Mill and Hume (1984) also formulated their influential political ideologies based on clear assumptions about human nature. Democracy and populism (although both mean rule by the people) make very different predictions about human psychology. Liberal democracy is an individualist credo and assumes a human ability for rational decision-making. Populism is a collectivist tribal ideology that subordinates the individual to the group and exploits our vulnerability to fear, insecurity, and uncertainty.

This chapter reviews evidence that populist movements weaponize feelings of insecurity by offering followers epistemic certainty, moral absolutism, and tribal identification. Populism emphasizes the necessity of conflict and struggles between the 'ingroup' and its enemies (class conflict in Marxism, racial conflict in Nazism, critical race theory, etc.; Forgas, 2020). This kind of group conflict ideology is incompatible with the individualism and progressivism of liberal democracy and its emphasis on due process, the rule of law, minority rights, and checks and balances.

Left- and right-wing populism share such a collectivist group conflict ideology, but employ different narratives (Cooper & Avery, 2021; see also Jussim et al., this volume). Right-wing populism invokes nativist values, emphasizing ethno-nationalism, order, structure, and predictability. Left-wing populism focuses on social justice and economic, ethnic, and racial equality (e.g., cultural Marxism, critical race theory, and woke activism). Liberal democracy, now under attack by both left- and right-wing populists, has historically withstood the earlier collectivist challenges posed by fascism and communism – but its current fragility gives great cause for concern.

An Evolutionary Perspective

Populism offers security by appealing to the tribal mentality of the Paleolithic, when group cohesion was the key requirement for survival (see also Fiske; Hirschberger; von Hippel & Merakovsky, this volume). The prodigious intellectual capacities of humans were shaped by the cognitive demands of maintaining group cohesion, the foundation of our evolutionary success, rather than the rational demands of understanding reality (Buss, 2019; Mercier & Sperber, 2017; von Hippel, 2018; see also von Hippel, this volume). Populism benefits from the tendency that much of human thinking is automatic and intuitive rather than analytical (Bachara, Damasio, Tranel & Damasio, 1997; Gigerenzer, Todd & Group, 1999; Kahneman, 2011; see also Krueger & Gruening, this volume).

Humans have a 'social brain' rather than a 'rational brain', shaped by the demands of group coordination, an idea that is supported by evidence showing that brain capacity in primates is closely correlated with group size (Dunbar, 1998). Insecurity may often trigger a reflexive return to the time-honored safety of archaic communalism, and populist movements offer security by appealing to the tribal mentality of the paleolithic (see also Hirschberger; van den Bos, this volume).

In contrast, the idea of the confident and self-sufficing individual is a radical Enlightenment invention, a fragile cultural fiction that emerged after centuries of horrific religious and tribal bloodshed in Europe. Despite the explosive success of individualist Western liberal systems (Pinker, 2018), our mass societies continue to be marked by endemic patterns of insecurity, loneliness, isolation, and anomie (Durkheim, 1964, 1966; Toennis, 1956; Zimbardo, 1977). Populist movements are psychologically attractive because they appeal to our ingrained stone-age tribal mentality, offering a heady mix of simplicity, certainty, moral absolutism, identity, and utopian promises.

The Quest for Simplicity and Certainty

Humans instinctively seek simplicity and certainty in their narratives (Kruglanski et al., 2021), and religions, creation stories, and mythologies throughout history fulfilled this epistemic need (Harari, 2014), offering simple but incorrect explanations rather than complex but correct ones (Crano & Gaffney, 2001; Kahneman, 2011; see also Crano & Hohman, this volume). Such cognitive bias also makes us vulnerable to political manipulation (Neal et al., 2022). Classic studies in empirical social psychology offer convincing evidence that human nature is marked by conformity, obedience, norm-following, and a readiness for inter-group conflict and discrimination (von Hippel & Merakovsky; Crano & Hohman; Hirschberger; Hogg & Gaffney, this volume). Our thinking was also shaped by evolutionary pressures and is similarly marked by cognitive habits such as categorization, confirmation bias, heuristics, fluency effects, and illusory correlations. These habits promote cognitive efficiency, but can also be exploited by populist propaganda (Krueger & Gruening, 2021; this volume).

For example, the overestimation of the correlation between negative behaviors and unfamiliar people (the *illusory correlation* bias) may well have served an adaptive alerting function in our ancestral environment (Hamilton & Rose, 1980). In a similar way, *cognitive fluency* effects produce an overestimation of the reliability of information that happens to be easy to process (Fiedler & McCaughey, this volume). Our experiments also found that people overestimate the truthfulness of statements that are easy to read and simple to understand, one of the defining features of populist communication (Forgas, 2013; Koch & Forgas, 2012).

Fake news and conspiracy theories spread precisely because they offer simplicity and a sense of privileged knowledge (van Prooijen, 2019; see also this volume). The *need for cognitive* closure (Webster & Kruglanski, 1994) promotes populist gullibility, and paradoxically, a lack of expertise increases confidence in erroneous beliefs (the *Dunning-Krueger effect*). Some Trump followers believe that their champion won the election. Left-wing woke activists may see science as a patriarchal conspiracy privileging white men and question scientific evidence for the heritability of human qualities (Myers, 2019).

Populist epistemic certainty *is promoted by confirmation bias*, the tendency to seek information supporting pre-existing beliefs. History is replete with enduring fictional beliefs: witch hunts, religious wars, creation myths, paranormal beliefs, and the QAnon cult all benefit from the confirmation bias (see also Kreko; van Prooijen; Fiske, this volume). On the political right, autocrats in power like Putin, Orban, and Erdogan use media control and propagate false certainties (Albright, 2014; Myers, 2020; Temelcuran, 2019). On the left, false Marxist ideas of class conflict continue to flourish (Popper, 1945). Once Marxist certainties were shaken by the horrors and collapse of communism, new and even more obscure 'certainties' were provided by incomprehensible postmodern ideologies promoted by the likes of Foucault and Derrida, offering ready refuge and new-found certainty to true believers. Critical social justice theories deny the possibility of truth; instead, tales about 'lived experience' offer epistemic comforts. Post-modernist conflict ideologies can 'colonize' entire academic fields such as critical literary studies, social anthropology, gender studies, or sociology to the exclusion of all other forms of thought (Davison, 2020; Jussim et al., this volume). Such postmodern critical theories can thrive in protected domains where empirical proof and real-life relevance are not even expected (Davison, 2020).

Epistemic certainty denies the value of rational discourse and sees debate as superfluous. Some Marxists still believe in the elusive proletarian revolution, and woke ideas about 'oppressive patriarchy' or universal white guilt reduce complex phenomena to simple-minded explanations. Claims that *all* white people are racist are transparently racist themselves and have no explanatory or discriminatory utility. The idea that group-based discrimination must be rectified by insisting on more group-based discrimination is manifestly false, yet such beliefs survive by satisfying the deeply felt human need for moral certainty and security.

Insecurity, Tribalism, and the Power of Belonging

Group loyalty offered a significant survival advantage to both individuals and their groups in our evolutionary past, and humans show a spontaneous preference for the ingroup and discrimination against outgroups (Hogg, 2007; Kruglanski et al., 2021; Tajfel & Forgas, 2000; see also von Hippel & Merakovsky;

Hirschberger, this volume). Seeking refuge in tribal certainties may be especially attractive to people who feel disappointed with their achievements and expectations (see also Crano & Gaffney, 2022). Insecurity can drive tribal affiliations, "a natural and nearly ineradicable feature of human cognition that no group – not even one's own – is immune" from (Clark et al., 2019, p. 587). Hogg's (2007; Hogg & Goetze-Astrup, 2021) uncertainty-identity theory explores how insecurity leads to group identification and political extremism, driving attachment to more radical and extremist groups (see also Mikulincer and Shaver, this volume).

Hatred of outsider 'elites' is a common populist rallying cry, and some Western 'elites' have indeed become captives to the ideological left-wing bias (Scruton, 2000), triggering right-wing populist reactions in the US (Trump), Germany (AfD), Austria (FPO), France (LePen), Britain (Brexit), and Italy (Salvini). However, once populist autocrats become the new elite, anti-elitism ceases to be a rallying cry, and the movement then survives on the tribal allegiances of its followers alone (e.g., Orban, Kaczinsky, Erdogan, Chavez; see also Kreko, 2021).

Insecurity also promotes authoritarianism (Albright, 2018; Feldman & Stenner, 1997) and a preference for tighter group norms (Feldman, 2021; Gelfand & Lorente, 2021). In two elections, in the US in 2016 and in France in 2017, insecure voters showed a greater preference for tight social norms and support for autocratic politicians (Gelfand & Lorente, 2021). Insecurity has driven ethno-nationalist 'my country first' movements in the US, Britain, France, Italy, and Hungary. In the US, General Social Survey data in 1996, 2004, and 2014 showed that nationalism was related to anti-immigrant attitudes and Republican identification among Whites (Huddy & DelPonte, 2021).

Tribalism becomes even more attractive when lack of personal achievement or traumatic group experiences require narrative explanation (Hogg & Goetze-Astrup, 2021). Narcissistic themes of injustice, betrayal, powerlessness, and victim mentality (e.g., 'Make America Great Again', 'Take Back Control', and 'We Are Sacred') are common narrative features to bolster group identity. Insecurity and collective narcissism are significant predictors of populist politics in countries such as the US, Britain, Poland, and Hungary (Cichochka, 2016; Forgas & Lantos, 2021; Goles de Zavala et al., 2020; Lantos & Forgas, 2021).

The recent COVID experience offers a striking example of how insecurity can promote tribalism and autocracy (Atlas, 2022; Frijters, Foster & Baker, 2021; see also Kreko, this volume). Otherwise, tolerant and liberal societies accepted unprecedented restrictions (Feldman, 2021; Foster, 2021), sometimes based on confusing and often misleading information. Media fearmongering, suppression of alternative opinions, and demands for autocratic restrictions prevailed. Important characteristics of the pandemic (such as posing very little extra danger to younger people) were ignored. "Reminiscent of other legendary frenzies in history, like the tulip bulb mania or the tech stock bubble, hypothetical extreme-risk scenarios went seemingly unchallenged and were given absolute credence" (Atlas, 2022, p. 1). Insecurity magnified common shortcomings of inductive reasoning. Some liberal democracies like Australia went so far as to prohibit citizens from *leaving* the country – the last and most basic human right an individual has when disagreeing with their government. Demonizing dissenters and imposing vaccine mandates in countries like Austria illustrate how fear and insecurity can lead to extreme authoritarianism and intolerance.

Moral Certitude and Virtue Signaling

When facing insecurity, great comfort can be gained by seeking moral certainty and virtue signaling. Claiming social status is a crucial adaptive resource for humans, and moral posturing is a common strategy for status seeking (Petersen, Osmundsen & Bor, 2021). Moral certitude offers personal significance (Kreko, 2021; Kruglanski et al., 2021). Many atrocities are committed by people with an unfailing belief in their moral cause (e.g., fascists, communists, religious fanatics, political ideologues such as Antifa, BLM, Proud Boys, woke activists, etc.). Moral certitude denies the legitimacy of any opposition and rejects the need for discussion, and followers of populist movements often try to outdo each other in righteous displays of moral fervor. Moral posturing was on display when Hilary Clinton called Trump's voters 'deplorables', or when Trump declared the press enemies of the people.

Moral absolutism also drives coercive authoritarian practices in many universities, organizations and institutions. It seems that tyrannical practices can now flourish in academic institutions in the absence of a one-party dictatorship (see Jussim et al., this volume). Slogans like 'silence is violence', seen at BLM rallies, claim that *not* having an opinion can be deplorable. While fascism has few remaining credible adherents, cultural Marxism and woke activism still retain a puzzling attraction for many intellectuals. Moral absolutism is often linked to *utopistic and millennial* narratives, promising a perfect future that vindicates every sacrifice and even violence. The dramatic promise of a 'thousand-year empire' (Nazism) or a perfect communist utopia (Marxism) has a powerful emotional appeal that liberal incrementalist ideologies cannot match (Scruton, 2000). In the face of insecurity, populism can tap into the all-too-human millennial quest for a perfect utopia.

The Role of Leaders

Populist leaders can alleviate insecurity by becoming the symbolic embodiments of their movements. Personality cults are endemic in populist regimes (Hiter, Mussolini, Stalin, Orban, Putin, Trump), mobilizing the common human tendency to personalize complex issues and ideas (Albright, 2018;

Myers, 2020). Populist leaders need to be strong, consistent, and uncompromising to satisfy their followers craving for simplicity and certainty. The more extreme the group, the more likely it is that the leader exhibits these qualities and satisfies followers' need for certainty (Hogg & Gaffney; Crano & Hohman, this volume). Repetition is a common propaganda tactic refined by the master propagandist Goebbels (Albright, 2018). Psychological research showed that repetition increases message credibility almost as much as hearing the same message from several independent sources.

As autocracy takes hold, populist leaders are increasingly characterized by their disrespect for the truth. In Fascist and Marxist dictatorships, truth is always secondary to propaganda, and obvious lies remain unchallenged. Trump told countless untruths, and Hungary's Orban is sometimes characterized as a 'spin-dictator' who won four elections by employing shamelessly dishonest propaganda. The invocation of moral absolutism allows populist leaders to ignore normal standards of honesty and suffer no censure for lying 'in the good cause'. In a sense, populist leaders come to embody the absurdities and fallacies of their narratives, as the recent histories of countries like Russia and Hungary also illustrate.

A Case Study: The Death of Democracy in Hungary

Hungary is the one country within the EU that has progressed perhaps furthest toward dismantling democracy and establishing an authoritarian illiberal regime, and the only country within the EU that Freedom House (2020) classifies as no longer a democracy. Psychological insecurity played a key role in this dramatic transformation that was accomplished without a political coup or military takeover, relying solely on psychological manipulation to secure the electoral support of a portion of the population (Beauchamp, 2018; Forgas, Kelemen & Laszlo, 2015). Surveys and empirical analyses demonstrate how fear, grievance, and a damaged sense of national identity are linked to the rise of autocracy (Ditto & Rodriguez, 2021; Keller, 2010), using propaganda to manipulate collective narcissism and self-uncertainty (Forgas & Lantos, 2020; Golec de Zavala, Lantos & Keenan, 2021; Lantos & Forgas, 2021).

Hungary is not an important country and barely provides .08% of the EU's economy, but its descent into one-party autocracy is widely regarded as a warning of how populist propaganda can exploit fear and insecurity. Hungary's ruler, Viktor Orbán made a great impression on Trump and his followers and now has admirers in many countries (Borger & Walker, 2019). In 12 years, he completely reshaped the country's political culture and institutions, illustrating how fake news, propaganda, conspiracy theories, and identity politics can be harnessed to destroy democracy – the same populist strategies that have been routinely employed by autocratic regimes since the 1930s (Albright, 2018).

From Democracy to Autocracy

Since 2010, Orbán has built a de facto one-party system he calls the "System of National Cooperation" (sic). He introduced a new constitution supported only by his own party, dismantled democratic institutions, abolished the system of checks and balances, and placed loyal party apparatchiks at the helm of most public institutions. A new electoral law guarantees the power of his party, giving him a two-third majority with barely 30% of eligible voters (Krekó & Enyedi, 2018, p. 42). In an infamous speech in 2014, Orbán announced that Hungary is turning its back on liberal democracy and sees autocratic Eastern states such as Russia and Turkey as its new role models, "because liberal values today mean corruption, sex, and violence" (Orbán, 2014). In the World Justice Project Rule of Law Index 2019, Hungary is now ranked dead last in the EU and North America region. In the 2019 edition of the Sustainable Governance Indicators, Hungary and Turkey occupy the two bottom places out of 40 countries when it comes to the rule of law. In the 2019 edition of The Global State of Democracy (International Institute for Democracy and Electoral Assistance, 2019), Hungary was listed as a country that has seen the greatest democratic erosion in the past five years.

The success of Orban's autocratic system rests on manipulating fear and insecurity. His party dominates most of the media, including the public broadcasting system that now functions as a party propaganda outlet. Hungary fell from 23rd to 87th in the international list of press freedom, and most public officials belong to the prime minister's loyal personal network. Some writers define Hungary as a post-communist mafia state (Forgas, Kelemen & Laszlo, 2015; Magyar, 2016), focusing on the all-encompassing corruption and godfather-like hierarchical power structures. Orbán's childhood friend, until 2010 a barely literate gasfitter has become the richest man in Hungary, and Orbán's son-in-law is now also a multi-billionaire, despite being accused of racketeering and corruption by the EU. Yet the number of significant corruption prosecutions has dropped to almost zero (Krekó & Enyedi, 2018, p. 44). This autocratic one-party regime has now been re-elected four times in elections commonly described as not fair and only partly free (Garton-Ash, 2019). This was largely achieved by manipulating and harnessing insecurity using state-controlled propaganda.

The Role of Insecurity

Manipulative and dishonest government propaganda exploiting the endemic insecurities of Hungarians played a critical role in legitimizing autocracy. Conspiracy theories, fake news, narcissism, the creation of fictitious enemies (migrants, foreigners, Jews, gays, the EU), historical fictions, and moral absolutism are key strategies (Forgas & Lantos, 2020). Hungary has had a traumatic history for over 500 years. Negative group experiences require narrative explanation, often built around themes of injustice, betrayal, powerlessness, and victim mentality (Bibó, 1948/2004; Lendvai, 2012). Populism becomes a truly dynamic political force when autocratic leaders can exploit the insecurity and collective narcissism associated with compromised group identity (Albright, 2018; Bar-Tal & Magal, 2021; Marcus, 2002; 2021; Lendvai, 2012). Surveys confirm that Hungarian national identity today is characterized by a deep sense of insecurity, inferiority, and lack of self-confidence, compensated by an overly unrealistic, grandiose, and narcissistic evaluation of the ingroup's imaginary virtues and entitlements. This predisposes many Hungarians to a kind of 'political hysteria', seeking comfort in nationalism and autocracy (Bibó, 1948/2004; Kelemen, 2010).

Verbal narratives. Numerous studies using the quantitative analysis of linguistic narratives in school history texts, historical novels, and everyday conversations documented the deep sense of insecurity and damaged national identity in Hungary (László, 2005, 2014; László & Ehman, 2013). The linguistic analyses looked at three domains: (1) descriptions of *causes* of historical events; (2) *emotional reactions*; and (3) national *self-evaluation*. The results showed Hungarians saw themselves as helpless victims with little causal influence over events, they blamed outsiders for their defeats, and their emotional reactions were dominated by sadness, fear, frustration, helplessness, and self-pity (László, 2005, 2014).

This deeply insecure view was compensated by an unrealistically grandiose and narcissistic *self-evaluation* claiming moral superiority and virtue compared to external groups who are blamed for failures (László, 2014, p. 96). This insecurity and vulnerability are actively exploited by Orbán's propaganda machine, which promotes fake 'historical' narratives of past greatness.

National insecurity was also documented in an ingenious study by Csepeli (2019), who compared the language and imagery of the Hungarian national anthem with other anthems. While neighboring nations' anthems feature words such as 'beauty, splendor, life, dawn, freedom, glory, love, fortune, joy, wealth, pride, victory, happiness, strength, the Hungarian anthem is replete with words such as 'misfortune, sin, punishment, sadness, moan, slavery, beaten, war, thundering sky, mounds of bones, ashes of your fetus, sea of flames, death growl, mourning, blood of the dead, torment'. This pattern of self-pity, victimhood, and insecurity offers fertile ground for political manipulation by an unscrupulous autocrat.

The Hungarian language also helps isolate Hungarians from international information. According to Eurostat (2016), very few Hungarians speak foreign languages, yet the Hungarian language functions as a key symbol of national pride and uniqueness. Many Hungarians believe that Hungarian is the most beautiful language in the world, yet speak no other language themselves. Naïve claims that ancient Hungarians invented runic writing (!) led many localities proudly displaying their names in runic writing – although no one can actually read it (Figure 17.2).

FIGURE 17.2 Examples of Hungarian government propaganda: Top left: 'Message to Brussels: Respect for Hungarians!' Top right: 'Soros can't have the last laugh!' Middle left: 'Resist (EU) blackmail: Defend Hungary' Middle right: 'You have a right to know what Brussels is planning for you'. Bottom left: Runic writing of locality names. Bottom right: 'Hungary will not give in!'

Insecurity in survey data. Representative national surveys further confirm the sense of insecurity demonstrated in linguistic analyses (Forgas et al., 2015; Forgas & Lantos, 2020; Kelemen, 2010; Kelemen et al., 2014; Szilágyi & Kelemen, 2019). In 2019, the majority of respondents expressed pessimism and insecurity endorsing statements such as 'a strong political leader is needed to solve the country's problems' (80%), 'democracy in Hungary will not function as it should for many decades' (72%), 'political parties do not really represent the interests of the people' (69%), 'people lived better before the change of regime' (55%), 'the average person has no influence on public life' (55%), 'the Hungarian economic and social structure should be radically transformed' (70%), 'the state of our society is getting worse every year' (63%), 'not everyone in Hungary has the opportunity to get rich and prosper' (54%), and 'most domestic political decision do not serve the public good' (55%) (Szilágyi & Kelemen, 2019, pp. 192–193). These attitudes were strongest among rural, poorer, and less educated respondents, who also provide the regime's main electoral support.

Other representative surveys by the Friedrich Ebert Stiftung in March 2020 also found that the majority of respondents believe that corruption (60%), public education (58%), health (63%), democracy and freedom of the press (50%), poverty (54%), and the international perception of the country (52%), have all become worse in the last ten years (Friedrich Ebert Stiftung, 2020, p. 86). Perceptions of the rule of law are equally negative – most respondents think that 'the law is applied differently to influential people than to the average person' (82%), 'not all people are equal before Hungarian courts' (65%), 'law and justice in court judgments often separated' (76%), 'the outcome of cases largely depends on the person of the judge' (75%), 'it is not worth litigating because it only favors lawyers' (58%), and 'the Hungarian judiciary is not independent of politics' (71%) (Szilágyi and Kelemen, 2019).

Consistent with the narrative analyses, these strongly negative opinions coexist with an unrealistically narcissistic national evaluation, as most voters felt that 'for me, Hungary is the most beautiful place in the world' (80%; Szilágyi & Kelemen, 2019). The sense of insecurity is also exploited by the manipulative, state-sponsored cult of allegedly world-beating 'hungaricums'. Official government committees announce which foods, practices, and inventions are suitable to confirm the unique genius of Hungarians. Obviously, a country with a realistic sense of self-confidence would not need their government to confirm which sausages, soups, spices, or drinks they should be collectively proud of from now on.

This paradoxical sense of vulnerable national identity is easily mobilized for political purposes by an autocratic regime that completely dominates the media (Albright, 2018; Bar-Tal, 2020; Golec de Zavala et al., 2021). Insecure voters who embrace political propaganda now provide the mainstay of electoral support for Orban's autocratic regime.

The Role of Political Propaganda

Emotional manipulation has long been part of political practice (Brader, 2006), and simple, endlessly repeated political messages glorifying the ingroup and creating external enemies and conspiracies is a well-established strategy also used by Mussolini, Hitler, and aspiring dictators ever since (Albright, 2018; Myers, 2019; see also Hirschberger; Hogg & Gaffney, this volume). In Hungary, state propaganda during the past ten years has variously portrayed the EU, refugees, gays, the opposition, or George Soros as mortal threats to national survival. Controlling the media is crucial for populist success (Krekó & Enyedi, 2018). In 2017 alone, about US$250 million was spent by the Hungarian regime on propaganda and fake 'national consultations', first used by Hitler's regime.

Hungary's propaganda expenditure before the 2018 election was several times the official amount spent by both sides on the Brexit campaign in the United Kingdom. And this propaganda exploiting insecurity seems to be working: Hungarians today are among the most xenophobic people in Europe, they are among the most habituated about corruption, and they fear Russia less than they fear Brussels and George Soros. In one recent survey, 51% of Fidesz voters said they would prefer Russia to the United States when choosing a strategic partner, and Vladimir Putin was more popular than Angela Merkel or Donald Trump (Krekó & Enyedi, 2018, p. 47).

In an even more astonishing recent survey, 47% of government voters held the US responsible for the Ukrainian war, but only 3% blamed the Russians (!). Among opposition voters, 61% blame Russia and only 7% the US. This dramatic contrast confirms that media dominance and unchecked misinformation can indeed produce a deeply warped perception of reality (see also Kreko; van Prooijen; this volume). Rather than physical oppression, mental manipulation has become the mainstay strategy of contemporary 'spin dictators' like Orban (Guriev & Treisman, 2022).

The Relationship between Insecurity and Populism

Insecurity is a strong feature of Hungarian political culture – but how does this translate into political support? In several studies, we explored the links between insecurity and populist political behavior. In our first study in 2017 (Lantos & Forgas, 2021), insecurity was assessed in 284 participants using a collective narcissism scale (Golec de Zavala et al., 2021) as a predictor of their support for Fidesz. Results showed that insecurity (collective narcissism) predicted populist voting in the previous, 2014 election, $r_{pb} = .21$, $p = .004$, as well as voting intentions in next, 2018 election $r_{pb} = .32$, $p < .001$. Subsequent mediational analyses also found that conservatism was a significant mediator between insecurity fuelled by collective narcissism and populist voting (Figure 17.3), consistent with the idea that insecure national identity predicts populist political preferences.

How Uncertainty Promotes Populism and Tribalism **321**

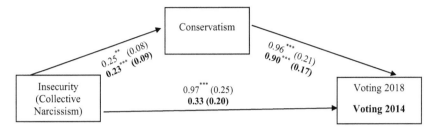

FIGURE 17.3 Insecurity as measured by collective narcissism predicts political preferences in 2014 ($N = 194$) and in 2018 ($N = 240$) with conservatism as a significant mediator (After Lantos & Forgas, 2021).

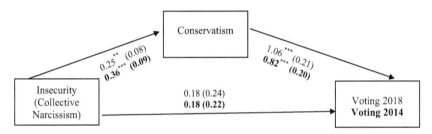

FIGURE 17.4 Mediation analysis showing that insecurity (collective narcissism) predicts populist voting in 2014 ($N = 265$) and in 2018 ($N = 155$), as well as relative deprivation, mediated by conservatism (After Lantos & Forgas, 2021).

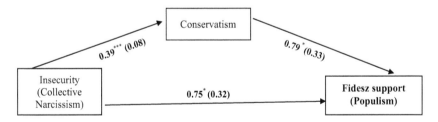

FIGURE 17.5 The direct and indirect effects of collective narcissism on Fidesz support ($N = 137$) and populism scores ($N = 440$), mediated by conservatism (after Lantos & Forgas, 2021).

In the next study (Lantos & Forgas, 2021; $N = 217$), we explored the relationship between insecurity and conservatism and again found a significant link between insecurity (collective narcissism) and populist voting (Figure 17.4).

In our third study carried out in 2020, 440 volunteer Hungarians completed an online questionnaire measuring populist politics, insecurity (collective narcissism), self-esteem, just world beliefs, and a variety of individual difference measures. Fidesz's support was again significantly predicted ($R^2 = .34$) by insecurity (*collective narcissism*, $\beta = .23$, $p = .04$), and conservatism was again a significant mediator (Figure 17.5).

These studies offer consistent evidence that experiences of insecurity have a major influence of populist politics, in our case mediated by conservatism. This pattern broadly supports the prediction that psychological insecurity makes people especially vulnerable to populist appeals that offer simplicity, certainty, tribal identification, and moral certitude, as suggested in the theoretical review in the first half of this chapter.

Conclusions

This chapter sought to highlight the important role that psychological insecurity plays in shaping political processes and the rise of populist movements in particular. We used the recent history of Hungary as an illustration of how such a shift to dictatorship can occur when unscrupulous rulers dominate the media and do not hesitate to use lies to manipulate fear and insecurity (Bibo, 1948/2004; 1991). In the first half of the chapter, we reviewed the recent literature in political psychology, offering a range of interesting hypotheses about how insecurity might motivate people to seek epistemic simplicity and certainty by joining collectivist, tribal movements that promise moral superiority (Marcus, 2021; Bar-Tal & Magal, 2021; Hogg & Goetze-Ostrup, 2021; see also Hogg & Gaffney; Hirschberger; von Hippel & Merakovsky, this volume). Insecure national identity, aching need for group identification, and collective narcissism offer fertile ground for the spread of populist ideologies. In the case of Hungary, propaganda was able to successfully exploit endemic insecurity by emphasizing external threats, employing conspiracy theories and fake news, and playing to the moral superiority of the ingroup – a strategy the opposition was unable or unwilling to adopt (Kelemen, 2010; Kreko, 2021; László, 2014; see also Kreko; van den Bos; van Prooijen, this volume).

The use of propaganda to manipulate insecurity is neither new nor particularly creative. Populist leaders such as Mussolini, Hitler, Goebbels, Putin, Erdogan, and now Orbán regularly use this method to strengthen their political legitimacy (Albright, 2018; Temelcuran, 2020). Orbán proved uniquely successful in disseminating such dishonest propaganda exploiting narcissistic feelings and promoting the deliberate falsification of history (Kelemen, 2010; Kreko, 2021; László, 2014). The empirical results broadly support the idea that feelings of insecurity and collective narcissism significantly predict support for the autocratic one-party state, which is consistent with the view that populism offers a collectivist pre-enlightenment value system based on tribalism and group identification (Cichochka, 2016; Marchlewska et al., 2017).

Populism represents a danger for liberal democracy because it has a deep affinity with the archaic, stone-age characteristics of the human mind, which evolved to serve the demands of group cooperation rather than the rational discovery of truth. Political movements succeed or fail depending on their ability to mobilize basic psychological needs, and both left-wing and right-wing populism exploit the human need for positive identity, epistemic certainty, simplicity, significance and belonging, and moral virtue. Both ascendant and in-power populist movements (Fascism, Marxism, Trumpism, cancel culture, Proud Boys, Antifa, and woke-ism) benefit from manipulating these evolutionary vulnerabilities.

The possibility that 'human nature' as shaped by evolution is ill-suited to the psychological requirements of liberal democracy (individualism, tolerance, and rationality) echoes Plato's age-old reservations (Brennan, 2017; Caplan, 2007). However, it may be encouraging that liberal democracies have survived for some hundreds of years now despite our underlying paleolithic inclinations. The role of insecurity in driving populist politics has now been convincingly demonstrated in many countries, including the US, Britain, Poland, and others (Federico & Golec de Zavala, 2018; Lantos & Forgas, 2021; Marchlewska et al., 2018; Myers, 2020).

It is also possible that the radical individualism and secularism of our age and the disappearance of genuine primary group experiences from our lives have left people particularly vulnerable to the siren calls of tribal ideologies. The growth of the internet and social media also contributes to undermining the once dominant public voice once reserved for enlightened liberalism (Haidt, 2022). People may now find their own 'tribe' in the metaverse, promoting consensual delusions, fake news, conspiracy theories, and sectional group ideologies (see also van den Bos; van Prooijen, this volume).

How can liberal individualism best respond to the populist challenge? Rational argument has limited utility in fostering security and convincing 'true believers' who reject the value of discussion. Totalitarian ideologues can acquire undue influence over once-liberal institutions (universities, media, law, education, and corporations; see Jussim et al., this volume). This is only possible as long as the silent majority remains silent. Understanding how populism operates should be the first step toward standing up to populist tyranny. Liberal democracies successfully rose to the challenge against both fascism and communism. New external threats like Chinese and Russian authoritarianism may yet produce a re-affirmation of our foundational values. In combating the dangers of populism, we certainly need a more thorough understanding of the psychological processes that underlie populist support.

References

Albright, M. (2018). *Fascism: A Warning*. New York: Harper Collins Press.
Atlas, S. (2022). How Panic Spread: COVID in the Early Days. 26th April, Brownstone, https://brownstone.org/articles/how-panic-spread-covid-in-the-early-days/
Brader, T. (2006). *Campaigning for Hearts and Minds: How Emotional Appeals in Political Ads Work*. Chicago: University of Chicago Press.

Bar-Tal, D., & Magal, T. (2021). Socio-psychological Analysis of the Deterioration of Democracy and the Rise of Authoritarianism: The Role of Needs, Values, and Context. In: Forgas, J. P., Crano, W. D., & Fiedler, K. (eds.). *The Psychology of Populism: The Tribal Threat to Liberal Democracy* (pp. 42–61). New York: Psychology Press.

Beauchamp, Z. (2018). It Happened Here: How Democracy Died in Hungary. Vox, September 13. https://www.vox.com/policy-and-politics/2018/9/13/17823488/hungary-democracy-authoritarianism-trump

Bechara, A., Damasio, H., Tranel, D., & Damasio, A. R. (1997). Deciding Advantageously Before Knowing the Advantageous Strategy. *Science*, *175*(28 February 1997), 1293–1295.

Beck, E. M., & Tolnay, S. E. (1990). The Killing Fields of the Deep South: The Market for Cotton and the Lynching of Blacks, 1882–1930. *American Sociological Review*, *55*(4), 526–539. https://doi.org/10.2307/2095805

Bibó, I. (1948/2004). Eltorzult magyar alkat, zsákutcás magyar történelem. *Válogatott tanulmányok*. Budapest, Corvina.

Bibó, I. (1991). *Democracy, Revolution, Self Determination*. Budapest: Eastern European Monographs.

Borger, J., & Walker, S. (2019). *Trump Lauds Hungary's Nationalist PM Orbán for 'Tremendous Job'*. Retrieved from: https://www.theguardian.com/us-news/2019/may/13/trump-latest-viktor-orban-hungary-prime-minister-white-house

Brennan, J. (2017). *Against Democracy*. Princeton, NJ: Princeton University Press.

Burke, B. L., Kosloff, S., & Landau, M. J. (2013). Death Goes to the Polls: A Meta-Analysis of Mortality Salience Effects on Political Attitudes. *Political Psychology*, *34*(2), 183–200.

Buss, D. (2019). *Evolutionary Psychology: The New Science of the Mind*. New York: Taylor & Francis.

Caplan, B. D. (2007). *The Myth of the Rational Voter: Why Democracies Choose Bad Policies*. Princeton: Princeton University Press.

Castano et, E., Leidner, B., Bonacossa, A., Nikkah, J., Perruli, R., Spencer, B. et al. (2011). Ideology, Fear of Death, and Death Anxiety. *Political Psychology*, *32*(4), 601–621.

Cichocka, A. (2016). Understanding Defensive and Secure In-group Positivity: The Role of Collective Narcissism. *European Review of Social Psychology*, 27, 283–317.

Cooper, J., & Avery, J. (2021). Value Framing and Support for Populist Propaganda. In: Forgas, J. P., Crano, W. D., & Fiedler, K. (eds.). *The Psychology of Populism: The Tribal Threat to Liberal Democracy* (pp. 319–331). New York: Psychology Press.

Clark, C. J. Liu, B. Winegard, B., & Ditto, P. (2019). Tribalism Is Human Nature. *Current Directions in Psychological Science*, *28*, 587–592.

Crano, W. D., & Gaffney, A. (2021). Social Psychological Contributions to the Study of Populism: Minority Influence and Leadership Processes in the Rise and Fall of Populist Movements. In: Forgas, J. P., Crano, W. D., & Fiedler, K. (eds.). *The Psychology of Populism: The Tribal Threat to Liberal Democracy* (pp. 158–173). New York: Psychology Press.

Csepeli, G. (2018). TedX Talks. 2018, December 20. *Új Hitel, Dr. Csepeli György, TEDxDanubia*. https://www.youtube.com/watch?v=Up9ry4AHbdY

Davies, W. (2019). *Nervous States: Democracy and the Decline of Reason* (First American edition. ed.). New York: W.W. Norton & Company.

Davison, A. (2020). A Darwinian Approach to Postmodern Critical Theory: Or How Did Bad Ideas Colonize the Academy? *Society, 57,* 417–424.
Ditto, P., & Rodriguez, C. G. (2021). Populism and the Social Psychology of Grievance. In: Forgas, J. P., Crano, W. D., & Fiedler, K. (eds.). *The Psychology of Populism: The Tribal Threat to Liberal Democracy* (pp. 23–41). New York: Psychology Press.
Doty, R. M., Peterson, B. E., & Winter, D. G. (1991). Threat and Authoritarianism in the United States, 1978–1987. *Journal of Personality and Social Psychology, 61*(4), 629–640.
Dunbar, R. I. M. (1998). The Social Brain Hypothesis. *Evolutionary Anthropology, 6,* 178–190.
Durkheim, E. (1964). *The Division of Labor in Society.* New York: Free Press.
Durkheim, E. (1966). *Suicide: A Study in Sociology.* London: routledge.
Eurostat. (2016). *Foreign Language Skills Statistics.* Retreived from: https://ec.europa.eu/eurostat/statistics-explained/index.php/Foreign_language_skills_statistics
Federico, C. M., & Golec de Zavala, A. (2018). Collective Narcissism and the 2016 United States Presidential Vote. *Public Opinion Quarterly, 82*(1), 110–121.
Feldman, S. (2021). Authoritarianism, Education, and Support for Right-Wing Populism. In: Forgas, J. P., Crano, W. D., & Fiedler, K. (eds.). *The Psychology of Populism: The Tribal Threat to Liberal Democracy* (pp. 348–364). New York: Psychology Press.
Feldman, S., & Stenner, K. (1997). Perceived Threat and Authoritarianism. *Political Psychology, 18*(4), 741–770.
Forgas, J. P. (2013). Don't Worry, Be Sad! On the Cognitive, Motivational, and Interpersonal Benefits of Negative Mood. *Current Directions in Psychological Science, 22*(3), 225–232.
Forgas, J. P., & Lantos, D. (2020). Understanding Populism: Collective Narcissism and the Collapse of Democracy in Hungary. In J. P. Forgas, W. D. Crano, & K. Fiedler (eds.). *Applications of Social Psychology: How Social Psychology Can Contribute to the Solution of Real-World Problems* (pp. 267–291). New York, NY: Routledge.
Forgas, J. P., & Lantos, D. (2021). When Populism Triumphs: From Democracy to Autocracy. In: Forgas, J. P., Crano, W. D., & Fiedler, K. (eds.). *The Psychology of Populism: The Tribal Threat to Liberal Democracy* (pp. 219–239). New York: Psychology Press.
Forgas, J. P., Kelemen, L., & László, J. (2014). Social cognition and democracy: An eastern European Case Study. In J. P. Forgas, B. Crano, & K. Fiedler (eds.). *Social Psychology and Politics.* New York: Psychology Press.
Forgas, J. P., Crano, W. D., & Fiedler, K. (eds.). *The Psychology of Populism: The Tribal Threat to Liberal Democracy.* New York: Psychology Press.
Freedom House. (2020). Nations in Transit 2020. Retrieved from https://freedomhouse.org/report/nations-transit/2020/dropping-democratic-facade
Friedrich Ebert Stiftung. (2020). *Az elmúlt tíz évtized a magyar társadalom szemével.* Budapest: Policy Solutions.
Frijters, P., Foster, G., & Baker, M. (2021). *The Great Covid Panic: What Happened, Why, and What To Do Next.* Austin, TX.: Brownstone Publications.
Fukuyama, F. (2022). *Liberalism and Its Discontents.* New York: Profile books.
Garton Ash, T. (2019). Europe Must Stop This Disgrace: Viktor Orban Is Dismantling Democracy. *The Guardian,* 20th June 2019. https://www.theguardian.com/commentisfree/2019/jun/20/viktor-orban-democracy-hungary-eu-funding

Gelfand, M., & Lorente, R. (2021). Threat, Tightness, and the Evolutionary Appeal of Populist Leaders. In: Forgas, J. P., Crano, W. D., & Fiedler, K. (eds.). *The Psychology of Populism: The Tribal Threat to Liberal Democracy* (pp. 297–318). New York: Psychology Press.

Gigerenzer, G., Todd, P. M., & Group, A. B. C. R. (1999). *Simple Heuristics That Make Us Smart*. New York: Oxford University Press.

Golec de Zavala, A., Dyduch-Hazar, K., & Lantos, D. (2019). Collective Narcissism: Political Consequences of Investing Self-Worth in the Ingroup's Image. *Political Psychology*, 40, 37–74.

Golec de Zavala, A. Lantos, D., & Keenan, O. (2021). Collective Narcissism and the Motivational Underpinnings of the Populist Backlash. In: Forgas, J. P., Crano, W. D., & Fiedler, K. (eds.). *The Psychology of Populism: The Tribal Threat to Liberal Democracy* (pp. 105–122). New York: Psychology Press.

Guriev, S., & Treisman, D. (2022). *Spin Dictators*. Princeton University Press. doi:10.1515/9780691224466. ISBN 9780691224466. S2CID 247903789.

Haidt, J. (2022). Why the Past 10 Years of American Life Have Been Uniquely Stupid. *The Atlantic*, April 11

Hamilton, D. L., & Rose, T. L. (1980). Illusory Correlation and the Maintenance of Stereotypic Beliefs. *Journal of Personality and Social Psychology*, 39(5), 832–845. https://doi.org/10.1037/0022-3514.39.5.832

Harari, Y. N. (2014). *Sapiens: A Brief History of Humankind*. London, UK: Random House.

Hobbes, T. (1968). *Leviathan*. Harmondsworth: Penguin.

Hogg, M. A. (2007). Uncertainty-identity Theory. *Advances in Experimental Social Psychology*, 39, 69–126. doi:10.1016/s0065-2601(06)39002-8

Hogg, M. A., & Gøtzsche-Astrup, O. (2021). Self-Uncertainty and Populism: Why we Endorse Populist Ideologies, Identify with Populist Groups, and Support Populist Leaders. In: Forgas, J. P., Crano, W. D., & Fiedler, K. (eds.). *The Psychology of Populism: The Tribal Threat to Liberal Democracy* (pp. 197–218). New York: Psychology Press.

Huddy, L., & Del Ponte, A. (2021). Rise of Populism in the USA: Nationalism, Race, and American Party Politics. In: Forgas, J. P., Crano, W. D., & Fiedler, K. (eds.). *The Psychology of Populism: The Tribal Threat to Liberal Democracy* (pp. 258–275). New York: Psychology Press.

Hume, D. (1984). *A Treatise of Human Nature*. London: Penguin Books.

International Institute for Democracy and Electoral Assistance. (2019). *The Global State of Democracy 2019: Addressing the Ills, Reviving the Promise*. https://www.idea.int/sites/default/files/publications/the-global-state-of-democracy-2019.pdf

Kahneman, D. (2011). *Thinking Fast and Slow*. Farrar, Straus and Giroux: New York.

Kelemen, L. (2010). *Miként vélekedjünk a jogról?* Budapest: Line Design. Muco66)%

Kelemen, L. Szabó, Zs. P. Mészáros, N. Zs. László, J., & Forgas, J. P. (2014). Social Cognition and Democracy. *Journal of Social and Political Psychology*, 2, 197–219.

Keller, T. (2010). Hungary on the world values map. *Review of Sociology*, 20, 27–51.

Koch, A. S., & Forgas, J. P. (2012). Feeling good and feeling truth: The interactive effects of mood and processing fluency on truth judgments. *Journal of Experimental Social Psychology*, 48(2), 481–485. https://doi.org/10.1016/j.jesp.2011.10.006

Krekó, P., & Enyedi, Z. (2018). Orbán's laboratory of illiberalism. *Journal of Democracy*, 29, pp. 39–51. General review of Orbán's methods.

Kreko, P. (2021). Populism in Power: The Tribal Challenge. In: Forgas, J. P., Crano, W. D., & Fiedler, K. (eds.). *The Psychology of Populism: The Tribal Threat to Liberal Democracy* (pp. 240–257). New York: Psychology Press.

Krueger, J., & Gruening, D. J. (2021). Psychological Perversities and Populism. In: Forgas, J. P., Crano, W. D., & Fiedler, K. (eds.). *The Psychology of Populism: The Tribal Threat to Liberal Democracy* (pp. 125–142). New York: Psychology Press.

Kruglanski, A. Molinario, E., & Sensales, G. (2021). Why Populism Attracts: On the Allure of Certainty and Dignity. In: Forgas, J. P., Crano, W. D., & Fiedler, K. (eds.). *The psychology of populism: the tribal threat to liberal democracy* (pp. 158–173). New York: Psychology Press.

Lantos, D., & Forgas, J. P. (2021). The Role of Collective Narcissism in Populist Attitudes and the Collapse of Democracy in Hungary. *Journal of Theoretical Social Psychology, 65*, 65–78.

László, J. (2014). *Historical Tales and National Identity*. London: Routledge.

László, J., & Ehmann, B. (2013). Narrative Social Psychology. In: J. P. Forgas, O. Vincze, & J. László (eds.). *Social Cognition and Communication*. New York: Psychology Press.

Lendvai, P. (2012). *Hungary: Between Democracy and Authoritarianism*. Columbia University Press.

Magyar, B. (2016). *Post-communist Mafia State*. Central European University Press.

Marchlewska, M., Cichocka, A., Panayiotou, O., Castellanos, K., & Batayneh, J. (2018). Populism as Identity Politics: Perceived In-Group Disadvantage, Collective Narcissism, and Support for Populism. *Social Psychological and Personality Science, 9*(2), 151–162. https://doi.org/10.1177/1948550617732393

Marcus, G. E. (2002). *The Sentimental Citizen: Emotion in Democratic Politics*. University Park, PA: Pennsylvania State University Press.

Marcus, G. E. (2021). The Rise of Populism: The Politics of Justice, Anger, and Grievance. In: Forgas, J. P., Crano, W. D., & Fiedler, K. (eds.). *The Psychology of Populism: The Tribal Threat to Liberal Democracy* (pp. 81–104). New York: Psychology Press.

Mercier, H., & Sperber, D. (2017). *The Enigma of Reason*. Cambridge, MA: Harvard University Press.

McCann, S. J. H. (1997). Threatening Times, "Strong" Presidential Popular Vote Winners, and the Victory Margin, 1924–1964. *Journal of Personality and Social Psychology, 73*(1), 160–170.

Myers, D. G. (2019). Psychological Science Meets a Post-truth World. In: J. P. Forgas & R. Baumeister (eds.). *Gullibility: Fake News, Conspiracy Theories and Irrational Beliefs* (pp. 77–101). New York: Psychology Press.

Neal, T. M. S., Lienert, P., Denne, E., & Singh, J. P. (2022). A General Model of Cognitive Bias in Human Judgment and Systematic Review Specific to Forensic Mental Health. *Law and Human Behavior, 46*(2), 99–120. http://dx.doi.org/10.1037/lhb0000482

Norris, P., & Inglehart, R. (2018). *Cultural Backlash: Trump, Brexit, and the Rise of Authoritarian-populism*. New York, NY: Cambridge University Press.

Oesch, D. (2008). Explaining Workers' Support for Right-Wing Populist Parties in Western Europe: Evidence from Austria, Belgium, France, Norway, and Switzerland. *International Political Science Review, 29*(3), 349–373.

Onraet, E., Alain, V. H., & Cornelis, I. (2013). Threat and Right-Wing Attitudes: A Cross-National Approach. *Political Psychology, 34*(5), 791–803.

Orbán, V. (2014). *Tusnádfürdői beszéd*, July 26th, 2014.

Petersen, M. B., Osmundsen, M., & Bor, A. (2021). Beyond Populism: The Psychology of Status-Seeking and Extreme Political Discontent. In: Forgas, J. P., Crano, W. D., & Fiedler, K. (eds.). *The Psychology of Populism: The Tribal Threat to Liberal Democracy* (pp. 62–80). New York: Psychology Press.

Pinker, S. (2018). *Enlightenment Now: The Case for Reason, Science, Humanism, and Progress.* New York, NY: Viking Penguin.

Plato. (1974). *The Republic.* Harmondsworth: Penguin.

Popper, K. (1945). *The Open Society and Its Enemies.* Routledge: London.

Riotta, C. (2019). *Trump 'Would Love to Have Orban's Situation in Hungary', US Ambassador Says.* Retreived from: https://www.independent.co.uk/news/world/americas/us-politics/trump-viktor-orban-hungary-david-cornstein-fidesz-illiberal-democracy-a8907246.html

Robin, C. (2004). *Fear: The History of a Political Idea.* New York: Oxford University Press.

Scruton, R. (2000). *The Meaning of Conservatism.* London: St. Augustine's Press.

Sustainable Governance Indicators. (2019). *Do Institutions Act in Accordance with the Law? Do They Check and Balance Each Other? Is Corruption Prevented?* Retreived from https://www.sgi-network.org/2019/Democracy/Quality_of_Democracy/Rule_of_Law

Szilágyi, I., & Kelemen, L. (2019). *Miként vélekedünk a jogról?* Budapest: HVGOrac.

Tajfel, H., & Forgas, J. P. (2000). Social Categorization: Cognitions, Values and Groups. In C. Stangor (ed.), *Key Readings in Social Psychology* (pp. 49–63). New York: Psychology Press.

Temelcuran, E. (2019). *How to Lose a Country.* New York: Harper Collins.

Toennies, F. (1956). *Community and Association.* London: Routledge.

Van Prooijen, J-W. (2019). Belief in Conspiracy Theories: Gullibility or Rational Scepticism? In: J. P. Forgas & R. Baumeister (eds.). *Gullibility: Fake News, Conspiracy Theories and Irrational Beliefs* (pp. 319–333). New York: Psychology Press.

von Hippel, W. (2018). *The Social Leap.* New York: Harper Collins.

Webster, D. M., & Kruglanski, A. W. (1994). Individual Differences in Need for Cognitive Closure. *Journal of Personality and Social Psychology,* 67(6), 1049–1062. https://doi.org/10.1037/0022-3514.67.6.1049

World Justice Project. (2019). *Rule of Law Index: 2019.* Washington: The World Justice Project.

Zimbardo, P. (1977). *Shyness: What It Is and What To Do about It.* New York: Da Capo Lifelong Books.

18
UNCERTAINTY, ACADEMIC RADICALIZATION, AND THE EROSION OF SOCIAL SCIENCE CREDIBILITY

Lee Jussim, Danica Finkelstein, and Sean T. Stevens

RUTGERS UNIVERSITY, NEW JERSEY, USA

Abstract

This chapter discusses the role of uncertainty and distress in the radicalization of the social sciences, including social psychology and the consequences of that radicalization. Political extremism often emerges as a response to uncertainty. Ironically, extremism often constitutes an embrace of simplistic dogmas and myopic certainties about answers to complex questions. After reviewing these processes, how they might apply to and explain the radicalization of the American social sciences is discussed. Evidence is reviewed documenting the extent to which such radicalization has occurred. The chapter then reviews evidence regarding ways in which ideological extremism can threaten the validity of some of the most highly canonized conclusions in the social sciences. We conclude the chapter by specifying some clearly falsifiable hypotheses that have emerged from this perspective, including the prediction that the continuing radicalization of the social sciences will erode public trust and undermine the credibility given to the claims that emerge on politicized topics.

Extremism can help mitigate uncertainty through the embrace of simplistic dogmas (see also Hogg & Gaffney; van den Bos, this volume). Ironically then, uncertainty can produce a particularly poisonous form of certainty. Whether it is fascists, White supremacists, Marxists, or anti-fascists, extremists are often characterized by dogmatic certainties about truth and morality that brooks no dissent. This psychological witch's brew, a key component of authoritarian movements everywhere, renders extremists particularly vulnerable to delusional scientific beliefs (e.g., the purging of "Jewish" science from Nazi

DOI: 10.4324/9781003317623-22

Germany and the embrace of Lysenkoism in the Soviet Union), terrible policy decisions, and violations of human rights, including large-scale mass murder.

This chapter presents a perspective on the role of uncertainty and distress in the radicalization of the social sciences. We restrict our discussion to the U.S. because we are most familiar with it. This review documents the extent to which the radicalization of the social sciences has occurred and provides evidence regarding ways in which this can threaten the validity of conclusions on politicized topics. We conclude the chapter by specifying some falsifiable hypotheses that have emerged from this perspective.

The Psychology of Uncertainty and Political Extremism

People can feel uncertain about many things such as their beliefs, values, relationships, careers, their future, and their place in the world. This can make it difficult to plan for the future and execute decisions, so people can become highly motivated to reduce uncertainty. In this section, we review the evidence showing that uncertainty can prompt a path of radicalization leading to extremist behaviors. To be clear, not all uncertainty leads to extremism, and uncertainty is not the only cause of extremism. Researchers have demonstrated that the path to extremism satisfies a multifaceted palette of needs, such as those for control (Hogg & Adelman, 2013) and belonging (Leary, Twenge & Quinlivanschiml., 2006; see also Hirschberger; Mikulincer & Shaver, this volume). It can also be a response to perceived unfairness or injustice (Moghaddam, 2005). Thus, processes producing extremism but unrelated to uncertainty are not the focus of the present chapter.

How Does Uncertainty Lead to Extremism?

Negative life events often produce feelings of insignificance, humiliation, or helplessness and create uncertainty. These events can generate distrust in institutions and elites, motivate tribal affiliations, and increase vigilance and intergroup animosity (van den Bos; van Prooijen; Kreko; Kruglanski & Ellenberg; Hirschberger; Hogg & Gaffney, this volume). Kruglanski and colleagues outlined a radicalization model that identified an individual's extremist behaviors stemming from a need to rectify a loss of significance in their life, now known as Significance Quest Theory (SQT). The quest for significance is a means to relieve the turmoil and distress that result from uncertainty.

Research examining SQT has found that perpetrators of ideologically motivated violence often had previously experienced significant economic and social loss (Jasko et al., 2017). Consequently, this triggers a quest to regain what has been lost and, as the environment becomes more complex, increases the demand on cognitive resources to mitigate uncertainty. Because people are limited in their processing capacities, they often apply shortcuts to arrive at acceptable solutions to loss of significance and regain a feeling of satisfaction or closure (Kruglanski et al., 2012). These shortcuts reflect a need for simplicity and can render individuals vulnerable to the simple answers provided by extremist narratives. For instance, when individuals are uncertain, they tend to seek out like-minded groups (Fiske & Taylor, 2008; Forgas, this volume; Hogg & Gaffney, this volume; Kruglanski & Ellenberg, this volume), have a preference for strong autocratic leadership (Rast et al., 2013), and more strongly identify with radical groups, increasing their intentions to behave in more extreme, active, and mobilized ways to protect and promote ingroups (Hogg et al., 2010).

This concoction of extremist beliefs, identities, and attitudes can cause a **myopic effect** as people become overly committed to the focal goal and the suppression of alternative considerations (Shah et al., 2002). This commitment is a key to radicalization because it identifies specific activities (e.g., crushing one's opponents) as the means to enhance personal significance (Shah et al., 2002). This myopic effect can be seen in a recent analysis of *monomania* (Haidt, 2021), defined as an unhealthy obsession with one thing. That one thing, in far-right circles, is often nationalism, and in far-left circles, is often how the powerful create and sustain injustice.

Regardless of what they are obsessed over, left- and right-wing extremists consider their political beliefs to be absolute and unquestionable on a range of topics, including health care, immigration, and affirmative action, compared with moderates (Toner et al., 2013). As a result, they are overconfident in their beliefs (van Prooijen et al., 2018) and despise their opponents (Iyengar et al., 2019; Westfall et al., 2015). Extremists provide simplistic answers to complex social and political problems. This is consistent with the psychology of the uncertain mind: a need to manage distress results in simple and certain solutions at the cost of accuracy. In short, extremists on both sides of the political spectrum engage in authoritarian behavior (Altemeyer, 1981; Costello et al., 2022; van Prooijen, this volume).

According to the model upon which this review is based, uncertainty about issues fundamental to self and identity, as well as uncertainty leading to loss of significance, is disturbing and motivates people to alleviate the distress (Forgas, this volume; Hogg & Gaffney, this volume; Kruglanski & Ellenberg, this volume). Thus, uncertainty leads to radicalization, dogmatism, and monomania. Later, we present evidence suggesting that such processes can be disastrous for the supposed truth-seeking commitments of social science.

Uncertainty and Extremism in the Academy

In this section, we first review empirical evidence for the radicalization of the American social sciences, focusing on survey research and real-world events. The second portion of this review is speculative. Given the wealth of evidence documenting a link between uncertainty and extremism, we apply the insights gleaned from that work to generate testable explanations for how uncertainty contributed to the leftward shift in the American social sciences and humanities.

The Politics of Academia Compared to the American Mainstream

The majority of college students and faculty identify as politically left-of-center (Gross, 2014; Honeycutt & Freberg, 2017; Kaufmann, 2021; Stevens, 2022; Stevens & Schwictenberg, 2020, 2021). The social sciences and humanities skew even more left, as confirmed by multiple methodologies (Buss & von Hippel, 2018; Honeycutt & Freberg, 2017; Langbert & Stevens, 2021), and some surveys show that roughly 40% of social science and humanities professors self-describe as activists, radicals, or Marxists (Gross & Simmons, 2014; Honeycutt, 2022; Kaufmann, 2021). The general public is a bit more moderate. According to Gallup, in January 2022, 37% of Americans identified as conservative, 36% as moderate, and 25% as liberal. In 2011, Pew Research Center found that the "progressive left" – very liberal, highly educated, and majority White – made up roughly 6% of the American general public and 7% of all registered voters. Donald Trump received almost 75 million votes in 2020, more than the almost 63 million he received in 2016. So, suffice it to say, what is politically "normal" in the academy is not "normal" outside of it.

The Radicalization of the Social Sciences and Humanities

The academy in general and the social sciences and humanities in particular did not always skew so left, although they have all long had a left-wing tilt (see Duarte et al., 2015). Since the 1990s, however, the number of professors identifying as liberal or far left has increased, while the number identifying as conservative has decreased. The number of professors identifying as far right has held roughly steady at about one half of one percent (Stoltzenberg, Eagan, Zimmerman, Berdan Lozano, Cesar-Davis, Aragon & Rios-Aguilar, 2019). This trend will likely continue as younger faculty and current graduate students are more liberal than their older counterparts (Kaufmann, 2021). This kind of environment is ripe for a spiral of silence (Noelle-Neumann, 1974), where minority viewpoints are expressed less and less often over time and come to be seen as "fringe" or "extreme." Finally, they are only expressed by their staunchest believers, who are typically perceived as extremists.

Figure 18.1 presents the Activist to Academia to Activism Pipeline Model of Academic Self-Radicalization. Although a full review of the evidence for the model can be found elsewhere (Honeycutt & Jussim, in press), we briefly summarize its main features next.

Step 1 captures the now-abundant evidence that activists, radicals, and extremists are overrepresented in academia. In Step 2, a hostile work

Uncertainty, Academic Radicalization, and the Erosion of Social Science Credibility **333**

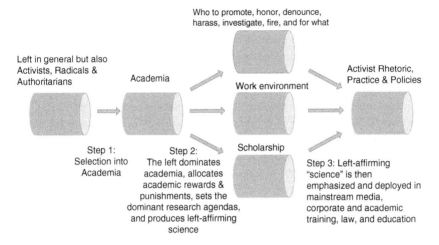

FIGURE 18.1 Activist to academia to activism pipeline model of radicalization.

environment is created not merely for political opponents but for almost any academic who expresses dissent against prevailing orthodoxies and shibboleths. Even after controlling for measures of actual academic achievement (such as publication), faculty holding more left positions on social attitudes end up at more prestigious institutions (Rothman & Lichter, 2009). Faculty (regardless of their personal politics) are vastly more likely to be derogated for expressing views skeptical of diversity, equity, and inclusion initiatives, arguing for merit-based policies rather than affirmative action, or views critical of the concepts commonly used in academia to craft narratives about the power and oppression (implicit bias, microaggressions, stereotype threat, systemic racism, White supremacy, etc.) than for expressing support for those initiatives or views (German & Stevens, 2021, 2022; Honeycutt & Jussim, in press).

Substantial numbers of academics *endorse* discriminating against their political opponents (e.g., Honeycutt & Freberg, 2017; Inbar & Lammers, 2012). Given the base rates of faculty political affiliation (Gross & Simmons, 2014; Honeycutt, 2022; Kaufmann, 2021) and spiral of silence theory (Noelle-Neumann, 1974), we propose the *academic political purity spiral hypothesis:* An academic field will move ever further to the left because of groupthink and political ingroup favoritism until either saturation is reached or some external factor disrupts the process. Consistent with the academic political purity spiral hypothesis, recent surveys of over 1,800 faculty and nearly 4,000 graduate students from over 350 U.S. colleges and universities from across the disciplines found that 40% of faculty and 58% of graduate students self-identify as radicals, activists, Marxists, socialists, or some combination of all four (Honeycutt, 2022).

One can also find evidence of censorship and aggression directed toward political opponents among faculty members. Surveys and real-world reports both document academics seeking to punish their colleagues for transgressing against left shibboleths (see Stevens et al., 2020 for a review; see also German & Stevens, 2021, 2022). Punishing apostates has occurred through much of human history (see Mchangama, 2022), and in modern American parlance, these are often referred to as "cancellation" attacks. For example, Kaufmann (2021) found that substantial minorities of faculty and nearly half of graduate students in social science and humanities fields *endorse* ousting academics who report research findings that challenge cherished left-wing beliefs, such as that diversity is not a net benefit, non-traditional parenthood is actually worse for children, or that women and minorities actually perform worse in some work or school context.

Far more liberals than conservatives abandon friendships over political differences (Cox, 2021), and academic outrage mobs are far more likely to call for punishing someone who violates left-wing values and beliefs (German & Stevens, 2021, 2022; Honeycutt & Jussim, in press; Stevens et al., 2020). Academic outrage mobs tend to rise up from positions of less authority in a chorus of denunciation of their target in order to get that target punished (deplatformed, disinvited, investigated, suspended, or fired). Mobs seeking to punish their ideological opponents are a classic manifestation of authoritarianism. For example, academic outrage mobs have sought to get papers retracted that have contested the sensibilities of transgender activists or criticized affirmative action, and they have sought to get faculty fired for expressing skepticism about microaggressions or publishing a book arguing that there was no evidence for a polar bear crisis resulting from climate change (Jussim, 2020; Stevens et al., 2020).

The Foundation for Individual Rights and Expression (FIRE) maintains a Scholars Under Fire database (see also German & Stevens, 2021, 2022). A scholar is deemed to be targeted for sanction when there is "…a campus controversy involving efforts to investigate, penalize, or otherwise *professionally* sanction a scholar for engaging in constitutionally protected forms of speech" (German & Stevens, 2021, p. 6). In 2021 alone, FIRE tracked over 200 such targeting incidents, with almost half involving scholars targeted from the left. Given what is known about faculty political beliefs, it is likely that most of these scholars were also on the political left, but just not as far left as the people targeting them. FIRE's database provides ample evidence of cancellation attacks from the right (more than half), but these are almost entirely from outside of academia rather than from inside, and, as such, are not relevant to this chapter on extremism *within* academia. Nearly all the attacks from the left were from *within* the academy. We conclude that there is converging evidence from surveys, news reports, and FIRE's database to indicate a substantial pattern of authoritarianism in the American academy.

Speculations on the Role of Uncertainty in the American Academy's Shift to the Left

The social sciences and humanities have long focused on understanding prejudice, oppression, and inequality (Adorno et al., 1950s; Allport, 1954/1979; La Piere, 1934). However, in the 1950s and 1960s, most of the legal structure supporting racial discrimination (school segregation, Jim Crow, redlining) was repealed and prohibited, and sweeping civil rights laws were passed. Nonetheless, few racial and ethnic inequalities have been eradicated (USA Today, 2020).

This is where uncertainty comes in. There are three broad classes of potential explanations for disparities: Current discrimination, historical discrimination, and something about the groups themselves. The explanatory power of current discrimination is obvious. However, past discrimination can also produce inequality in the absence of present discrimination by creating enduring socio-economic differences. For example, relatively few Black American soldiers received the benefits of the 1944 G.I. Bill (Ibrahim, 2021) to attend college and reduced rate mortgages. If poverty motivates some crime, then the poverty induced by historical discrimination can also create disparities in imprisonment. Past discrimination can cause present disparities, even in the absence of present discrimination.

Last, there is the possibility that something about the groups causes inequality. At least since the publication of *Blaming the Victim* (Ryan, 1972), even raising the possibility that something about groups may cause inequality has inspired outrage (e.g., Jussim, 2019). Nonetheless, *some* disparities are not readily attributable to either present or historical discrimination. For example, Asian Americans have higher education and income levels than do any other racial/ethnic group in the U.S. Census (Guzman, 2017; Ryan & Bauman, 2016). Yet no one has argued that America is an "Asian Supremacy." Indeed, America has a history of anti-Asian discrimination, and the spike in anti-Asian hate crimes during the COVID-19 pandemic indicated that current discrimination against Asians still occurs (d for the Study of Hate & Extremism, 2020). Nonetheless, neither present nor past anti-Asian discrimination can explain the *higher* educational and income status of Asians. The idea that *there are no differences between different cultures* is absurd. If there are differences between cultures, then it is reasonable to think that some differences in characteristics, behaviors, and norms between different cultures may produce different outcomes (Henrich, Heine, & Norenzayan, 2010).

Although outcome disparities may result from current and past discrimination and group differences, research has rarely compared the three explanations against one another. Indeed, given the risk of denunciation (e.g., German & Stevens, 2021, 2022), it is likely that research comparing discrimination accounts has been suppressed (Honeycutt & Jussim, in press; Stevens et al., 2020; Zigerell, 2018). Thus, claims about what causes observed disparities are unlikely to come from sound research and often show a "flagrant disregard for truth" (Frankfurt, 2005), ignoring potential alternatives. In other words, the nature and sources of the inequalities that academics care about are often not scientifically well established.

Uncertainty and Distress-Induced Authoritarianism: The Rise of Cancellation Attacks in the Aftermath of George Floyd

George Floyd's murder heightened general distress over racism and injustice in America, especially among those on the left, who, according to some studies, experienced ambient distress over these issues long before Floyd's murder (Napier & Jost, 2008). Core characteristics of authoritarianism include efforts to punish one's political opponents and deprive them of basic human rights. (Because we are discussing the social justice protests and the radicalization of the academy, we refer readers to others who have reviewed right-wing authoritarianism (e.g., Altemeyer, 1981).) Thus, we speculate that the stress and uncertainty produced by Floyd's murder manifested as increased extremism, intolerance, and authoritarianism in the academy.

Evidence of an increase in intolerant, hostile, censorious behavior post-Floyd in the academy is observable in the wave of cancellation attacks since the summer of 2020. For example, FIRE's Scholars Under Fire database currently identifies 414 scholars who are subject to cancellation attacks from their left between 2015 and 2022. Table 18.1 below only shows data through 2021, although we note that as of September 30, 2022, there have already been 41 attempts to sanction scholars from their left. Four things are clear: (1) Cancellation attacks have been generally increasing; (2) There was a spike in attempts from the left during 2020 that continued through 2021; (3) Attempts to sanction scholars from within the academy (other scholars and graduate students)

TABLE 18.1 Cancellation attacks from within academia in FIRE's scholars under fire database

Year	Scholars Under Fire from their left and targeted by other scholars	Scholars Under Fire from their right and targeted by other scholars	Scholars Under Fire from their left and targeted by graduate students	Scholars Under Fire from their right and targeted by graduate students
2015	4	0	8	0
2016	4	0	3	0
2017	10	2	6	0
2018	12	2	6	1
2019	11	1	5	0
2020	34	2	13	0
2021	34	7	22	0

have also increased; and (4) Attacks from the right within the academy have also occurred, but at less than 10% the frequency of attacks from the left.

In addition to cancellation attacks, there has also been a rise in suppressing views that oppose far-left ideology. An excellent example of the left-authoritarian response to Floyd's murder was the Princeton Faculty Letter (2020), which calls for an egregious abrogation of academic freedom and investigations of "racist behaviors, incidents, research, and publication on the part of faculty." This Orwellian call for Big Brother-like oversight of faculty intellectual activities sets a new standard for limiting expression. The fact that so many faculty, graduate students, and alumni signed an open letter advocating limits to free speech and expression documents the prevalence of the radical academic left at Princeton.

How Radicalization and Political Activism Corrupts Scholarship

Detecting Politically Biased Social Science

Political biases have no effect on nonpolitical scientific topics. They can only distort science on politicized topics (Crawford & Jussim, 2018; Duarte et al., 2015; Honeycutt & Jussim, 2020, in press; Jussim et al., 2015; Martin, 2016; Redding, 2013, in press; Zigerell, 2019). Of course, just because some conclusions vindicate an ideological narrative does not mean the work is biased. Just as one can reach a valid conclusion regarding whether judgments are racially or gender biased (e.g., Darley & Gross, 1983; Fiske & Neuberg, 1990), one can also determine whether a paper or an area of research is politically biased.

Tests for Political Bias

The social sciences are overwhelmingly populated by people on the left, so the tests we describe next focus on left-wing bias. Nonetheless, they could also be used to test for right-wing biases. Below we describe four tests that can be used to determine whether a claim is politically biased, with examples to illustrate how the tests operate.

Test 0: Does the study vindicate some left narrative? Test 0 is necessary because if the claim does not vindicate a left narrative, it cannot possibly be left-biased. It is not sufficient because the claim or study may be valid. To answer this question, one needs to be able to identify major left narratives. Although there are others, here are some common ones:

1 Equalitarianism (Clark & Winegard, 2020), which includes denying biological differences between groups, claiming that prejudice and discrimination are the only sources of group differences and claiming that society has a moral obligation to arrange that all groups are equal on socially valued outcomes.

2 Claims advancing "social justice," or the programs designed to increase it (such as diversity, equity, and inclusion programs, diversity training, implicit bias training, etc.).
3 Claims that liberals are more competent, informed, and morally superior to conservatives.
4 Attitudes supporting environmentalism are good.

Many more examples can be found in the broader literature on manifestations of political biases in social science (e.g., Clark & Winegard, 2020; Crawford & Jussim, 2018; Duarte et al., 2015; Honeycutt & Jussim, 2020, in press; Martin, 2016; Redding, 2013, in press; Zigerell, 2019).

Test 1: Did they misinterpret or misrepresent their results in ways that unjustifiably vindicate a left narrative? If the results show X and the claims based on the study are left of X, then the paper is politically biased. If the results are muddled or mixed and the conclusions emphasize left-affirming narratives, then the paper is politically biased. A common example is when a demographic gap is interpreted as reflecting discrimination, ignoring other alternatives, including group differences in behavior or preferences. A famous gender discrimination lawsuit in graduate admissions in the 1970s failed because, although it was true that women were admitted at lower rates than were men, this occurred because women disproportionately applied to programs with lower admissions rates (Bickel et al., 1975).

Test 2: Do the authors systematically ignore papers and studies inconsistent with their left-affirming conclusions? One can usually present a "compelling narrative" simply by ignoring results that disconfirm that narrative. A paper may appear "scientific," having all the trappings and form of science – but may not deserve to be called "science" if it ignores evidence inconsistent with its narrative (e.g., Gelman, 2017; Schimmack, 2021). One example is work on gender bias in peer review. The left position is to "smash the patriarchy" by exposing the power of biases favoring men. But what happens when there is evidence of bias favoring men, favoring women, or no bias?

Honeycutt & Jussim (2020) evaluated whether claims about gender bias in peer review qualify as science or pseudoscience. They found that papers showing gender bias against women had far smaller sample sizes (and thus smaller reliability) yet were cited far more frequently than papers reporting no bias or bias against men that had far greater sample sizes (and reliability). Sample size is one of the main markers of more credible research (Fraley & Vazire, 2014). The irony here is that the literature on gender bias in peer review cites less credible papers more than four times as often as those that, by scientific standards, are more credible.

Test 3: Leaping to left-affirming conclusions based on weak data. Sometimes, results actually do vindicate a left narrative but should be considered tentative and preliminary (e.g., if based on small samples or results emerging from a single research lab or group). If the conclusions are expressed with triumphant certainty, the paper is politically biased. The poster child for this is implicit bias. The Implicit association test (IAT) is the workhorse method for assessing implicit bias and has been used in thousands of studies. If claims based on the IAT are often of dubious credibility, then much of the field of "implicit bias" has feet of clay (see Jussim et al., in press, for a review of those criticisms; see the list of over 40 such articles (with new additions added regularly) critical of the IAT and implicit bias in this online repository (Jussim et al., 2022)). In brief, claims based on the IAT labeling 70% of Americans or more as implicit racists are wrong; the IAT, long claimed to measure "unconscious" racism, does no such thing; it is riddled with measurement error and statistical artifacts; it has weak predictive validity; changing IAT scores has no effect on discrimination; its effect sizes are overstated; and it has never been shown to cause any gap (see Jussim et al., in press, for a review).

Future Directions: Some Falsifiable Hypotheses

This analysis argued that the academic social sciences have *already radicalized*, so the hypotheses below would be confirmed. If uncertainty/distress increase radicalization, then these political biases should *increase* after events that increase feelings of distress. Such events need to be widely publicized to undercut what Smith (2014) has referred to as "sociology's sacred mission", "…exposing, protesting, and ending through social movements, state regulations, and government programs all human inequality, oppression, exploitation, suffering, injustice, poverty, discrimination, exclusion, hierarchy, constraint, and domination by, of, and over other humans" (Smith, 2014, p.6). The analysis suggests much of social psychology shares that sacred mission. Smith (2014, p.4) seems to believe so, too: "…What I describe here about sociology is obviously also embedded in the intellectual and moral culture of American higher education and elite, knowledge-class culture more broadly."

Smith's (2014) analysis helps inform what types of events might increase social science radicalization in the U.S. They would be events that contravene this sacred mission, events that are widely seen as increasing inequality, oppression, injustice, etc. Such events would include elections of Republican Presidents (e.g., Donald Trump); appointment of conservative Supreme Court justices (e.g., Neil Gorsuch, Brett Kavanaugh, Amy Coney Barret); activities of far-right extremists (e.g., the events of January 6th); and news emblematic of oppression of sacred victim groups (e.g., the George Floyd killing). Our prediction is that, *if* such an event occurs, it will likely increase academic radicalization.

These are to be distinguished from "negative" predictions, which constitute phenomena expected *not* to occur and which, if they did occur, would also falsify the theoretical perspective at least in part. A good theory not only generates hypotheses about what will happen, but it also excludes certain possibilities (e.g., Roberts & Pashler, 2000). Therefore, we present both positive and negative predictions. As with most social psychological mini theories, although no single disconfirmation would falsify the entire theory, consistent disconfirmation would.

Falsifiable Positive Predictions

Citation bias. Citation biases have been called "unscientific" (Schimmack, 2021) because systematically ignoring research that contests one's preferred narrative or conclusions corrupts the purpose of scientific research. This requires acknowledging uncertainty. Citation biases occur for all sorts of reasons and are not restricted to politics (see, e.g., De Vries et al., 2018, for an example involving interventions for depression). The radicalization of academia perspective, however, predicts that when new, otherwise similar studies (in topic and methods) are published either vindicating or contesting left narratives, the vindicating studies will be cited at much higher rates than those that contest those narratives. This will hold true even when other aspects of the publications are held constant (such as outlet, impact factors, methodological quality factors such as sample size/sample representativeness/number of replications/consistency across different methods, etc.). As most papers are mostly ignored, this hypothesis becomes disconfirmable only when two papers on the same left-hot button topic are published with conflicting findings, and at least one is cited at least 50 times. In such cases, the hypothesis is that most of the papers with left-affirming findings will be cited at higher rates than those that contest left narratives.

Trapped priors. A trapped prior is a belief or expectation (in the Bayesian sense) that cannot be updated, no matter how much data or how high quality the data that conflicts with it (Siskind, 2021). The trapped prior concept can be exploited to generate falsifiable hypotheses regarding the extent to which dogmatic certainties have corrupted the scientific mission of the social sciences. One can test for the presence of trapped priors by providing evidence contesting left-hot button issues to social scientists and then evaluating the extent to which it changes their prior. For example, how many social scientists who believe stereotypes are inaccurate will have that belief changed by the actual evidence of moderate to high levels of stereotype accuracy from over 50 studies demonstrating it (Jussim, Crawford & Rubinstein, 2015)? Can social scientists who believe stereotypes are inaccurate identify *any data* that would lead them to believe, or publicly admit, that the stereotypes people actually endorse are often not inaccurate? One could ask researchers whether there is any point (5 studies? 10? 100?) at which they would agree that something they hold sacred to be true (Smith, 2014) is actually false (inaccuracy of stereotypes, power of implicit bias, racism in the present causing most racial inequality, LGBTQ parenting being just as good as heterosexual parenting, affirmative action being a good solution to inequality, etc.).

Cancelation attacks. The present perspective also predicts that denunciation, ostracism, and punishment of scholars by other scholars will be primarily from the left for violating left shibboleth's and Smithian (2014) sacred missions. These would manifest as firings, suspensions, loss of positions, investigations, and forced retractions of articles triggered by academic outrage mobs who fail to identify data fraud or unusually high levels of data errors or irregularities. The failure to identify data errors is key here because articles are justifiably retracted when the underlying data are shown to be fraudulent or so riddled with errors as to lose all credibility.

Declining credibility among the public. We doubt that the public is paying rapt attention to the day-to-day events inside academia. However, people have a reasonably good intuitive sense that fields with large minorities (or more) of radicals, activists, and Marxists (Honeycutt, 2022) do not deserve the same credibility ascribed to fields studying apolitical topics or with better representation from across the political spectrum. For example, Marietta & Barker (2019) found that not only do Republicans distrust academia more than do Democrats, but regardless of personal politics, the more people viewed academia as left-skewed, the less they trusted it. Thus, another prediction is that the credibility the public ascribes to the social sciences will decline until these political trends reverse or practices emerge to limit unjustified claims on politicized topics.

Falsifiable Negative Predictions

Conservative underrepresentation. Conservatives are one of the most underrepresented groups in social psychology (Garcia et al., 2019). We use "underrepresented" in its descriptive sense to mean "represented in numbers lower than in the population." For example, the Black membership of The Society for Personality and Social Psychology (SPSP) is only about 3%. Compared to the Black portion of the U.S. population (U.S. Census, 2021), Black people are underrepresented at SPSP by over 75%. By comparison, conservatives constitute about 4% of the SPSP membership (Garcia et al., 2019) and about 36% of the U.S. population (Saad, 2022). Thus, conservatives are underrepresented by almost 90% in SPSP.

Nonetheless, the present perspective predicts that professional social psychological associations (e.g., SPSP, SPSSI, and SESP) will not acknowledge this state of affairs in these terms. Specifically, none will publicly describe conservatives as "underrepresented." In modern academic parlance, "underrepresented" is generally used *prescriptively*, as a rhetorical springboard to justify directing extra initiatives and resources toward including such groups as the progressive left considers protected or oppressed. Because our perspective predicts a process of increasing left-wing radicalization and extremism, because extremists usually despise their political opponents, and because "underrepresented" is a term suggesting more support is required, characterizing conservatives as "underrepresented" in this sense is specifically predicted not to occur among the major social science organizations. Should this change, i.e., should one of the main professional organizations in social psychology start even *referring* to conservatives as "underrepresented" as if it were a problem requiring attention, it would disconfirm this prediction.

Male underrepresentation. The 2019 SPSP diversity survey found that men are now underrepresented (36.5%) in the SPSP (Garcia et al., 2019), a pattern that gets more extreme as the stage of their career becomes earlier. As of 2019, men made up 20% of undergraduate members, 33% of graduate students, 41% of associate professors, and 47% of full professors (SPSP, 2019). The present perspective predicts that SPSP will not propose initiatives to increase the representation of men, who are not usually considered oppressed or "underrepresented" by progressives. This prediction would be disconfirmed if, for example, SPSP officers started raising alarms about male underrepresentation and/or started creating initiatives to stem the drain of men from the field.

Conclusions

In this chapter, we have developed a perspective on ways in which uncertainty and distress have contributed to the radicalization of the social science professoriate and how that has undermined some of its scholarship on politicized topics. As a theoretical perspective, we readily acknowledge that there is no data (yet) directly establishing some of the phenomena proposed herein. For example, there is no data directly linking uncertainty/distress to academic radicalization; and there have been no studies assessing the prevalence of left-wing authoritarianism in the academy.

Nonetheless, we did review and summarize some of the key conclusions in the now-extensive literature linking uncertainty and distress to political extremism. These links seem quite well established by sound empirical social science.

Second, we reviewed evidence of the radicalization of the social science professoriate. The far left (self-described radicals, activists, Marxists, and socialists) is massively overrepresented among social science professors. We speculated on the presence of an inordinate number of authoritarians among those professors, primarily on the basis of three well-established facts: (1) People high in left-wing authoritarianism actively endorse censoring and punishing their political opponents; (2) Academics actually attempting to censor and punish their opponents has been on the rise for some time now; and (3) There was a spike in such efforts shortly after George Floyd's highly publicized and disturbing murder, a pattern consistent with the uncertainty/distress linkage to extremism.

Third, we reviewed some of the ways that common, highly touted, and canonized claims have gone wrong, consistently in such a manner as to vindicate left narratives without sufficient evidence that justifies doing so. Although for brevity we merely *illustrated* this problem with examples about gender and implicit biases, many more examples can be found in various reviews (Crawford & Jussim, 2018; Duarte et al., 2015; Honeycutt & Jussim, 2020, in press; Jussim et al., 2015; Martin, 2016; Redding, 2013, in press; Zigerell, 2019).

Fourth, we have generated clear, falsifiable hypotheses about future behavior and practices in the social sciences. We look forward to empirical tests of those hypotheses. It is possible that the social sciences will stem the tide of increasing political radicalization. It is more likely that this will get more extreme before it moderates. Whereas social science as a club for progressives exploiting scientific skills to advance political agendas will work well for those inside the club (it probably will facilitate getting papers published, grants funded, awards, promotions, and jobs), it risks severely undercutting the credibility of the social sciences outside of progressive circles and among the wider public. But this may be a price many see as worth paying in order to have a clearer and easier path toward professional accomplishments such as tenure and promotions. It is also easier to advance progressive political goals on the basis of peer-reviewed social science if there are few social scientists willing to contest those goals by debunking unjustified claims and skeptically evaluating dubious research. A culture of uniformity of values, open hostility to opponents, and exclusion is, for far-left academics, a professional-win-political-win situation.

Acknowledgment

This chapter was completed with support from the Institute for Humane Studies.

References

Adorno, T. W., Frenkel-Brunswik, E., Levinson, D., & Sanford, R. N. (1950). *The authoritarian personality*. New York: Harper & Row.

Allport, G. W. (1954/1979). *The nature of prejudice* (2nd edn.). Cambridge, MA: Perseus.

Altemeyer, B. (1981). *Right-wing authoritarianism*. Winnipeg: University of Manitoba Press.

Bickel, P. J., Hammel, E. A., & O'Connell, J. W. (1975). Sex bias in graduate admissions: Data from Berkeley: Measuring bias is harder than is usually assumed, and the evidence is sometimes contrary to expectation. *Science*, *187*(4175), 398–404.

Buss, D. M., & Von Hippel, W. (2018). Psychological barriers to evolutionary psychology: Ideological bias and coalitional adaptations. *Archives of Scientific Psychology*, *6*(1), 148.

Center for the Study of Hate & Extremism (2020). Fact Sheet: Anti-Asian prejudice March 2020. Retrieved 3/6/22 from: https://www.csusb.edu/sites/default/files/FACT%20SHEET-%20Anti-Asian%20Hate%202020%203.2.21.pdf

Clark, C. J., & Winegard, B. M. (2020). Tribalism in war and peace: The nature and evolution of ideological epistemology and its significance for modern social science. *Psychological Inquiry*, *31*(1), 1–22.

Costello, T. H., Bowes, S. M., Stevens, S. T., Waldman, I. D., Tasimi, A., & Lilienfeld, S. O. (2022). Clarifying the structure and nature of left-wing authoritarianism. *Journal of Personality and Social Psychology*, *122*(1), 135.

Cox, D. (2021, June 8). The state of American friendship: Change, challenges, and loss. The Survey Center on American Life. https://www.americansurveycenter.org/research/the-state-of-american-friendship-change-challenges-and-loss/

Crawford, J. T., & Jussim, L. J. (Eds.). (2018). *The politics of social psychology*. New York: Psychology Press.

Darley, J. M., & Gross, P. H. (1983). A hypothesis-confirming bias in labeling effects. *Journal of Personality and Social Psychology*, *44*(1), 20.

De Vries, Y. A., Roest, A. M., de Jonge, P., Cuijpers, P., Munafo, M. R., & Bastiaansen, J. A. (2018). The cumulative effect of reporting and citation biases on the apparent efficacy of treatments: The case of depression. *Psychological Medicine*, *48*, 2453–2455.

Duarte, J. L., Crawford, J. T., Stern, C., Haidt, J., Jussim, L., & Tetlock, P. E. (2015). Political diversity will improve social psychological science 1. *Behavioral and Brain Sciences*, *38*.

Fiske, S. T., & Neuberg, S. L. (1990). A continuum of impression formation, from category-based to individuating processes: Influences of information and motivation on attention and interpretation. In M. P. Zanna (Ed.), *Advances in experimental social psychology* (Vol. 23, pp. 1–74). San Diego, CA: Academic Press.

Fiske, S. T., & Taylor, S. E. (2008). *Social cognition: From brains to culture*. New York: McGraw-Hill.

Fromm, E. (1947). *Man for himself: An inquiry into the psychology of ethics*. New York: Rinehart.

Gelfand, M. J., LaFree, G., Fahey, S., & Feinberg, E. (2013). Culture and extremism. *Journal of Social Issues*, *69*, 495–517.

Frankfurt, H. (2005). *On Bullshit*. Princeton, NJ: Princeton University Press.

Fraley, R. C., & Vazire, S. (2014). The N-pact factor: Evaluating the quality of empirical journals with respect to sample size and statistical power. *PloS One*, *9*(10), e109019.

Friedersdorf, C. (2020, November 10). Why California Rejected racial preferences, again. *The Atlantic*. https://www.theatlantic.com/ideas/archive/2020/11/why-california-rejected-affirmative-action-again/617049/

Garcia, J., Sanchez, D., Wout, D., Carter, E., & Pauker, K. (2019). SPSP diversity survey: Final report. Retrieved on 3/38/22 from: https://spsp.org/sites/default/files/SPSP_Diversity_and_Climate_Survey_Final_Report_January_2019.pdf

Gelman, A. (2017, October 18). From perpetual motion machines to embodied cognition: The boundaries of pseudoscience are being pushed back into the trivial. | Statistical Modeling, Causal Inference, and Social Science. Statmodeling.stat.columbia.edu. https://statmodeling.stat.columbia.edu/2017/10/18/perpetual-motion-machines-embodied-cognition-boundaries-pseudoscience-pushed-back-trivial/

German, K. T., & Stevens, S. T. (2021). Scholars under fire: The targeting of scholars for ideological reasons from 2015 to present. The Foundation for Individual Rights in Education. Available online: https://www.thefire.org/research/publications/miscellaneous-publications/scholars-under-fire/

German, K., & Stevens, S. T. (2022). Scholars under fire: 2021 year in review. *The Foundation for Individual Rights in Education*. Available online at: https://www.thefire.org/research/publications/miscellaneous-publications/scholars-under-fire-2021-year-in-review/scholars-under-fire-2021-year-in-review-full-text/

Guzman, G. G. (2017). *Household income: 2016*. US Department of Commerce, Economics and Statistics Administration, US Census Bureau.

Haidt, J. (2021). Monomania is illiberal and stupefying. *Persuasion*. Retrieved March 28, 2022, from https://www.persuasion.community/p/haidt-monomania-is-illiberal-and?s=r

Henrich, J., Heine, S. J., & Norenzayan, A. (2010). The weirdest people in the world?. *Behavioral and Brain Sciences, 33*(2–3), 61–83.

Hogg, M. A., & Adelman, J. (2013). Uncertainty–identity theory: Extreme groups, radical behavior, and authoritarian leadership. *Journal of Social Issues, 69*(3), 436–454.

Hogg, M. A., Meehan, C., & Farquharson, J. (2010). The solace of radicalism: Self-uncertainty and group identification in the face of threat. *Journal of Experimental Social Psychology, 46*(6), 1061–1066.

Honeycutt, N. (2022). *Manifestations of political bias in the academy*. Unpublished doctoral dissertation, Rutgers University.

Honeycutt, N., & Freberg, L. (2017). The liberal and conservative experience across academic disciplines: An extension of Inbar and Lammers. *Social Psychological and Personality Science, 8*(2), 115–123.

Honeycutt, N., & Jussim, L. (2020). A model of political bias in social science research. *Psychological Inquiry, 31*(1), 73–85.

Honeycutt, N., & Jussim, L. (in press). Political bias in the social sciences: A critical, theoretical, and empirical review. To appear in C. L. Frisby, R. E. Redding, W. T. O'Donohue, & S. O. Lilienfeld (Eds.). *Ideological and political bias in psychology: Nature, scope and solutions*. New York: Springer.

Horowitz, J. M. (2019). Americans see advantages and challenges in country's growing racial and ethnic diversity. Pew Research Center. Retrieved on 3/13/22 from: https://www.pewresearch.org/social-trends/2019/05/08/americans-see-advantages-and-challenges-in-countrys-growing-racial-and-ethnic-diversity/

Ibrahim, N. (2021, 6/25). Were Black World War II veterans excluded from G.I. Bill benefits? Retrieved on 3/29/22 from: https://www.snopes.com/fact-check/black-world-war-ii-vets-gi-bill/

Inbar, Y., & Lammers, J. (2012). Political diversity in social and personality psychology. *Perspectives on Psychological Science, 7*(5), 496–503.

Iyengar, S., Lelkes, Y., Levendusky, M., Malhotra, N., & Westwood, S. J. (2019). The origins and consequences of affective polarization in the United States. *Annual Review of Political Science, 22*, 129–146.

Jasko, K., LaFree, G., & Kruglanski, A. (2017). Quest for significance and violent extremism: The case of domestic radicalization. *Political Psychology, 38*(5), 815–831.

Jussim, L. (2017). Why brilliant girls tend to favor non-STEM careers. *Psychology Today*. Retrieved on 3/8/22 from: https://www.psychologytoday.com/us/blog/rabble-rouser/201707/why-brilliant-girls-tend-favor-non-stem-careers

Jussim, L. (2019). The threat to academic freedom ... from academics. Retrieved on 3/13/22 from: https://psychrabble.medium.com/the-threat-to-academic-freedom-from-academics-4685b1705794

Jussim, L., Cain, T. R., Crawford, J. T., Harber, K., & Cohen, F. (2009). The unbearable accuracy of stereotypes. *Handbook of Prejudice, Stereotyping, and Discrimination*, 199, 227.

Jussim, L., Careem, A., Goldberg, Z., Honeycutt, N., & Stevens. S. T. (in press). IAT scores, racial gaps, and scientific gaps. In Krosnick, J. A., Stark, T. H & Scott, A. L. (Eds.). *The future of research on implicit bias*. New York: Cambridge University Press.

Jussim, L., Crawford, J. T., & Rubinstein, R. S. (2015). Stereotype (in)accuracy in perceptions of groups and individuals. *Current Directions in Psychological Science*, 24(6), 490–497.

Jussim, L., Crawford, J. T., Anglin, S. M., & Stevens, S. T. (2015). Ideological bias in social psychological research. In J. Forgas, W. Crano, & K. Fiedler (Eds.). *Social psychology and politics* (pp. 91–109). New York, NY: Taylor & Francis.

Jussim, L., Stevens, S. T., & Honeycutt, N. (2018). Unasked questions about stereotype accuracy. *Archives of Scientific Psychology*, 6(1), 214.

Jussim, L., Stevens, S. T., Honeycutt, N., Anglin, S. M., & Fox, N. (2019). Scientific gullibility. In J. Forgas & R. Baumeister (Eds.), *The social psychology of gullibility: Conspiracy theories, fake news and irrational beliefs* (pp. 279–303). New York: Routledge.

Jussim, L., Thulin, E., Fish, J., & Wright, J. D. (2022, March 14). Articles Critical of the IAT and Implicit Bias. Retrieved from osf.io/74whk

Kaufman, E. (2021). Academic freedom in crisis: Punishment, political discrimination, and self-censorship. *Center for the Study of Partisanship and Ideology*, 2, 1–195.

Kruglanski, A. W., Bélanger, J. J., Chen, X., Köpetz, C., Pierro, A., & Mannetti, L. (2012). The energetics of motivated cognition: A force-field analysis. *Psychological Review*, 119(1), 1.

Langbert, M., & Stevens, S. (2021). Partisan registration of faculty in flagship colleges. *Studies in Higher Education*, 47(8), 1–11.

Leary, M. R., Twenge, J. M., & Quinlivan, E. (2006). Interpersonal rejection as a determinant of anger and aggression. *Personality and Social Psychology Review*, 10(2), 111–132.

Marietta, M., & Barker, D. C. (2019). *One nation, two realities: Dueling facts in American democracy*. New York: Oxford University Press.

Martin, C. C. (2016). How ideology has hindered sociological insight. *The American Sociologist*, 47(1), 115–130. https://doi.org/10.1007/s12108-015-9263-z

Mchangama, J. (2022). *Free speech: A history from Socrates to social media*. New York: Basic Books.

Moghaddam, F. M. (2005). The staircase to terrorism: A psychological exploration. *American Psychologist*, 60(2), 161.

Napier, J. L., & Jost, J. T. (2008). Why are conservatives happier than liberals. *Psychological Science*, 19(6), 565–572.

Noelle-Neumann, E. (1974). The spiral of silence a theory of public opinion. *Journal of Communication*, 24(2), 43–51.

Pew Research center (2021, Nov. 9). 11. Progressive left. https://www.pewresearch.org/politics/2021/11/09/progressive-left/

Princeton Faculty Letter. (2020, July 4). GoogleDocs. https://docs.google.com/forms/d/e/1FAIpQLSfPmfeDKBi25_7rUTKkhZ3cyMICQicp05ReVaeBpEdYUCkyIA/viewform

Rast III, D. E., Hogg, M. A., & Giessner, S. R. (2013). Self-uncertainty and support for autocratic leadership. *Self and Identity*, 12(6), 635–649.

Redding, R. E. (2013). Politicized science. *Society, 50*(5), 439–446. https://doi.org/10.1007/s12115-013-9686-5.

Roberts, S., & Pashler, H. (2000). How persuasive is a good fit? A comment on theory testing. *Psychological review, 107*(2), 358.

Rothman, Stanley, & Lichter, S. Robert. (2009). The vanishing conservative—is there a glass ceiling? In Maranto, Robert, Redding, Richard, and Hess, Frederick, (Eds.). *The politically correct university: Problems, scope and reforms* (pp. 60–76).

Ryan, C. L., & Bauman, K. (2016). Educational attainment in the United States: 2015. *Current Population Reports*. US Census Bureau. https://vtechworks.lib.vt.edu/bitstream/handle/10919/83682/EducationalAttainment2015US.pdf?sequence=1&isAllowed=y

Ryan, W. (1972). *Blaming the victim*. New York: Vintage Books.

Saad, L. (2022). U.S. political ideology steady; conservatives, moderates tie. https://news.gallup.com/poll/388988/political-ideology-steady-conservatives-moderates-tie.aspx

Schimmack, U. (2021). Invalid claims about the validity of implicit association tests by prisoners of the implicit social-cognition paradigm. *Perspectives on Psychological Science, 16*(2), 435–442.

Shah, J. Y., Friedman, R., & Kruglanski, A. W. (2002). Forgetting all else: On the antecedents and consequences of goal shielding. *Journal of Personality and Social Psychology, 83*(6), 1261.

Siskind, S. (2021). Trapped priors as a basic problem of rationality. Retrieved from Astral Codex 10 on 3/28/22 at: https://astralcodexten.substack.com/p/trapped-priors-as-a-basic-problem?s=r

Smith, C. (2014). *The sacred project of American sociology*. Oxford University Press.

SPSP (2019). Member diversity statistics. Retrieved on 3/30/22 from: https://spsp.org/sites/default/files/Member-Diversity-Statistics-December-2019.pdf

Stevens, S. (2022). 2022–2023 College Free Speech Rankings: What's the Climate for Free Speech on America's College Campuses. The Foundation for Individual Rights in Education. Available online: https://www.thefire.org/research/publications/student-surveys/2022-college-free-speech-rankings/2022-college-free-speech-rankings-full-text/

Stevens, S. T., Jussim, L., & Honeycutt, N. (2020). Scholarship Suppression: Theoretical Perspectives and Emerging Trends. *Societies, 10*(4), 82.

Stevens, S., & Schwictenberg, A. (2021). 2021 College free speech rankings: What's the climate for free speech on America's college campuses. *The Foundation for Individual Rights in Education*. Available online: https://www.thefire.org/research/publications/student-surveys/2021-college-free-speech-rankings/

Stevens, S., & Schwictenberg, A. (2020). 2020 College free speech rankings: What's the climate for free speech on America's college campuses. *The Foundation for Individual Rights in Education*. Available online: https://www.thefire.org/research/publications/student-surveys/2020-college-free-speech-rankings/

Stolzenberg, E. B., Eagan, M. K., Zimmerman, H. B., Berdan Lozano, J., Cesar-Davis, N. M., Aragon, M. C., & Rios-Aguilar, C. (2019). *Undergraduate teaching faculty: The HERI Faculty Survey 2016–2017*. Los Angeles: Higher Education Research Institute, UCLA.

Toner, K., Leary, M. R., Asher, M. W., & Jongman-Sereno, K. P. (2013). Feeling superior is a bipartisan issue: Extremity (not direction) of political views predicts perceived belief superiority. *Psychological Science, 24*(12), 2454–2462.

U.S. Census (2021). Quick facts, U.S. Retrieved on 3/29/22 from: https://www.census.gov/quickfacts/US

USA Today (2020, June 18). 12 charts show how racial disparities persist across wealth, health, education, and beyond. Retrieved on 3/13/22 from: https://www.usatoday.com/in-depth/news/2020/06/18/12-charts-racial-disparities-persist-across-wealth-health-and-beyond/3201129001/

van Prooijen, J. W., & Douglas, K. M. (2018). Belief in conspiracy theories: Basic principles of an emerging research domain. *European Journal of Social Psychology, 48*(7), 897–908.

Westfall, J., Van Boven, L., Chambers, J. R., & Judd, C. M. (2015). Perceiving political polarization in the United States: Party identity strength and attitude extremity exacerbate the perceived partisan divide. *Perspectives on Psychological Science, 10*(2), 145–158.

Zigerell, L. J. (2019). Left unchecked: Political hegemony in political science and the flaws it can cause. *PS: Political Science & Politics, 52*(4), 720–723. https://doi.org/10.1017/S1049096519000854

Zigerell, L. J. (2018). Black and White discrimination in the United States: Evidence from an archive of survey experiment studies. *Research & Politics, 5*(1), 2053168017753862.

19
ESCAPE FROM UNCERTAINTY
To Conspiracy Theories and Pseudoscience

Péter Krekó

EOTVOS LORAND UNIVERSITY, HUNGARY

Abstract

In times of high anxiety, uncertainty, and existential fears, people tend to embrace any – even false – promises of social and epistemic security. The advantage of conspiracy theories in crises in general, and during the COVID-19 pandemic in particular, was that they provided – scary but robust deductive theories – causal explanations on who and why created, let free and controlled the spread of the coronavirus. This way, conspiracy theories can help in the symbolic coping with new threats. At the same time, conspiracy theories as a form of *collectively motivated cognition* are emerging on the basis of group identities, and these "tribal myths", bound to group membership, can be psychologically reassuring when survival is at stake. But the price of this psychological comfort can be high: many studies (including our empirical research) have found that conspiracy theories can undermine rational individual responses to the pandemic. Beliefs in malevolent and secret plots by scientists, politicians, and background powers are undermining people's willingness to vaccinate themselves – putting their lives at much greater risk.

Crises, Insecurity, and Conspiracy Theories[1]

Uncertainties, crises, and insecurities are always providing a fertile ground for conspiracy theories. People often look for hidden causes that match the magnitude of the cataclysm they are facing. It is simply hard to accept, for instance, that great historical figures can die for banal causes – for example, because of a drunken driver, as it happened with Lady Diana. Many find it difficult to grasp that a virus that caused more than half a billion infections and more than 6 million deaths in the world by August 2022 might be the result of an undercooked bat on a kitchen table. Or some other unfortunate and extraordinary encounter of an inhabitant of Wuhan with an animal carrying the virus – as the official, most likely explanation for the origin of the virus (see, for example, NIAID, 2022) tells us.

The more tragic the event and the deeper its impact, the more we suspect an underlying conspiracy (van Prooijen and E. Van Dirk, 2014; see also van Prooijen, this volume). We feel the need to establish a symmetry between the significance of an event and the gravity of its cause: extraordinary events must be precipitated by extraordinary, huge plots. An experimental research project conducted in the 1970s (McCauley, Jacques, 1979) already showed that a successful assassination attempt on a US president was far more likely to give rise to an American conspiracy theory than an unsuccessful one. The authors argued that this is not an irrational lunacy, but instead, a logical conclusion based on a Bayesian logic: a conspiracy to kill a president is simply more likely to succeed than an assassination attempt conducted by a lone wolf. Coming up with a large-scale explanation on the events of the World help to reduce insecurity (see also Fiske, this volume).

In times of economic, political, and social crisis, the temptation to fall back on conspiracy theories to rationalize the unexpected is especially high (Moscovici and Graumann, 1981). Anxiety and feelings of insecurity triggered by adverse and unexpected events play a crucial role in this. When people feel they cannot influence happenings that pose a threat to their daily lives, they may try to recreate an illusion of control by overzealously seeking explanations (Park, 2010; see also Cooper & Pearce; van den Bos; Hirt, Eyink & Heiman; Krueger & Gruening, this volume).

Extraordinary events require extraordinary interpretations that reach beyond the well-known, official, and trivial stories. Furthermore, in historical times riddled with frustration and uncertainty, the need to identify an enemy or culprit can be overwhelming (see also von Hippel & Meraqkovsky; Fiske, this volume). Conspiracy theories provide an outlet for aggressive impulses pointing to specific foes in ways that seem to explain abstract, impersonal problems or complex social processes, which can otherwise appear impossible to understand (Volkan, 1985; Abalakina-Paap et al., 1999). The pandemic provides a textbook case for all these ills.

Some empathic researchers of rumors and conspiracy theories point to their function as collective problem-solving mechanisms that can help in the social adjustment to changing circumstances (see Shibutani, 1965). Rumors surface when the previous world order is turned on its head and former reference points lose their validity or cogency, triggering new collective explanations. Sudden unexpected events, with a powerful social impact that demands rationalization, tend particularly to trigger conspiracy theory narratives – even if the individual is not personally affected by the event in question. Festinger, for example, (1957) observed that, after an earthquake, scaremongering visions of an apocalyptic future rather spread in communities not directly affected by the disaster – as people who experienced no tragedy personally will look to justify their strong – but unfounded – anxieties (see Cooper & Pearce, this volume).

Healthcare crises usually go hand in hand with economic ones, and COVID-19 was no exception. Economic crises have all the features that make them good targets for conspiracy theorizing. Explanations and available analyses are contradictory, complex, and difficult for the lay public to understand. The information environment is heavily ambiguous, and public opinion turns to simplistic explanations as a consequence. People end up naming names. The biggest advantage of conspiracy theories is that they give "face" to complex, abstract, multi-actor, and multi-causal processes and give some epistemic certainty: a familiar enemy to blame (see also Kruglanski & Ellenberg, this volume).

Narratives about sinister and powerful groups typically multiply after terrorist attacks or major accidents that impact the community as a whole. Events like 9/11 or the 7/7 London bombings and the murder of Osama bin Laden all spawned widespread conspiracy theories (see for example: Swami et al., 2010). Of course, this is hardly surprising given that terror attacks are conspiratorial by nature.

Conspiracy theories are tales expressing absolute mistrust of established authority and power. All institutions and individuals associated with the "system" are suspects: politicians, media, experts, scientists, healthcare institutions, pharmaceutical companies, and doctors. In recent years, anti-science conspiracy theories have multiplied at an astonishing rate (Lewandowsky Oberauer, 2016), posing a considerable challenge to scientists and decision-makers (Goertzel, 2010). Trust can serve as an important factor in reducing the feeling of insecurity both socially and epistemologically (see also van den Bos, this volume).

Conspiracy theories about healthcare can be particularly dangerous. They endanger those who believe them, their families, and the broader social environment. The coronavirus resulted in a bizarre, ambivalent response regarding trust in science and scientists. While scientists are often mentioned as the most reliable source of COVID-related information according to, for example, Eurobarometer surveys, we could see a cascade of anti-scientific conspiracy theories as a consequence of the pandemic, for example, more than one-fourth of the European citizens think that viruses are developed by governments to control the freedom of the citizens.

This endemic paranoiac thinking is not a new phenomenon. A spreading narrative that birth control is part of a genocidal conspiracy against black Americans also led to a setback in efforts to encourage condom use. It contributed to a spike in unwanted pregnancies and the spread of sexually transmitted diseases (see for example: Bogart and Burn, 2013). Other stories that may appear less damaging, such as those offering ways of losing weight, have also provoked irrational behavior and responses. In Brooklyn's African American circles, for example, rumors were spread that sodas and fast food produced by certain chains (Snapple and Church's Chicken) contain added chemicals that sterilize black men (Fletcher, 1996). Research has also shown that about a quarter of African Americans believe that the HIV virus was released into the population as part of a conspiracy to eradicate black people. This belief contributed to a significant reluctance among many black men to use condoms or see a doctor to help prevent or treat HIV/AIDS. Doctors felt that they could be part of the conspiracy. The risk of HIV infection in their community was thereby greatly increased.

The belief among African Americans that vaccines may be potential weapons of genocide was obviously greatly amplified by acute experiences of genuine racial injustice and subjugation (Quinn et al., 2017). Their distrust must be considered alongside the history of black oppression in the US and unethical research programs, like the *Tuskegee Syphilis Study* between the 1930s and 1970s, in which black American men were subjected to human experimentation and mistreatment. (Freimuth ez al., 2001). Ultimately, though, faith in conspiracy narratives ends up harming the believers, regardless of their historical memory.

Anti-vaccination attitudes reach far beyond the boundaries of the black community. The anti-vaxx movement has spread worldwide, and hundreds of thousands have fallen victim to its prejudices (Poland and Jacobson, 2012).

Anti-vaxx and Conspiracy Theories

Many expected at the beginning of the epidemic that the coronavirus would take away all the credibility of the anti-vaccination movements. What could be a bigger defeat for the anti-vaxxers than a global virus where everybody is looking for vaccines as a panacea? Just the opposite happened: the insecurity caused by the pandemic gave momentum to anti-vaxxers (Ball, 2020). Conspiracy theories proved to be shockingly efficient in reducing the vaccination rate, as many studies suggested (Pertwee et al., 2022). The anti-vaxxer movement could build on two layers of uncertainties during COVID: First, about the pandemic as such, and second, about the newly developed vaccines which proved to be an easy target for fearmongering, claiming that humans only serve as guinea pigs in a huge social lab of virologists and politicians.

The anti-vaccination movement has been most successful in spreading conspiracy theories. According to data produced by the World Health Organization, measles remains one of the leading causes of child death globally. This is true despite the existence of the MMR vaccine, widely used in Europe, which protects against measles, mumps, and rubella. The inoculation is affordable and accessible to everyone. Nonetheless, in 2016, there were 89,780 fatal measles infections worldwide, and in 2015, 367 people died of the disease every day. The good news is that, between 2000 and 2016, mandatory vaccination lowered the number of cases with a fatal outcome by 84%, saving over 20 million lives. But in the same period, several million people died because they were not vaccinated (WHO, 2019).

However, because of adverse publicity from the anti-vaccination movement, which declared MMR its arch-enemy, measles has returned to countries where it had previously been eradicated: Germany and the United States, for instance. According to a project that provides lists of vaccine hesitancy victims based on public reports, failure to vaccinate caused 9028 deaths between 2007 and mid-2015 in the US alone (Anti-vaccine Body Count, 2015).

In 2017, measles resurfaced repeatedly in Europe, most of all in Germany, Italy, France, and Romania. According to a study of several European countries, there is a widespread belief that the negative and harmful effects of vaccines outweigh their advantages (Karafilikis and Larson, 2017). In Hungary, the anti-measles vaccine was introduced in 1969, so here the population is relatively well protected. After the emergence and proliferation of anti-vaxx views over the border in Romania, however, the disease resurfaced in Hungary as well (WHO, 2019).

Thanks to recent research, we know increasingly more about how anti-vaccination movements operate, what arguments they present, and the kind of healthcare conspiracy theories they promote. Misconceptions about healthcare are astonishingly widespread. According to the results of one US study, almost half of Americans believe in at least one medical conspiracy theory (Douglas et al., 2015). Those who do so are also more likely to seek recourse from "alternative therapies" which may have little or no effect and may prove harmful if the patient skips conventional medical therapy (Oliver, Wood, 2014). Rather than turning to health experts and doctors, this group will be inclined to take advice from people they trust – usually family members, friends, or acquaintances with no corresponding qualifications.

Medical conspiracy theories also often correlate with insecurity and anxieties about the impact of technological change on human health – most typical of the elderly, religious believers, and right-wingers (Lachbach and Furnham, 2017). An analysis of anti-vaccination tweets in the US between 2009 and 2015 has shown that the number of tweets was greater in states with a higher proportion of women who had recently given birth, of people aged between 40 and 44, of men with no higher education, or of households with a higher income (2017). As these results indicate, the anti-vaccination movement can gain popularity among very different groups at the same time: young mothers, men with little education, and people with high earnings. In the US, anti-vaxx tweets are by no any means the most popular in economically less developed states, as might be expected.

Anti-vaccination beliefs, medical conspiracy theories, and faith in alternative medicine all appear to correlate with political conspiracy theories (Galiford and Furnham, 2017). According to a study based on data from numerous countries, those who believe in conspiracy theories about the Kennedy assassination, Princess Diana's death, or 9/11 tend to be more skeptical about vaccines (APA, 2018). The role of misconceptions in dismissing the H1N1 Swine Flu vaccine (driven to some extent by political considerations) has been proven by Hungarian studies as well (Nguyen Luu et al., 2010).

Experiment-based research has shown that conspiracy theories can undermine faith in vaccines and diminish people's inclination to get vaccinated (Jolley and Douglas, 2014a). It also seems that such beliefs remain stubborn in the face of challenges. Prevention – based on promoting the logic of vaccination – is therefore more effective in the fight against anti-vax conspiracy theories than subsequent salvaging. Another investigation had shown that arguments against conspiracy theories were only compelling when they were presented to subjects in advance of the conspiracy narrative itself. It is easier to immunize those who refuse to succumb to such theories generally than to correct anti-vaccination attitudes that have already been formed (Jolley and Douglas, 2017).

Anti-scientific conspiracy narratives can do harm well beyond anti-vaccine sentiments as well, however. One experiment has shown, for example, that conspiracy theories targeted at climate change skeptics substantially diminished people's willingness to observe their carbon footprint or actively do something for the environment. The narratives promote the view that global warming is a mischievous invention aimed at weakening developed countries. They intensify public feelings of disappointment and insecurity while strengthening the sense that the individual is helpless (Jolley and Douglas, 2014b).

Insecurity and Conspiracy Theories: Empirical Findings

In the section below, I will briefly introduce a few pieces of my own empirical findings that connect conspiracy theories to uncertainty and insecurity from different aspects, including novel, preliminary findings from an international comparative database.

Conspiracy Theories and Welfare Concerns about the Future

From 2008 onwards, Hungary has been hit especially hard by the economic crisis, also combined with political instability and a social crisis (see, for example, Krekó and Enyedi, 2018). In representative survey research conducted in 2010, we found that generic conspiracy beliefs were the highest among people who saw a darker future. As Figure 19.1 illustrates, "prospective optimism" is a general feature among the respondents, who tend to evaluate the next 12 months of their household more positively, than the past 12 months. But this gap between the past and the future is closing as conspiracy beliefs are on the rise and become statistically insignificant in groups with high levels of conspiracy beliefs.

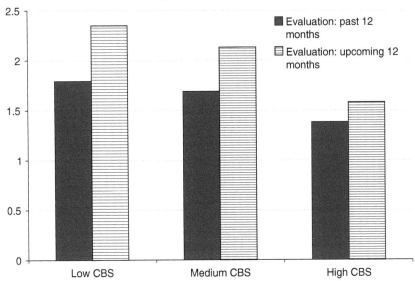

FIGURE 19.1 People with high-conspiracy beliefs are significantly less optimistic about the future and less certain that the next year will be better than the previous year, compared to people with low-conspiracy beliefs. CBS= Conspiracy Beliefs Scale.
Source: Krekó (2015).

Conspiracy Theories and Political Insecurity

Deprivation of political power is a hotbed for conspiracy theorizing. Hofstadter's classical (1965) essay on the paranoid style in American politics – one of the earliest social science papers on this topic – explained the rise of conspiracy theories on the American Populist Right with a feeling of powerlessness – both in political (lack of institutions) and cultural (lack of normative influence) terms. When the institutions are perceived as hostile and unreliable, the feeling of insecurity can become endemic in the community. This can be exploited by political leaders in ways that promote a "siege mentality" (see also Forgas, Ch. 17, this volume). In a comparative research encompassing 26 countries, Imhoff et al. (2022) found that conspiracy mentality is higher (a) among opposition voters, than among governmental voters, (b) among voters on the extreme left or extreme right of the political spectrum, than among voters on the center, and (c) among voters on the right than among voters of the left. The first two findings are obviously related to the feeling of being excluded from the power (opposition as well as the extremes are typically out of power). The right-leaning conspiracy theorizing might also be related to the higher levels of existential insecurity (see, for example, Van Der Linden et al., 2021).

Conspiracy Theories and Insecurity in the Context of COVID in Central and Eastern Europe

As Van Prooijen (2018, p. 8.) argues, *"Conspiracy theories are a natural defensive reaction to feelings of uncertainty and fear, blaming dissimilar outgroups for the distressing circumstances that one has to deal with"* (see also van Prooijen, this volume). In line with this argument, we aimed to identify the insecurity- and uncertainty-related predictors of conspiracy theories using the set of variables available in the European Social Survey. While doing so, we relied on the systematic review of Van Mulukom et al. (2022), who identified nine types of antecedents of belief in COVID-19 conspiracy theories – and several consequences, among which we only focused on the vaccination intention. Our research aimed at (a) revealing the relative explanatory power of different uncertainty-related attitudes in a robust, multi-country representative survey and (b) comparing the explanatory power of conspiracy attitudes against the respondents' own experiences on vaccination intentions. While common sense would suggest that one's own experiences with COVID (getting infected or having a relative get infected) are more important when deciding whether to get vaccinated or not, our hypothesis was the opposite that conspiracy theories – these tribal myths that offer a crutch among the feelings of epistemic and social uncertainty – will be more robust predictors of vaccination intentions than one's own experiences.

In the first, exploratory part of our research, we identified four different kinds of insecurity-related attitude clusters that can have an impact on conspiracy beliefs.

1 Personal insecurity. As the "conspiracy theories are for losers" theory suggests (Uscinski and Parent, 2014), the feeling of deprivation of material assets, political power, or fear of losing other important assets such as physical safety can be an important driver for belief in conspiracy theories. We assumed that the higher the level of these uncertainties were, the stronger the conspiracy beliefs.

2 Social insecurity and social identity. Generally, self-categorization and embracing group identities can be attempts to reduce subjective uncertainties (see, for example, the uncertainty-identity theory, Hogg, 2000; see also Hogg & Gaffney, this volume). On the other hand, some forms of strong and politically heated beliefs can be a hotbed for conspiracy theories (see for example: Sternisko et al., 2020). Nationalism and belief in conspiracy theories came hand in hand as a response to the pandemic (Jutzi et al., 2020), and stronger association with the political right (Imhoff et al., 2022) is associated with a stronger conspiracy mentality. Stronger religious

affiliation is also found to be positively correlated with conspiracy beliefs (Frenken et al., 2022), and stronger partisan beliefs can also be associated with stronger conspiracy beliefs (Enders et al., 2019). Individualism as an ideology has been found to be strongly related to conspiracy theories about COVID (Briddlestone, Green, and Douglas, 2020) and also to vaccination intentions. We expected that stronger identification with the attitude categories under discussion (nation, religion, right-wing ideology, and party) would be associated with stronger levels of conspiracy beliefs.

3   Insecurity of identity: Prejudices and authoritarianism. The classics of the Frankfurt School, such as Erich Fromm, have already referred to authoritarianism and prejudice as a way to "escape from freedom" – and, of course, the insecurity and uncertainty that come along with this freedom (Fromm, 1994). The feeling of uncertainty about one's self and group belonging can lead to prejudiced perception, radical political reaction, and the embracement of authoritarian leadership (see also Hirschberger, this volume). Previous studies have found a relationship between conspiracy beliefs and authoritarianism (see, for example, Gzeziak-Feldman, 2015), as well as between conspiracy theories and prejudices towards minority groups. Conspiracy theories can serve as an "outlet for hostility" (Abalakina-Paap et al., 1999) that can help express hostile feelings against outgroups – that we are less familiar with, therefore making us feel insecure and uncertain. In the context of the pandemic, this insecurity becomes even more important, as reminding people of their mortality encourages negative reactions to others whose behavior or values deviate from the cultural worldview (Pyszczynski & Sundby, this volume).

4   Mistrust: epistemic uncertainty and social insecurity. Trust is the best way for reducing epistemic and social-relational insecurity. When we trust a person or an institution, it means that we regard them as a credible source of information – and also as competent problem solvers (see also van den Bos, this volume). Conspiracy theories are theories of endemic mistrust (see for example: Inglehart, 1987, Kramer and Jost, 2002, Pierre, 2020); so it is probably not surprising that institutional (vertical) and interpersonal (horizontal) trust (Van Prooijen, 2022), satisfaction with democracy (Pantazi et al., 2022), and attitudes toward science (Douglas et al., 2020) are all related to conspiracy theories according to the empirical research so far. Mistrust has its evolutionary advantage as well, though. It helps us avoid interactions with hostile, and potentially dangerous individuals, groups, or even institutions (see also von Hippel & Merakovsky; Fiedler & McCaughey, this volume).

Examining all these four types of uncertainties on the open-access database of the European Social Survey can be a valuable contribution to the already existing research for at least three reasons. (1) Most of the research so far could only focus on one or two factors at once, while this rich database allows us to examine all of them at the same time. (b) We could cover seven countries in our analysis with sizeable (N = 1000–3000) databases. (c) All the samples were representative and based on a rigorous sampling design that increases the validity and comparability of the data from different countries.

Methodology

In order to examine the role of insecurities in conspiracy beliefs, we have conducted a two-step analysis on the freshly published open-access European Social Survey database (ESS Round 10, 2020). We focused on eight Central and Eastern European and Baltic countries in our analysis; all of them are members of both the European Union and NATO: Bulgaria (N = 2718), Czechia (N = 2476), Estonia (N = 1542), Finland (N = 1577), France (N = 1977), Croatia (N = 1592), Hungary (N = 1849), Lithuania (N = 1660), Slovenia (N = 1252), and Slovakia (N = 1418). All of the samples were representative, and the survey was conducted with the same methodology (CAPI). The data collection period was the autumn of 2021, at a time when vaccines were already available for these populations (except for part of the sample in Slovenia).[2]

The aim of the analysis was to (a) examine the strength and influence of the above-mentioned four types of uncertainty and insecurity on conspiracy beliefs, and (b) examine the impact that conspiracy beliefs have on the behavioral intention – the respondents' willingness to vaccinate themselves – compared to their own experiences with the virus. As indicated above, the hypothesis was that conspiracy beliefs would be more important determinants of vaccination intentions than the respondents' own experiences.

We created a conspiracy scale from three items using principal component analysis. One was a conspiracy statement on the pandemic itself ("COVID-19 is result of deliberate and concealed efforts of some government or organization"); while the other two were more general assumptions on the existence of grand political ("Small secret group of people responsible for making all major decisions in world politics") and scientific ("Groups of scientists manipulate, fabricate, or suppress evidence in order to deceive the public") conspiracies.

While we used the combined data pool of eight post-communist countries in broader Central and Eastern Europe when investigating the links between insecurities, conspiracy beliefs, and vaccination intentions, the scale also allows us to compare the prevalence of conspiracy beliefs between the countries. As Figure 19.2 shows, countries in south-eastern Europe (or the Western Balkans) are the most enthusiastic believers in conspiracy theories, while conspiracy beliefs were the lowest in the more "Nordic" countries within this branch: Chechia, Lithuania, and Estonia. Finland was not included in our analysis as it is not a post-socialist country- but left in this comparison as an interesting odd-one out in this group with the remarkable lack of conspiracy beliefs (Finnish

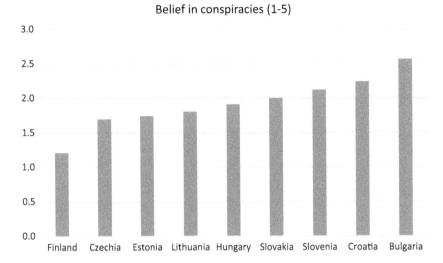

FIGURE 19.2 Conspiracy beliefs in nine European countries.

respondents had only an average of 1.2 points on the conspiracy beliefs scale of 1–5.) – a reminder of how much conspiracy beliefs are determined by the social-political environment and the quality of the educational system.

First, we used a linear regression model in order to examine the predicting value of the four different kinds of insecurity- and uncertainty-related variables. Second, we used a multinominal regression model to decide if personal experiences or conspiracy beliefs had a bigger impact on vaccination intentions.

The list of variables we put in the model is as follows:

1 Personal insecurity and uncertainty	
a Material uncertainty and insecurity	Feeling about household's income nowadays
b Physical uncertainty and insecurity	Feeling of safety of walking alone in local area after dark
2 Social insecurities and group identities	
a Vertical trust	A principal component of six institutional trust items
b Horizontal trust	The principle component of three items: Most people can be trusted or you can't be too careful; Most people try to take advantage of you, or try to be fair; Most of the time people helpful or mostly looking out for themselves
c Satisfaction with democracy	How satisfied with the way democracy works in country
d Attitudes towards science	Trust in scientists

3 Social insecurity – Group identities (response to social insecurity)
a National identity How emotionally attached to [country]
b Religious identity How religious are you
c Ideological identity (left-right scale) Placement on left-right scale
e Political identity How close do you feel yourself to this particular party
f individualism A principal component from three items: "More important for governments to prioritize public health or economic activity when fighting a pandemic
More important for governments to monitor and track the public or to maintain public privacy when fighting a pandemic
More important for governments to prioritize public health or economic activity"

Identity insecurity: prejudices & authoritarianism
Anti-gay attitudes principle component of three items:
1 Gays and lesbians free to live life as they wish
2 Ashamed if a close family member is gay or lesbian
3 Gay and lesbian couples' right to adopt children
Anti-immigrant attitudes principle component of three items:
Immigration bad or good for country's economy
Country's cultural life is undermined or enriched by immigrants
Immigrants make a country worse or a better place to live
Authoritarianism Principal component of three items:
Acceptable for a country to have a strong leader above the law; obedience and respect for authority most important virtues children should learn
Country needs most loyalty towards its leaders

Our linear regression model had a very good explanatory value: the set of insecurity-related variables (put in the model individually) could explain 30% of the variance ($R^2 = 0,301; p < 0,01$) (Table 19.1).

In the graph below, we indicated the seven most important predictors – in the order of their beta values. Mistrust (in scientists, in democracy, with beta values around 0.2 in political institutions, but also, in other people) clearly proved to be the most important factor predicting conspiracy beliefs. Authoritarianism, concerns about the future finances of the household, and COVID-individualism as a strange (anti-identity) identity category also played some role with lower beta values of 0,08–0,12 (Figure 19.3).

TABLE 19.1

Model B	Unstandardized coefficients		Standardized coefficients	t	Sig
		Std error	Beta		
1 (Constant)	,423	,175		2417	,16
Feeling about households income nowadays	106	,032	,085	3,336	,001
Feeling of safety of walking alone in a local area after dark	−,028	,032	−,021	−,863	,388
Verticaltrust	−,159	,034	−,151	−4,619	,000
Horizontaltrust	−,121	,025	−,127	−4,891	,000
Trust in scientists	−,084	,011	−,202	−7,534	,000
How satisfied with the way democracy works in a country	−0,71	,011	−,199	−6,190	,000
How emotionally attached to [country]	,021	,011	,046	1,869	,062
How religious are you	,013	,008	,041	1,685	,092
Placement on left-right scale	,021	,008	,067	2,603	,009
How close to party	,012	,037	,008	,318	,751
COVID-individualism	,118	,024	,120	5,000	,000
Homophobic discrimination	−,035	,028	−,034	−1,260	,208
Pro-immigrationism	,004	,026	,004	,168	,866
Authoritarianism	−,117	,025	−,124	−4,753	,000

[a]Dependent variable; conspiracy beliefs.

It is also interesting to see which variables have *not* proved to be important predictors of belief in conspiracies: prejudice, nationalism, and religiousness, for example, proved to be insignificant, at least in this set of data. Also, fears over public safety and security played no role in conspiracy theorizing. The direction of the relations was in line with what we expect: authoritarianism, mistrust, COVID-individualism, and worries about finances were all positively correlated with the conspiracy beliefs.

Our multinominal regression model has also confirmed our hypothesis: conspiracy beliefs proved to be a powerful predictor of vaccination intention: Rating high on the conspiracy scale (1 standard deviation from average on the scores of the principal component, a standardized variable) reduces the chance of having been vaccinated by 58%.

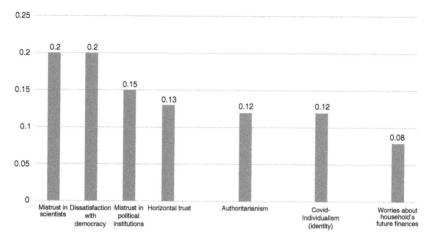

FIGURE 19.3 The seven most important predictors of conspiracy beliefs (Standardized Betas).

At the same time, a person's own experiences with COVID (own infection, family member's infection, and changing working conditions, for example, lost job played almost no role in the respondent's tendency to vaccinate themselves). The only significant result we found here is that having been infected with COVID and being tested positive is positively correlated with not having been vaccinated but planning it by 20%, which makes sense, as the infection delayed the vaccination process for many.

Conclusion: Conspiracy Theories as a Collective Form of Self-Handicapping: The Only Thing We Know Is Uncertainty

In times of endemic uncertainty with a lot of challenges to predict and adapt, the psychological stability and security that a strong attitude can provide in explaining the world, driving emotions, and directing behavior simply cannot be overstated (see also Hirt et al., this volume). Conspiracy beliefs and identity-based attitudes paradoxically provide this subjective certainty and security – even if, according to their conclusions, nothing is reliable, certain, or secure. But this fragile psychological "comfort" is dangerous, as it makes people avoid adaptive responses that help their survival – such as vaccination. The "safety" of this attitude is more important for the individuals than their own experiences with the virus and how it may affect their families – a strong reminder of how powerful notions of conspiracy can be in shaping socio-political reality.

As we could see throughout the chapter, uncertainty and insecurity are providing fertile ground for conspiracy theorizing. Unexpected events, crises, and tragic situations are calling for atypical explanations for the world's ills, and conspiracy theories are ready to fill the epistemical vacuum when needed.

As some previous empirical research has clearly proven, fears over the future and a feeling of uncertainty because of a lack of power are both associated with conspiracy theories. It is not coincidental, then, that the pandemic served as a perfect opportunity for conspiracy theories to spread and gain popularity. The pandemic polarized the already polarized societies even further (Falyuna and Krekó, 2021). Our results show that conspiracy beliefs were more easily aroused on the basis of some ideologies, such as authoritarianism and individualism. Conspiracy theories are ambiguous and controversial in this sense as well: They are collective beliefs but can express some form of (anti-statist) individualism.

Citizens could feel that they lost political control and are just subjects of the decisions of authorities: governments and doctors (Krekó, 2021). Comparative research has shown that the "conspiracy Zeitgeist" is not balanced: Central and Eastern Europe seem to be particularly vulnerable to conspiracy theories than in Western and Northern Europe (Krekó, in press). This might be partially explained by the diverse histories of these regions, and especially by the general experience of lack of trust, control, and lack of sovereignty in Eastern Europe (see also Forgas, this volume).

Even if conspiracy theories can serve some motivational functions, their disadvantages for the self and the group often exceed their advantages (e.g. Fiedler & McCaughey; Van Prooijen, this volume), and for this reason, they can be mostly regarded as a self-defeating way of social cognition (Douglas, Sutton, Cichocka, 2017), or collective cognition (Krekó, 2015). Our investigations based on the ESS international database have clearly shown that mistrust – in scientists, in political authorities, but also in others – is by far the most important predictor behind conspiracy theorizing. But again, mistrust only adds to the social and epistemic uncertainties. Given their numerous negative consequences, and even self-fulfilling nature, they are also related to self-handicapping strategies (for an overview on self-handicapping, see, for example, Hirt et al, this volume), as they can increase the outgroup's hostility against the ingroup in a predictable way that further helps to explain the ingroup's disadvantaged position (for the negative stigma of the COVID-related conspiracy theories, see again Van Prooijen, this volume).

So, we can conclude that conspiracy theories provide a dangerous and potentially harmful – even if often entertaining – and fragile comfort in reaction to crises, feelings of uncertainty and unexpected developments. They should be understood more as tribal myths (Krekó, 2021, Forgas, this volume) that can exacerbate the problems rather than as adaptive tools to cope with epistemic and social challenges. This is why it is crucially important to have more systemic research on their cures (Krekó, 2020).

Notes

1 For a more detailed description of the topic, see: Why conspiracy theories soar in times of crises, Péter Krekó, 23 June 2020 (Krekó, 2020/a) https://www.eurozine.com/why-conspiracy-theories-soar-in-times-of-crises/.
2 Except for a part of the Slovenian sample that was collected before the vaccine roll-out, and these cases were excluded from the calculations.

References

Abalakina-Paap, M., Stephan, W. G., Craig, T., & Gregory, W. L. (1999). Beliefs in conspiracies. *Political Psychology, 20*(3), 637–647.
Anti Vaccine Body Count (2015): Number of Preventable Deaths https://www.jennymccarthybodycount.com/PreventableDeaths.html
APA (2018). Belief in Conspiracy Theories Associated with Vaccine Skepticism, American Psychological Association, 1 February 31, http://www.apa.org/news/press/releases/2018/02/vaccine-skepticism.aspx
Ball, P. (2020). 'Anti-vaccine movement could undermine efforts to end coronavirus pandemic, researchers warn.' *Nature, 581*.
Biddlestone, M., Green, R., & Douglas, K. M. (2020). Cultural orientation, power, belief in conspiracy theories, and intentions to reduce the spread of COVID-19. *British Journal of Social Psychology, 59*(3), 663–673.
Bogart, L. M., & Bird, S. T. (2003). Exploring the relationship of conspiracy beliefs about HIV/AIDS to sexual behaviors and attitudes among African-American adults. *Journal of the National Medical Association, 95*(11), 1057.
Douglas, K. M., Sutton, R. M., & Cichocka, A. (2017). The psychology of conspiracy theories. *Current Directions in Psychological Science, 26*(6), 538–542.
Douglas, K. M., Sutton, R. M., Jolley, D., & Wood, M. J. (2015). The social, political, environmental, and health-related consequences of conspiracy theories: Problems and potential solutions. In Bilewicz, Cichocka, Soral (Eds.): *The psychology of conspiracy* (pp. 201–218). London: Routledge.
Douglas, K. M., Uscinski, J. E., Sutton, R. M., Cichocka, A., Nefes, T., Ang, C. S., & Deravi, F. (2019). Understanding conspiracy theories. *Political Psychology, 40*, 3–35.
Enders, A. M., & Smallpage, S. M. (2019). Informational cues, partisan-motivated reasoning, and the manipulation of conspiracy beliefs. *Political Communication, 36*(1), 83–102.
ESS Round 10: European Social Survey Round 10 Data (2020). Data file edition 1.2. Sikt - Norwegian Agency for Shared Services in Education and Research, Norway – Data Archive and distributor of ESS data for ESS ERIC. doi:10.21338/NSD-ESS10–2020.
European Commission: Public Opinion in the Times of the Coronavirus.
Falyuna and Krekó (2022). Miért csábít minket az áltudományok diszkrét bája? In *Krekó-Falyuna (szerk): Sarlatánok Kora- Miért dőlünk be az áltudományoknak?* Athenaeum Kiadó, Budapest.
Fenster, M. (1999). *Conspiracy theories: Secrecy and power in American culture*. Minneapolis: University of Minnesota Press.
Festinger, L. (1957). *A theory of cognitive dissonance* (Vol. 2). Redwood City: Stanford University Press.
Fletcher, M. A. (1996). Conspiracy theories can often ring true: History feeds blacks' mistrust. *Washington Post*.

Forgas, J. P., Crano, W. D., & Fiedler, K. (2021). *The Psychology of Populism: The Tribal Challenge to Liberal Democracy*. New York: Routledge.

Freimuth, V. S., Quinn, S. C., Thomas, S. B., Cole, G., Zook, E., & Duncan, T. (2001). African Americans' views on research and the Tuskegee Syphilis Study. *Social Science & Medicine, 52*(5), 797–808.

Frenken, M., Bilewicz, M., & Imhoff, R. (2022). On the relation between religiosity and the endorsement of conspiracy theories: The role of political orientation. *Political Psychology. 44*, doi: 10.1111/pops.12822.

Fromm, E. (1994). *Escape from freedom*. London: Macmillan.

Galliford, N., & Furnham, A. (2017). Individual difference factors and beliefs in medical and political conspiracy theories. *Scandinavian journal of psychology, 58*(5), 422–428.

Goertzel, T. (2010). Conspiracy theories in science: Conspiracy theories that target specific research can have serious consequences for public health and environmental policies. *EMBO Reports, 11*(7), 493–499.

Grzesiak-Feldman, M. (2015). Are the high authoritarians more prone to adopt conspiracy theories? The role of right-wing authoritarianism in conspiratorial thinking. In Bilewicz, Cichocka, Soral (Eds.): *The psychology of conspiracy* (pp. 117–139). Routledge.

Hofstadter, R. (2012). *The paranoid style in American politics*. New York: Vintage.

Hogg, M. A. (2000). Subjective uncertainty reduction through self-categorization: A motivational theory of social identity processes. *European Review of Social Psychology, 11*(1), 223–255.

Hogg, M. A., & Adelman, J. (2013). Uncertainty–identity theory: Extreme groups, radical behavior, and authoritarian leadership. *Journal of Social Issues, 69*(3), 436–454.

Imhoff, R., Zimmer, F., Klein, O., António, J. H., Babinska, M., Bangerter, A.,... & Van Prooijen, J. W. (2022). Conspiracy mentality and political orientation across 26 countries. *Nature Human Behaviour, 6*(3), 392–403.

Imhoff, R., Zimmer, F., Klein, O., António, J. H., Babinska, M., Bangerter, A.,... & Van Prooijen, J. W. (2022). Conspiracy mentality and political orientation across 26 countries. *Nature Human Behaviour, 6*(3), 392–403.

Jolley, D., & Douglas, K. M. (2014). The effects of anti-vaccine conspiracy theories on vaccination intentions. *PloS One, 9*(2), e89177.

Jolley, D., & Douglas, K. M. (2014). The social consequences of conspiracism: Exposure to conspiracy theories decreases intentions to engage in politics and to reduce one's carbon footprint. *British Journal of Psychology, 105*(1), 35–56.

Jolley, D., & Douglas, K. M. (2017). Prevention is better than cure: Addressing anti-vaccine conspiracy theories. *Journal of Applied Social Psychology, 47*(8), 459–469.

Jutzi, C. A., Willardt, R., Schmid, P. C., & Jonas, E. (2020). Between conspiracy beliefs, ingroup bias, and system justification: how people use defense strategies to cope with the threat of COVID-19. *Frontiers in psychology, 11*, 578586.

Karafillakis, E., & Larson, H. J. (2017). The benefit of the doubt or doubts over benefits? A systematic literature review of perceived risks of vaccines in European populations. *Vaccine, 35*(37), 4840–4850.

Kramer, R. M., & Jost, J. (2002). Close encounters of the suspicious kind: Outgroup paranoia in hierarchical trust dilemmas. In D. Mackie, E Smits (Eds.): *From prejudice to intergroup emotions: Differentiated reactions to social groups* (pp. 173–189). London: Psychology Press/Taylor & Francis.

Krekó, P. (2015). Conspiracy theory as collective motivated cognition. In Bilewicz, Cichocka, Soral (Eds.): *The Psychology of Conspiracy* (pp. 80–94). London: Routledge.

Krekó, P. (2020/a): Why conspiracy theories soar in times of crises. 23 June 2020 (Krekó, 2020) Eurozine, https://www.eurozine.com/why-conspiracy-theories-soar-in-times-of-crises/.
Kreko, P. (2020). Countering conspiracy theories and misinformation. In M. Butter & P. Knight (Eds.), *Routledge handbook of conspiracy theories* (pp. 242–256).
Krekó, P. (2021). The mutations of science in the pandemic. *Eurozine*, 6 December 2021. https://www.eurozine.com/mutations-of-science-in-the-pandemic/
Krekó, P., & Enyedi, Z. (2018). Explaining Eastern Europe: Orbán's laboratory of illiberalism. *Journal of Democracy, 29*(3), 39–51.
Lahrach, Y., & Furnham, A. (2017). Are modern health worries associated with medical conspiracy theories? *Journal of Psychosomatic Research, 99*, 89–94.
Lewandowsky, S., & Oberauer, K. (2016). Motivated rejection of science. *Current Directions in Psychological Science, 25*(4), 217–222.
McCauley, C., & Jacques, S. (1979). The popularity of conspiracy theories of presidential assassination: A Bayesian analysis. *Journal of Personality and Social Psychology, 37*(5), 637.
Moscovici, S. (1987). The conspiracy mentality. In C.F. Graumann and S. Moscovici (eds.), *Changing conceptions of conspiracy*, Springer-Verlag, 151–169.
Nguyen Luu, L. A., Borsfay, K., & Kende, J. (2010). H1N1 influenzával és a vírus elleni oltással kapcsolatos attitűdök (Attitudes to Vaccines against the H1N1 Influenza and Virus), MPT Kongresszus, Pécs.
NIAID (2022). *Origins of Coronaviruses.* https://www.niaid.nih.gov/diseases-conditions/origins-coronaviruses.
Oliver, J. E., & Wood, T. (2014). Medical conspiracy theories and health behaviors in the United States. *JAMA Internal Medicine, 174*(5), 817–818.
Pantazi, M., Papaioannou, K., & van Prooijen, J. W. (2022). Power to the people: The hidden link between support for direct democracy and belief in conspiracy theories. *Political Psychology, 43*(3), 529–548.
Park, C. L. (2010). Making sense of the meaning literature: An integrative review of meaning making and its effects on adjustment to stressful life events. *Psychological Bulletin, 136*, 257–301.
Pertwee, E., Simas, C., & Larson, H. J. (2022). An epidemic of uncertainty: Rumors, conspiracy theories and vaccine hesitancy. *Nature Medicine, 28*(3), 456–459.
Pierre, J. M. (2020). Mistrust and misinformation: A two-component, socio-epistemic model of belief in conspiracy theories. *Journal of Social and Political Psychology, 8*(2), 617–641.
Poland, G. A., & Jacobson, R. M. (2012). The clinician's guide to the anti-vaccinationists' galaxy. *Human Immunology, 73*(8), 859–866.
Prooijen, J. W. (2018). *The psychology of conspiracy theories.* London: Routledge.
Quinn, S. C., Jamison, A., Freimuth, V. S., An, J., Hancock, G. R., & Musa, D. (2017). Exploring racial influences on flu vaccine attitudes and behavior: Results of a national survey of White and African American adults. *Vaccine, 35*(8), 1167–1174.
Shibutani, T. (1966). *Improvised news.* Ardent Media.
Sternisko, A., Cichocka, A., & Van Bavel, J. J. (2020). The dark side of social movements: Social identity, non-conformity, and the lure of conspiracy theories. *Current Opinion in Psychology, 35*, 1–6.

Swami, V., Chamorro-Premuzic, T., & Furnham, A. (2010). Unanswered questions: A preliminary investigation of personality and individual difference predictors of 9/11 conspiracist beliefs. *Applied Cognitive Psychology, 24*(6), 749–761.

Tomeny, T. S., Vargo, C. J., & El-Toukhy, S. (2017). Geographic and demographic correlates of autism-related anti-vaccine beliefs on Twitter, 2009–15. *Social science & medicine, 191,* 168–175.

Uscinski, J. E., & Parent, J. M. (2014). *American conspiracy theories.* Oxford: Oxford University Press.

Van der Linden, S., Panagopoulos, C., Azevedo, F., & Jost, J. T. (2021). The paranoid style in American politics revisited: An ideological asymmetry in conspiratorial thinking. *Political Psychology, 42*(1), 23–51.

van Mulukom, V., Pummerer, L. J., Alper, S., Bai, H., Čavojová, V., Farias, J.,... & Žeželj, I. (2022). Antecedents and consequences of COVID-19 conspiracy beliefs: A systematic review. *Social Science & Medicine,* 114912.

Van Prooijen, J. W., Douglas, K. M., & De Inocencio, C. (2018). Connecting the dots: Illusory pattern perception predicts belief in conspiracies and the supernatural. *European Journal of Social Psychology, 48*(3), 320–335.

Van Prooijen, J. W., Spadaro, G., & Wang, H. (2022). Suspicion of institutions: How distrust and conspiracy theories deteriorate social relationships. *Current Opinion in Psychology, 43,* 65–69.

Van Prooijen, J.-W., & Van Dijk, E. (2014). When consequence size predicts belief in conspiracy theories: The moderating role of perspective taking. *Journal of Experimental Social Psychology, 55,* 63–73; LeBoeuf, A., & Norton, M. I. (2012). Consequence-cause matching: Looking to the consequences of events to infer their causes. *Journal of Consumer Research, 39,* 128–141.

Volkan, V. (1985). The need to have enemies and allies: A developmental approach. *Political Psychology, 6,* 219–247.

Wagner-Egger, P., Bangerter, A., Gilles, I., Green, E., Rigaud, D., Krings, F.,... & Clémence, A. (2011). Lay perceptions of collectives at the outbreak of the H1N1 epidemic: heroes, villains and victims. *Public Understanding of Science, 20*(4), 461–476.

WHO (2019). Measles, Fact-sheet. https://www.who.int/news-room/fact-sheets/detail/measles.

20
FEELINGS OF INSECURITY AS A DRIVER OF ANTI-ESTABLISHMENT SENTIMENTS

Jan-Willem van Prooijen

VU AMSTERDAM, NSCR, AND MAASTRICHT UNIVERSITY, NETHERLANDS

Abstract

Many citizens in contemporary society hold strong anti-establishment sentiments, as reflected in conspiracy theories, populist attitudes, and support for radical political movements. The current contribution examines the role of insecurity in such anti-establishment sentiments. First, the chapter reviews evidence from various disciplines that feelings of insecurity are a main driver of these different manifestations of anti-establishment sentiments. Second, the chapter illuminates how anti-establishment sentiments may be emotionally and cognitively appealing when people feel insecure. Specifically, (1) anti-establishment beliefs satisfy a cognitive need for epistemic clarity, by making complex societal problems and their solutions easy to understand. Furthermore, (2) anti-establishment beliefs foster self-enhancement: Believing that oneself or one's group is part of a meaningful struggle against an immoral establishment increases feelings of importance. Finally, (3) anti-establishment beliefs stimulate a range of emotions that help people cope with insecurity. Anti-establishment movements glorify the past, and therefore, anti-establishment attitudes are associated with feelings of nostalgia; moreover, people may experience the struggle against powerful elites as exciting, highlighting the rewarding aspects of insecurity. I conclude that feelings of insecurity are key to understand anti-establishment sentiments.

Many citizens of democratic societies hold strong anti-establishment sentiments: Negative feelings and beliefs toward established power holders. This "establishment" most clearly includes politicians and political parties, but also businesspeople, bankers, scientists, journalists, and other people of influence. Anti-establishment sentiments can manifest themselves in various ways. Populist

DOI: 10.4324/9781003317623-24

movements at both the left and right often voice strong anti-establishment rhetoric, and indeed, anti-elitism is one of the defining dimensions of populism (Akkerman et al., 2014; Mudde, 2004; Mudde & Kaltwasser, 2017). Relatedly, many citizens support politically extreme movements that rigidly endorse a strong left-wing or right-wing ideology (see also Forgas; Jussim et al., this volume). These strong ideologies often imply antagonism toward a more moderate establishment and are empirically related to institutional distrust and EU skepticism (Kutiyski, Krouwel, & Van Prooijen, 2021). Finally, many citizens believe conspiracy theories that assume nefarious plots by societal power holders. While not all conspiracy theories necessarily are anti-establishment – as people also can hold conspiracy theories about low-power groups in society, such as ethnic minorities (Nera et al., 2021) – a substantial portion of conspiracy theories accuse legitimate power holders of deliberately causing a pandemic, exaggerating climate change, committing terrorist strikes against its own citizens, and rigging elections (for overviews, see Butter & Knight, 2020; Douglas et al., 2019; Van Prooijen, 2020, 2022; see also Kreko, this volume). Consistently, a general propensity to believe conspiracy theories is associated with prejudice toward high-power groups (Imhoff & Bruder, 2014).

One common assumption is that anti-establishment sentiments increase when people feel insecure, for instance, due to societal crisis situations (Douglas et al., 2019; Van Prooijen & Douglas, 2017; Van Prooijen, 2020), economic anxiety (Mols & Jetten, 2017), perceived injustice (Van den Bos, 2020), and personal or group-based humiliation (Kruglanski et al., 2014; see also Hogg & Gaffney; Hirschberger; Kruglanski & Ellenberg; van den Bos, this volume). Feelings of insecurity are broadly defined as the extent to which people do not feel safe in their social environment and are anxious about the future. Feelings of insecurity instigate an epistemic sense-making process that is focused on ascribing moral responsibility and blame for the predicament that perceivers find themselves in (Alicke, 2000; see also Fiske, this volume). Such moral responsibility and blame are often ascribed to societal power holders who have contributed to policies that citizens perceive as failing to resolve, or even causing, their problems. Moreover, power holders are an easy and convenient scapegoat, providing a compelling target to explain one's situation in cognitively simple terms. The present contribution will focus on the presumed link between insecurity and anti-establishment sentiments by pursuing two broad goals. First, it will review the empirical evidence that feelings of insecurity increase anti-establishment sentiments. Second, by integrating recent empirical findings, it will illuminate various different reasons why anti-establishment sentiments are appealing to citizens when they feel insecure.

To accomplish these goals, I broadly conceptualize anti-establishment sentiments as populist attitudes, anti-establishment conspiracy theories, and political extremism. It first needs to be acknowledged that these are three specific and conceptually different constructs. For instance, populist attitudes are a worldview (Mudde, 2004; Müller, 2016) or communication style (Jagers & Walgrave, 2007) that highlight a struggle between the "elites" and "the people". Conspiracy theories in turn include a concrete allegation of unethical or even criminal conduct by those elites (Van Prooijen, 2022). Moreover, populism is not necessarily the same as political extremism: Some populist leaders are not politically extreme (e.g., Berlusconi; see Mudde, 2004) and some populist movements combine traditionally "left-wing" and "right-wing" positions, making them difficult to classify as far-left or far-right (e.g., Italy's Five-Star Movement).

These differences notwithstanding, populism, conspiracy theories, and political extremism are closely related constructs and particularly converge in a psychological rejection of societal elites. Populist attitudes are robustly associated with conspiracy beliefs (Erisen et al., 2021; Silva et al., 2017), a finding that has replicated across 13 EU countries, albeit with strong variation in effect sizes (Van Prooijen, Cohen Rodrigues, et al., 2022). Moreover, even when the relationship between populism and political extremism is imperfect, the political reality is that populist parties in national parliaments often are at the far-left or far-right (e.g., Akkerman et al., 2014; Mudde & Kaltwasser, 2017; Müller, 2016). Finally, political extremism is associated with increased belief in conspiracy theories (Imhoff et al., 2022; Van Prooijen, Krouwel, & Pollet, 2015), particularly when those conspiracy theories are anti-establishment (i.e., conspiracy theories that target vulnerable groups in society are more common at the political right; Nera et al., 2021). The present contribution is primarily focused on what these constructs have in common – namely, anti-establishment sentiments – and how insecurity links these constructs. Also, the chapter addresses the question of what makes these anti-establishment sentiments appealing when people feel insecure.

Does Insecurity Increase Anti-Establishment Sentiments?

A common assumption is that anti-establishment sentiments are rooted in feelings of insecurity (e.g., Douglas et al., 2019; Mols & Jetten, 2017; Van den Bos, 2020; Van Prooijen, 2020; see also Forgas; Hogg & Gaffney; Kreko; and Van den Bos, this volume). What empirical evidence exists for this assumption? The current section will examine the effects of insecurity on populism, conspiracy theories, and political extremism separately. While theorizing and research differ across these fields, I will examine to what extent these fields converge in acknowledging the causal role of insecurity.

Populism

Populism is a complex and multilayered construct, and important differences exist between left- and right-wing populist movements (e.g., Akkerman,

Zaslove, & Spruyt, 2017). Yet, all populist movements share at least two underlying dimensions: *Anti-elitism*, defined as an aversion against ruling elites and power holders; and *people-centrism*, defined as a belief that the "will of the people" should be the leading principle in politics (e.g., Mudde, 2004; Müller, 2016). Most relevant for the present purposes is the anti-elitism dimension, which is also a large part of the appeal of populist movements. For instance, people who score low on agreeableness – making them generally distrustful and cynical in their interpersonal relationships – are drawn to populist movements, particularly due to the anti-establishment rhetoric of such movements (Bakker, Schumacher, & Rooduijn, 2021). Moreover, populist movements' tendency to blame pressing societal problems on the actions of the political and societal establishment provides people with a causal and simple explanation for their feelings of insecurity (Erisen et al., 2021): The problems underlying their insecurity are largely or entirely the result of mismanagement and corruption by societal elites.

Qualitative analyses have underscored that ontological insecurity – that is, feelings of uncertainty, anxiety, and fear – explains people's support for populist movements, particularly at the far-right (Kinvall, 2018; Forgas, this volume). Quantitative studies also support this notion, as many studies have established relationships between populist attitudes and dissatisfaction with society – a feeling that can be expected particularly among citizens who experience insecurity as a result of the way society is governed. For instance, societal pessimism is strongest among supporters of right-wing populist movements, closely followed by supporters of left-wing populist movements (Steenvoorden & Harteveld, 2018; see also Jussim et al., this volume). Protest attitudes also predict voting for populist leaders, suggesting that populist attitudes are rooted in dissatisfaction with the societal system (Schumacher & Rooduijn, 2013). Finally, one study has found that collective angst – a collective emotion signaling that a group to which perceivers attach their identity is under threat – predicts populist attitudes (Van Prooijen, Rosema, Chemke-Dreyfus, Trikaliti, & Hormigo, 2022).

Accumulating research also supports a causal effect of insecurity on populist attitudes. For instance, a large-scale longitudinal study (including six waves over a time period of five years) has revealed that societal discontent predicts a progressive increase in populist attitudes over time. Interestingly, this study also provides evidence for a reverse temporal order: Populist attitudes also fuel increased societal discontent over time. Apparently, populist movements are more appealing to citizens who are dissatisfied, but also populist communication increases feelings of dissatisfaction (Rooduijn, Van der Brug, & De Lange, 2016). Furthermore, experimental research suggests that insecurity increases people's preference for a populist leader (Sprong et al., 2019; see also Pyszczynski, this volume).

These effects depend on the subjective experience of insecurity, which holds complex implications for the link between objective life circumstances and populism. In their work on the "Wealth Paradox," Mols and Jetten (2017) have found that economic hardship is a poor predictor of populist attitudes. While in some countries populist sentiments are strong among the economically deprived segments of society, in other countries populist sentiments are quite strong, and sometimes even stronger, among the economically affluent. The explanation for this paradox is that economically affluent people also experience their own forms of insecurity, notably status anxiety – the fear of losing one's privilege. Instead, economic inequality is a good macro-level predictor of support for strong, populist leaders, as inequality promotes perceptions of anomie among citizens: A form of insecurity characterized by the belief that society is breaking down (Sprong et al., 2019). Altogether, the evidence supports the notion that subjective feelings of insecurity predict increased populist attitudes.

Conspiracy Theories

One key insight is that conspiracy theories flourish, particularly in societal crisis situations (Van Prooijen & Douglas, 2017; see also Kreko, this volume). For instance, the recent COVID-19 pandemic has stimulated many conspiracy theories asserting that the coronavirus was created by humans in the lab, that 5G radiation was the actual cause of the pandemic, that the coronavirus is a hoax, and that the mRNA vaccines to protect people from COVID-19 have been tampered with (Bierwiaczonek, Gundersen, & Kunst, 2022; Freeman et al., 2022; Imhoff & Lamberty, 2020; Pummerer et al., 2021; Van Prooijen et al., 2021). Theoretical models have highlighted that the link between societal crisis situations and conspiracy beliefs is due to the subjective insecurity that people experience during such situations. Specifically, in crisis situations, people often experience existential threat, which broadly entails the insecurities that people feel when they are harmed or suffer losses (Douglas et al., 2019; Van Prooijen, 2020; see also Fiske; Pyszczynski & Sundby, this volume). Such existential threats instigate an epistemic sense-making process that may lead people to blame the threat on the covert actions of salient antagonistic (and often powerful) outgroups. The resulting conspiracy theories, in turn, do not necessarily alleviate such threats and may even increase them by signaling the dangers of hostile conspiracies (see also Liekefett, Christ, & Becker, 2021; Van Prooijen & Van Vugt, 2018).

A wealth of research has supported the idea that feelings of insecurity are associated with increased conspiracy beliefs. Many empirical studies have established that conspiracy beliefs are correlated with psychological indicators such as trait anxiety (Grzesiak-Feldman, 2013), fear of death (Newheiser, Farias, & Tausch, 2011), powerlessness and anomie (Abalakina-Paap, Stephan, Craig, & Gregory, 1999), self-uncertainty (Van Prooijen, 2016), and lack of control (Kofta, Soral, & Bilewicz, 2020; Van Prooijen & Acker, 2015). Societal indicators also support a link between insecurity and conspiracy beliefs, as conspiracy beliefs are related to perceived economic inequality (Casara, Suitner, & Jetten, 2022), levels of corruption in a country (Alper, 2022), and the perception that core values in society are changing (Federico, Williams, & Vitriol, 2018).

These associations are further supported by experimental studies, which show causal effects of insecurity indicators on conspiracy beliefs. Conspiracy beliefs increase following experimental manipulations of the perception that one's societal system is under threat (Jolley et al., 2018; Mao et al., 2021), perceived economic inequality (Casara et al., 2022), and ostracism (Poon et al., 2020). Moreover, conspiracy beliefs increase when people experience a lack of political control (Kofta et al., 2020) or personal control (Van Prooijen & Acker, 2015; Whitson & Galinsky, 2008). These effects of lacking control seem restricted to specific conspiracy beliefs, however (i.e., specific allegations pointing to the concrete actions of an identifiable group) and do not generalize to more general beliefs that events in the world are caused by conspiracies, referred to as conspiracy mentality (Stojanov & Halberstadt, 2020). Empirical evidence supports the notion that feelings of insecurity increase conspiracy beliefs.

Political Extremism

Political extremism can be defined as the extent to which citizens polarize into, and strongly identify with, generic left- or right-wing ideological outlooks on society (Van Prooijen & Krouwel, 2019). This definition explicitly includes radical groups that use aggression against what they see as "the establishment" to support their ideological goals (e.g., physical violence toward the police; death threats toward politicians), but it also includes regular, law-abiding citizens that self-identify as extremely left-wing or right-wing, and vote for anti-establishment parties at the political fringes (e.g., at the left, for instance, "die Linke" in Germany; Podemos in Spain; the Socialist party in the Netherlands; at the right, for instance, "AfD" in Germany; "PVV" in the Netherlands; "Front National" in France). Arguably the most influential theoretical perspective accounting for political extremism is significance quest theory (Kruglanski et al., 2014; see also Kruglanski & Ellenberg, this volume). While originally designed to explain violent extremism, the processes stipulated by the significance quest theory can also explain relatively extreme political beliefs among regular citizens (Van Prooijen & Kuijper, 2020; Webber et al., 2018). The significance quest theory proposes that people become susceptible to radical ideologies when they experience significance loss – that is, feelings of insecurity through experiences of humiliation, deprivation, loss, or injustice, either suffered by themselves or their group. These experiences inspire a quest for significance, which people may satisfy by strongly committing to focal ideological goals. In particular, endorsing extreme political ideologies helps to restore a sense of significance through the feeling that one is supporting a meaningful cause.

Empirical research supports a link between feelings of insecurity and political extremism. One large-scale study conducted in the Netherlands assessed participants' socio-economic fear: That is, fear that the wellbeing of oneself or one's collective will be compromised by political or societal developments. Results revealed evidence for a quadratic effect, such that the left- and right extremes expressed stronger socio-economic fear than political moderates (Van Prooijen, Krouwel, Boiten, & Eendebak, 2015). Relatedly, feelings of societal discontent predict voting for extremely left-wing or right-wing political parties (Van der Bles, Postmes, LeKander-Kanis, & Otjes, 2018). Other research found a robust link between political extremism and anger through analyses of the emotional tone of Twitter feeds, texts by political organizations, rhetoric by members of the US Congress, and news media. This link was attributable to increased feelings of threat from political opponents among political extremists (Frimer, Brandt, Melton, & Motyl, 2019).

Experimental studies expand these correlational data by revealing the causal effects of insecurity on political extremism. Burke, Kosloff, and Landau (2013) have assessed the effects of mortality salience on political ideology through a meta-analysis: Does reminding people of their own mortality produce shifts to the political right or to both extremes? This analysis yielded support for both perspectives. Apparently, feelings of insecurity – as caused by a reminder that death is inevitable – increase political extremism, although this effect is likely context-dependent in light of the finding that exclusively right-wing shifts also occur (cf. Pyszczynski & Sundby, this volume). Additional experiments have shown that inducing self-uncertainty increases peoples' identification with radical groups (Hogg & Gaffney, this volume; Hogg, Meehan, & Farquharson, 2010). Moreover, Hales and Williams (2018) manipulated whether or not participants were ostracized by others, an aversive experience that is associated with feelings of insecurity. Such ostracism increased participants' willingness to attend a meeting of a radical activist group to reduce tuition fees at the university and participants' interest in being a member of a gang. Altogether, the evidence indicates that feelings of insecurity stimulate support for politically extreme ideas and movements.

Conclusion

The fields of populism, conspiracy theories, and political extremism are separate research domains that have their own theoretical perspectives and empirical research traditions. Yet, these fields are also empirically and conceptually related, particularly as high levels of populism, conspiracy theories, and political extremism often imply a psychological rejection of the political and societal establishment. For the present purposes, it is therefore of notable interest that these three research fields have all attributed a central causal role to feelings of insecurity, broadly defined. The findings reviewed here are therefore consistent with a model asserting that feelings of insecurity increase anti-establishment sentiments.

The Appeal of Anti-Establishment Sentiments When Feeling Insecure

Why do feelings of insecurity make anti-establishment sentiments so appealing? The answer to this question is relatively complicated due to the antagonistic nature of anti-establishment sentiments. Specifically, a natural desire when people feel insecure is to regulate such feelings in order to re-establish a sense of safety (e.g., Van den Bos, 2009; see also Hirschberger, this volume). Yet, the paradox here is that anti-establishment sentiments are not particularly effective in reducing feelings of insecurity (Kreko, this volume). Recall that in the long run, populist attitudes *increase* (instead of decrease) feelings of societal discontent (Rooduijn et al., 2016). Similar findings have been observed for conspiracy theories: Two longitudinal studies have revealed that over time, conspiracy theories do not decrease, and sometimes even increase, feelings of anxiety, uncertainty aversion, and existential threat (Liekefett et al., 2021). Believing that powerful societal actors have nefarious intentions apparently does not make people feel secure. Anti-establishment sentiments hence can be experienced as threatening in themselves, creating a self-reinforcing feedback loop that amplifies further anti-establishment sentiments (Van Prooijen, 2020).

Despite the ineffectiveness of establishing feelings of safety in the long run, anti-establishment sentiments contain a range of properties that make them appealing when experiencing insecurity. Specifically, in this section, I propose that anti-establishment sentiments provide people with a sense of meaning and purpose by (1) identifying a clear cause of their problems and a justification of perceived solutions (i.e., epistemic clarity), (2) exaggerating the importance of themselves and their groups, and (3) stimulating a range of emotional processes that are not solely negative. As such, anti-establishment sentiments provide people with psychological tools to cope with feelings of insecurity in the short run; however, insofar as these tools neither remove societal power holders from their positions nor restore trust in them, they are ineffective in establishing a sense of security in the long run. In the following, I will review these three short-term coping mechanisms in more detail. Figure 20.1 displays the links between insecurity, anti-establishment sentiments, and the factors that make anti-establishment appealing.

Epistemic Clarity

A basic evolutionary motivation when faced with insecurity is to understand and identify the causes of these aversive feelings (von Hippel & Merakovsky, this volume). Anti-establishment sentiments provide a reasonably straightforward conceptual framework to achieve such epistemic clarity by proffering a relatively simple explanation and solution for the personal or societal problems that gave rise to feelings of insecurity. In particular, anti-establishment

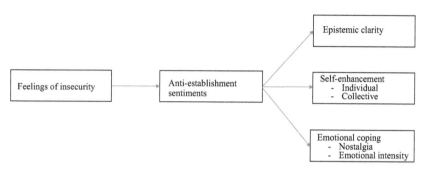

FIGURE 20.1 The relationships between feelings of insecurity, anti-establishment sentiments, and the factors that make anti-establishment sentiments appealing.

sentiments imply that personal or societal problems are externally attributed to the incompetence or immorality of powerful elites and can be solved by replacing these elites with populist or politically extreme leaders who acknowledge these problems and provide simple solutions for them. Anti-establishment sentiments hence provide people with a sense of meaning and purpose by giving them a clear scapegoat for their predicaments (i.e., societal elites), a belief to understand the causes of the situation they are insecure about, and a sense of direction by identifying and legitimizing the actions that are necessary to solve these problems.

Consistent with this perspective, populism, conspiracy beliefs, and political extremism are all empirically related to the belief that simple solutions exist for complex societal problems (Erisen et al., 2021; Van Prooijen et al., 2015), an apparently universal human evolutionary tendency that is present in many other areas of social judgment as well (see Krueger & Gruening, this volume). Relatedly, political extremism predicts a cognitively simplified perception of political stimuli (Lammers, Koch, Conway, & Brandt, 2017) and is associated with a need for closure (Webber et al., 2018). Populist attitudes are rooted not in critical thinking but in gullibility (Van Prooijen, Cohen Rodrigues, et al., 2022), and conspiracy beliefs are associated with lower education levels and decreased analytical thinking (Van Prooijen, 2017). These various findings are consistent with the broader assertion that a relatively simplistic information processing style, which ignores many of the complexities commonly associated with societal problems, predicts anti-establishment sentiments.

Somewhat paradoxically, such a simplistic information processing style is likely to increase the confidence that people have in their beliefs. Reducing complexity also reduces ambiguity, increasing certainty (cf. Jussim et al., this volume). This process may be partially compensatory. Feelings of insecurity in one domain (e.g., through experiences of humiliation, injustice, or economic anxiety) may stimulate feelings of certainty in a different, ideological domain (e.g., McGregor, Prentice, & Nash, 2013). Research findings, for instance, reveal that, as compared with political moderates, people at the political left and right extremes are more convinced of the objective correctness of their ideological beliefs (Toner, Leary, Asher, & Jongman-Sereno, 2013) and express more certainty about their knowledge of societal problems despite not having more knowledge (Van Prooijen, Krouwel, & Emmer, 2018). Relatedly, overclaiming one's own knowledge about political issues predicts subsequent anti-establishment voting during a referendum (Van Prooijen & Krouwel, 2020). Populist attitudes predict rejection of science due to an increased reliance on one's own "common sense" (Staerklé, Cavallero, Cortijos-Bernabeu, & Bonny, 2022), and the link between populist attitudes and conspiracy beliefs is mediated by the faith that people have in their own intuitions (Van Prooijen, Cohen Rodrigues, et al., 2022). Rejecting the scientific approach when it challenges left-wing ideology is even common in some academic institutions (see also Jussim et al., this volume). Apparently, through anti-establishment sentiments, people can compensate for their underlying feelings of insecurity through increased confidence in their subjective beliefs.

The epistemically clear worldview that anti-establishment sentiments offer also gives a sense of direction and provides a justification for actions against the establishment. These actions may range from voting behavior (Van der Bles et al., 2018), peaceful protest or civic disobedience (e.g., not complying with governmental regulations to contain the coronavirus; Bierwiaczonek et al., 2022), to criminal behavior, such as destruction of property (Jolley & Paterson, 2020), and other forms of extremist violence (Rottweiler & Gill, 2020). Conspiracy theories appear to accelerate the process through which extremist political groups turn violent (Bartlett & Miller, 2010). Hence, anti-establishment sentiments not only contribute to a sense of understanding the causes of personal or societal problems but also stimulate action that perceivers believe may solve these problems. If such actions are ineffective, however, the epistemic clarity that anti-establishment sentiments bring may be frustrating in the long run.

Self-enhancement

Feelings of insecurity are threatening to perceivers and their groups by implying that they are meaningless, marginalized, and insignificant. For instance, experiences of humiliation and injustice directly suggest that the perceivers and their group are considered to be of low worth by the perpetrators. Moreover, being anxious about one's economic future suggests that one is, or might become, of low status. One way to cope with insecurity may therefore be self-enhancement: An increased, and often exaggerated, perceived importance of oneself and one's groups. People can derive such importance from the notion that they and their groups are part of an important struggle in pursuit of a better world. Anti-establishment sentiments may therefore be appealing when faced with insecurity, by increasing self-enhancement: Believing that powerful societal actors are corrupt provides perceivers with the meaningful goal to oppose oppressors. This process corresponds to the notion that extremist beliefs are driven by a quest for significance (Kruglanski et al., 2014).

Empirical findings support a link between anti-establishment sentiments and self-enhancement. Consistent with significance quest theory (Kruglanski et al., 2014), both political and religious extremism are associated with the extent to which people derive a sense of significance from their ideological beliefs (Van Prooijen & Kuijper, 2020). Moreover, narcissism – the dispositional tendency to exaggerate one's own self-worth – predicts political extremism (Duspara & Greitemeyer, 2017) and conspiracy beliefs (Cichocka et al., 2016; see also Forgas, this volume). Finally, an individual-level predictor of conspiracy beliefs is the need to feel unique and special: Conspiracy beliefs often imply that perceivers have discovered truly important information about the malpractice of powerful societal actors and are in a unique position to expose the conspiracy to the public (Imhoff & Lamberty, 2017; Lantian, Muller, Nurra, & Douglas, 2017). At the individual level, anti-establishment sentiments are associated with an inflated sense of self-importance.

These observations generalize to the group level in that anti-establishment sentiments are associated with a perception of one's own group as superior. In fact, the quest for collective significance appears to be more impactful than the quest for personal significance in explaining violent extremism (Jasko et al., 2020; see also Hirschberger; Hogg & Gaffney, this volume). A frequently investigated construct in the context of anti-establishment sentiments is collective narcissism: A belief that one's ingroup is exceptional and superior, yet insufficiently recognized by others. Such collective narcissism is associated with intergroup hostility and is therefore often part of the rhetoric of politically extreme or populist movements that construe the societal or political establishment as an antagonistic outgroup (Golec de Zavala, Dyduch-Hazar, & Lantos, 2019; Golec de Zavala & Lantos, 2020). Collective narcissism indeed predicts support for populist movements (Federico & Golec de Zavala, 2018; Marchlewska, Cichocka, Panayiotou, Castellanos, & Batayneh, 2018; Forgas, this volume). Furthermore, collective narcissism is robustly associated with conspiracy beliefs, both in longitudinal studies (Golec de Zavala & Federico, 2018) and across many different nations (Sternisko, Cichocka, Cislak, & Van Bavel, 2021). Anti-establishment sentiments are associated with self-enhancement motives at both the individual and collective levels.

Emotional Coping

While anti-establishment sentiments have been mostly associated with negative feelings such as anxiety, dissatisfaction, and anger (e.g., Frimer et al.,

2019; Rooduijn et al., 2016; Van Prooijen et al., 2015), recent findings have suggested that anti-establishment sentiments are also associated with positive emotional experiences that help people cope with feelings of insecurity. Here, I will highlight two different forms of emotional coping. The first is that anti-establishment sentiments elicit feelings of nostalgia, which help to psychologically buffer against feelings of insecurity (e.g., Baldwin, White, & Sullivan, 2018). The second is that people experience anti-establishment sentiments with intense emotions, making them attractive to sensation-seekers and providing people with a form of entertainment.

Nostalgia. Anti-establishment rhetoric typically does not only criticize current societal leaders; it often also includes memories of a glorious past. This is evident, for instance, in the rhetoric of Trump during the 2016 election campaign ("Make America Great Again") as well as the arguments in favor of a "Leave" vote during the Brexit referendum ("We want our country back"). Populist leaders often accuse the political establishment of "ruining" their country, thus implicitly or explicitly suggesting that the country was in much better shape during the "good old days." Besides painting a bleak picture of current society, anti-establishment movements increase citizens' hope for the future by promising a return to the past. This suggests that anti-establishment sentiments are associated with feelings of nostalgia, defined as a sentimental longing for a better past. Nostalgia is a bittersweet feeling that contains both negative and positive emotions, although accumulating research suggests that nostalgia has stronger positive than negative effects. In particular, feelings of nostalgia help people cope with threats by increasing positive affect, self-esteem, a sense of belonging, and meaning in life (Sedikides, Wildschut, Arndt, & Routledge, 2008).

Empirically, feelings of nostalgia are particularly pronounced among anti-establishment parties the far-right that voice strong anti-immigration sentiments (Lammers & Baldwin, 2020). This is likely due to various specific features of the far-right, including a relatively strong focus on tradition (Jost, 2017) in combination with a link of nostalgia with rejection of immigrants who were not part of a country's glorious past (Smeekes et al., 2018). Also independent of the content of one's ideology, however, anti-establishment sentiments contain properties that are likely to increase feelings of nostalgia. The societal pessimism espoused by left- and right-wing populist movements usually pertains to present-day society, emphasizing how society has deteriorated as compared with the past (Steenvoorden & Harteveld, 2018). Consistent with this line of reasoning, the speeches of the Spanish left-wing populist party *Podemos* contain many references to a glorious national past that, according to this movement, has been lost through the policies of current political elites (Custodi, 2021). The same kind of appeal to populist nostalgia is also present in the propaganda used by right-wing autocrats such as Victor Orban in Hungary (see Forgas, this volume).

Three empirical studies have investigated the link between populist attitudes and feelings of nostalgia while empirically controlling for political orientation (Van Prooijen, Rosema, et al., 2022). These studies consistently revealed that, above and beyond the effects of political orientation, populist attitudes are uniquely associated with feelings of nostalgia. Also, experimental findings indicated a causal effect such that exposing people to populist rhetoric increases feelings of nostalgia. When people feel insecure, anti-establishment political movements are attractive as they stimulate feelings of nostalgia.

Emotional intensity. Anti-establishment sentiments imply that people perceive themselves as part of a competitive struggle with powerful others. To some extent, this may be frightening, but it might also be exciting. Perceiving a struggle with a powerful enemy can be considered adventurous, and presuming hidden evil plots by powerholders gives people the opportunity to uncover a mystery (see also Hirschberger, this volume). More generally, the common assumption throughout social psychology that insecurity is an exclusively aversive experience does not correspond well with the realities of everyday life, as people quite often make decisions that will make them insecure (for related arguments, see Fiedler & McCaughey; Kruglanski & Ellenberg, this volume). Sometimes people make short-term choices that increase insecurity merely for their own entertainment (e.g., watching scary movies; gambling; skydiving; traveling to unknown places) but sometimes people also make impactful choices that imply substantial uncertainty about whether they will ultimately change life for better or worse (e.g., quitting a stable and well-paying job to start a risky business; divorcing a reliable marriage partner to pursue a passionate affair). Various research domains in social psychology acknowledge these attractive features of insecurity, for instance through Prospect Theory's notion that people sometimes are risk-seeking (Kahneman & Tversky, 1979). Also, people differ in their craving for new, exciting, and sometimes risky experiences, a trait referred to as sensation seeking (Zuckerman, 1994).

Anti-establishment sentiments make these positive aspects of insecurity salient by highlighting a challenging but exciting and meaningful struggle with powerful societal actors. These positive aspects of insecurity can be summarized through people's desire for intense emotional experiences, which may be positive, negative, or both in valence (Van Boven, Kane, McGraw, & Dale, 2010). Such intense emotions distract people from boredom and stimulate the feeling that one is living life to its fullest. If the perceived struggle with the societal or political establishment is appealing due to their link with intense emotional experiences, one hypothesis that follows is that anti-establishment sentiments should be more pronounced among people high in sensation seeking. Various studies support this notion. Sensation seeking appears to be one possible motivation to join violent extremist groups that fight against the political establishment (Nussio, 2020). Also, survey studies reveal that sensation seeking predicts political extremism, as reflected in a stronger willingness to self-sacrifice and use violence in support of political goals (Schumpe, Bélanger, Moyano, & Nisa, 2020). Moreover, both dispositional susceptibility to boredom (Brotherton & Eser, 2015) and sensation-seeking more generally (Van Prooijen, Ligthart, Rosema, & Xu, 2022) predict increased belief in conspiracy theories.

Experimental studies have shown that one reason people believe conspiracy theories is because they find them entertaining. In two experiments, participants were exposed to conspiracy theories that the Notre Dame was set on fire deliberately and that the wealthy sex offender Jeffrey Epstein was murdered in his prison cell. In the control conditions, participants read the official reading of these events (i.e., the Notre Dame fire was accidental; Jeffrey Epstein committed suicide). Participants found the conspiratorial text more entertaining than the control text; moreover, these entertainment appraisals mediated the effects of the manipulation on the extent to which participants believed these conspiracy theories. Furthermore, entertainment appraisals were strongly correlated with emotional intensity but weakly with emotional valence (Van Prooijen, Ligthart, et al., 2022). Anti-establishment sentiments can be exciting, thus emphasizing the rewarding aspects of insecurity.

Discussion and Conclusion

The present chapter sought to examine the link between insecurity and anti-establishment sentiments by pursuing two broad goals. The first goal was to examine what empirical evidence exists for the common assumption that anti-establishment sentiments originate from underlying feelings of insecurity. Integrating the research literatures on populism, conspiracy theories, and political extremism reveals the central causal role of insecurity in explaining these different expressions of anti-establishment sentiments (e.g., Forgas; Hogg & Gaffney; Kreko, this volume). The second goal was to assess the appeal of anti-establishment sentiments when people feel insecure. A review of recent studies suggests that anti-establishment sentiments provide people with a sense of meaning and purpose in three complementary manners. First, anti-establishment sentiments increase epistemic clarity by suggesting a clear cause of and a clear solution for one's problems (Jussim et al.; Krueger & Dunning, this volume). Second, anti-establishment sentiments increase self-enhancement by providing people with the feeling that they are pursuing a meaningful cause, making them feel important. And third, anti-establishment sentiments help people cope with insecurity through positive emotional processes. Anti-establishment movements promise a return to a better past, thus creating hope for the future through feelings of nostalgia. Also, the struggle against political elites may, to some extent, be experienced as exciting, providing people with intense emotional experiences that they consider rewarding.

It should be emphasized that while these processes make anti-establishment sentiments appealing in the short run, they can likewise make these sentiments frustrating in the long run. Clearly recognizing the establishment as responsible for one's problems, seeing oneself and one's groups as important, feeling nostalgic, and feeling excited may all contribute to a sense of hope that society will change for the better in the near future. But as time progresses and nothing really seems to change, people may lose hope, thus further increasing feelings of insecurity (Liekefett et al., 2021; Rooduijn et al., 2016). A more effective coping strategy in the long run to reduce cognitive dissonance may therefore be attempts to restore trust with societal powerholders (see Cooper & Pearce, this volume). This can be quite challenging in practice, however, as many citizens experience a deep-rooted distrust toward power holders. Moreover, actual integrity violations and examples of mismanagement among power holders do occur and reinforce feelings of insecurity and anti-establishment sentiments.

The broad conceptualization of anti-establishment sentiments through populism, conspiracy theories, and political extremism implies both a strength and a limitation. It is a strength by enabling a broad analysis of research relevant to understand anti-establishment sentiments while also integrating these closely related research domains. It is also a limitation, however, as there are conceptual differences between these constructs, and for some of the findings reviewed here, it is yet unclear whether they generalize to other expressions of anti-establishment sentiments. Clearly, these are research fields in development, and the ideas presented in the current chapter may suggest novel research questions about anti-establishment sentiments. For instance, do people vote for anti-establishment politicians (e.g., Trump) simply because they find them more entertaining than "mainstream" candidates (e.g., Clinton; Biden)? Likewise, is belief in anti-establishment conspiracy theories associated with feelings of nostalgia?

Altogether, the present chapter provides a conceptual framework that assigns a pivotal role to feelings of insecurity in understanding anti-establishment sentiments. This framework may also help explain why anti-establishment political movements have been relatively successful across the US and the EU in the past decade. The world has faced (and is still facing) many realistic challenges, including refugee crises, wars, a pandemic, climate change, political polarization, and economic volatility. Also, we live in a time where society changes quickly, for instance due to rapid technological development and shifting norms (e.g., about gender roles, race, and sexual identity) that call long-standing traditions into question. Finally, through the Internet and social media, people have unlimited access to information, which may confuse them when different sources provide conflicting information. For better or worse, these societal developments are likely to fuel feelings of insecurity among large groups of citizens. It can hence be expected that widespread anti-establishment sentiments will continue to be a reality in the foreseeable future.

References

Abalakina-Paap, M., Stephan, W., Craig, T., & Gregory, W. L. (1999). Beliefs in conspiracies. *Political Psychology, 20,* 637–647.

Alicke, M. D. (2000). Culpable control and the psychology of blame. *Psychological Bulletin, 126,* 556–574.

Akkerman, A., Mudde, C., & Zaslove, A. (2014). How populist are the people? Measuring populist attitudes in voters. *Comparative Political Studies, 47,* 1324–1353.

Akkerman, A., Zaslove, A., & Spruyt, B. (2017). 'We the People' or 'We the Peoples'? A comparison of support for the populist radical right and populist radical left in the Netherlands. *Swiss Political Science Review, 23,* 377–403.

Alper, S. (2021). There are higher levels of conspiracy beliefs in more corrupt countries. *PsyArXiv.* doi: 10.31234/osf.io/2umfe

Bakker, B. N., Schumacher, G., & Rooduijn, M. (2021). The populist appeal: Personality and antiestablishment communication. *The Journal of Politics, 83,* 589–601.

Baldwin, M., White, M. H., & Sullivan, D. (2018). Nostalgia for America's past can buffer collective guilt. *European Journal of Social Psychology, 48,* 433–446.

Bartlett, J., & Miller, C. (2010). *The power of unreason: Conspiracy theories, extremism and counterterrorism.* London, UK: Demos.

Bierwiaczonek, K., Gundersen, A. B., & Kunst, J. R. (2022). The role of conspiracy beliefs for COVID-19 health responses: A meta-analysis. *Current Opinion in Psychology, 46,* 101346. https://doi.org/10.1016/j.copsyc.2022.101346

Brotherton, R., & Eser, S. (2015). Bored to fears: Boredom proneness, paranoia, and conspiracy theories. *Personality and Individual Differences, 80,* 1–5.

Burke, B. L., Kosloff, S., & Landau, M. J. (2013). Death goes to the polls: A meta-analysis of mortality salience effects on political attitudes. *Political Psychology, 34,* 183–200.

Butter, M., & Knight, P. (2020). *Routledge Handbook of Conspiracy Theories.* Oxon, UK: Routledge.

Casara, B. G. S., Suitner, C., & Jetten, J. (2022). The impact of economic inequality on conspiracy beliefs. *Journal of Experimental Social Psychology, 98,* 104245.

Cichocka, A., Marchlewska, M., & Golec de Zavala, A. (2016). Doe self-love or self-hate predict conspiracy beliefs? Narcissism, self-esteem, and the endorsement of conspiracy theories. *Social Psychological and Personality Science, 7,* 157–166.

Custodi, J. (2021). Nationalism and populism on the left: The case of Podemos. *Nations and Nationalism, 27,* 705–720.

Douglas, K. M., Uscinski, J. E., Sutton, R. M., Cichocka, A., Nefes, T., Ang, C. S., & Deravi, F. (2019). Understanding conspiracy theories. *Advances in Political Psychology, 40,* 3–35.

Duspara, B., & Greitemeyer, T. (2017). The impact of dark tetrad traits on political orientation and extremism: An analysis in the course of a presidential election. *Heliyon, 3,* e00425.

Erisen, C., Guidi, M., Martini, M., Toprakkiran, S., Isernia, P., & Littvay, L. (2021). Psychological correlates of populist attitudes. *Advances in Political Psychology, 42,* 149–172.

Federico, C. M., & Golec de Zavala, A. (2018). Collective narcissism and the 2016 US presidential vote. *Public Opinion Quarterly, 82*(1), 110–121. https://doi.org/10.1093/poq/nfx048

Federico, C. M., Williams, A. L., & Vitriol, J. A. (2018). The role of system identity threat in conspiracy theory endorsement. *European Journal of Social Psychology, 48,* 927–938.

Freeman, D., Waite, F., Rosebrock, L., Petit, A., Causier, C., East, A.,… Lambe, S. (2022). Coronavirus conspiracy beliefs, mistrust, and compliance with government guidelines in England. *Psychological Medicine, 52,* 251–263. doi:10.1017/S0033291720001890

Frimer, J. A., Brandt, M. J., Melton, Z., & Motyl, M. (2018). Extremists on the Left and Right Use Angry, Negative Language. *Personality and Social Psychology Bulletin, 45,* 1216-123, doi: 10.1177/0146167218809705

Golec de Zavala, A., Dyduch-Hazar, K., & Lantos, D. (2019). Collective narcissism: Political consequences of investing self-worth in the ingroup's image. *Advances in Political Psychology, 40,* 37–74.

Golec de Zavala, A., & Federico, C. M. (2018). Collective narcissism and the growth of conspiracy thinking over the course of the 2016 United States presidential election: A longitudinal analysis. *European Journal of Social Psychology, 48,* 1011–1018.

Golec de Zavala, A., & Lantos, D. (2020). Collective narcissism and its social consequences: The bad and the ugly. *Current Directions in Psychological Science, 29,* 273–278.

Grzesiak-Feldman, M. (2013). The effect of high-anxiety situations on conspiracy thinking. *Current Psychology, 32,* 100–118.

Hales, A. H., & Williams, K. D. (2018). Marginalized individuals and extremism: The role of ostracism in openness to extreme groups. *Journal of Social Issues, 74,* 75–92.

Hogg, M. A., Meehan, C., & Farqueharson, J. (2010). The solace of radicalism: Self-uncertainty and group identification in the face of threat. *Journal of Experimental Social Psychology, 46,* 1061–1066.

Imhoff, R., & Bruder, M. (2014). Speaking (un-)truth to power: Conspiracy mentality as a generalized political attitude. *European Journal of Personality, 28,* 25–43.

Imhoff, R., & Lamberty, P. (2017). Too special to be duped: Need for uniqueness motivates conspiracy beliefs. *European Journal of Social Psychology, 47,* 724–734.

Imhoff, R., & Lamberty, P. (2020). A bioweapon or a hoax? The link between distinct conspiracy beliefs about the coronavirus disease (COVID-19) outbreak and pandemic behavior. *Social Psychological and Personality Science, 11,* 1110–1118.

Imhoff, R., Zimmer, F., Klein, O., António, J. H. C., Babinska, M., Bangerter, A., … & Van Prooijen, J.-W. (2022). Conspiracy mentality and political orientation across 26 countries. *Nature Human Behaviour, 6,* 392–403.

Jagers, J., & Walgrave, S. (2007). Populism as political communication style: An empirical study of political parties' discourse in Belgium. *European Journal of Political Research, 46,* 319–345.

Jasko, K., Webber, D., Kruglanski, A. W., Gelfand, M., Taufiqurrohman, M., Hettiarachchi, M., & Gunaratna, R. (2020). Social context moderates the effects of quest for significance on violent extremism. *Journal of Personality and Social Psychology, 118,* 1165–1187. https://doi.org/10.1037/pspi0000198

Jolley, D., Douglas, K. M., & Sutton, R. M. (2018). Blaming a few bad apples to save a threatened barrel: The system-justifying function of conspiracy theories. *Political Psychology, 39,* 465–478.

Jolley, D., & Paterson, J. L. (2020). Pylons ablaze: Examining the role of 5G COVID-19 conspiracy beliefs and support for violence. *British Journal of Social Psychology, 59,* 628–640.

Jost, J. J. (2017). Ideological asymmetries and the essence of political psychology. *Political Psychology, 38,* 167–208.

Kahneman, D., & Tversky, A. (1979). Prospect theory: An analysis of decision under risk. *Econometrica, 47,* 263–291.

Kinvall, C. (2018). Ontological insecurities and postcolonial imaginaries: The emotional appeal of populism. *Humanity & Society, 42,* 523–543.

Kofta, M., Soral, W., & Bilewicz, M. (2020). What breeds conspiracy antisemitism? The role of political uncontrollability and uncertainty in the belief in Jewish conspiracy. *Journal of Personality and Social Psychology, 118,* 900–918.

Kruglanski, A. W., Gelfand, M. J., Bélanger, J. J., Sheveland, A., Hetiarachchi, M., & Gunaratra, R. (2014). The psychology of radicalization and deradicalization: How significance quest impacts violent extremism. *Advances in Political Psychology, 35*(S1), 69–93. https://doi.org/10.1111/pops.12163

Kutiyski, Y., Krouwel, A. P. M., & Van Prooijen, J.-W. (2021). Political extremism and distrust: Does radical political orientation predict distrust and negative attitudes towards European integration? *The Social Science Journal, 58,* 1–16. DOI: 10.1016/j.soscij.2019.03.004

Lammers, J., & Baldwin, M. (2020). Make America gracious again: Collective nostalgia can increase and decrease support for right-wing populist rhetoric. *European Journal of Social Psychology, 50,* 943–954.

Lammers, J., Koch, A., Conway, P., & Brandt, M. J. (2017). The political domain appears simpler to the political extremes than to political moderates. *Social Psychological and Personality Science, 8,* 612–622.

Lantian, A., Muller, D., Nurra, C., & Douglas, K. M. (2017). "I know things they don't know!": The role of need for uniqueness in belief in conspiracy theories. *Social Psychology, 48,* 160–173.

Liekefett, L., Christ, O., & Becker, J. C. (2021). Can conspiracy beliefs be beneficial? Longitudinal linkages between conspiracy beliefs, anxiety, uncertainty aversion, and existential threat. *Personality and Social Psychology Bulletin.* https://doi.org/10.1177/01461672211060965

Mao, J.-Y., Van Prooijen, J.-W., Yang, S.-L., & Guo, Y.-Y. (2021). System threat during a pandemic: How conspiracy theories help to justify the system. *Journal of Pacific Rim Psychology, 15,* 1–11.

Marchlewska, M., Cichocka, A., Panayiotou, O., Castellanos, K., & Batayneh, J. (2018). Populism as identity politics: Perceived in-group disadvantage, collective narcissism, and support for populism. *Social Psychological and Personality Science, 9,* 151–162.

McGregor, I., Prentice, M., & Nash, K. (2013). Anxious uncertainty and reactive approach motivation (RAM) for religious, idealistic, and lifestyle extremes. *Journal of Social Issues, 69,* 537–563.

Mols, F., & Jetten, J. (2017). *The wealth paradox: Economic prosperity and the hardening of attitudes.* Cambridge, UK: Cambridge University Press.

Mudde, C. (2004). The populist zeitgeist. *Government and Opposition, 39,* 541–563.

Mudde, C., & Kaltwasser, C. R. (2017). *Populism: A very short introduction.* New York, NY: Oxford University Press.

Müller, J.-W. (2016). *What is populism?* Philadelphia, PA: University of Pennsylvania Press.

Nera, K., Wagner-Egger, P., Bertin, P., Douglas, K., & Klein, O. (2021). A power-challenging theory of society, or a conservative mindset? Upward and downward conspiracy theories as ideologically distinct beliefs. *European Journal of Social Psychology, 51*, 740–757. https://doi.org/10.1002/ejsp.2769

Newheiser, A.-K., Farias, M., & Tausch, N. (2011). The functional nature of conspiracy beliefs: Examining the underpinnings of belief in the *Da Vinci Code* conspiracy. *Personality and Individual Differences, 51*, 1007–1011.

Nussio, E. (2020). The role of sensation seeking in violent armed group participation. *Terrorism and Political Violence, 32*, 1–19.doi: 10.1080/09546553.2017.1342633

Poon, K.-T., Chen, Z., & Wong, W.-Y. (2020). Beliefs in conspiracy theories following ostracism. *Personality and Social Psychology Bulletin, 46*, 1234–1246.

Pummerer, L., Böhm, R., Lilleholt, L., Winter, K., Zettler, I., & Sassenberg, K. (2021). Conspiracy theories and their societal effects during the COVID-19 pandemic. *Social Psychological and Personality Science.* https://doi.org/10.1177/19485506211000217

Rooduijn, M., Van der Brug, W., & De Lange, S. L. (2016). Expressing or fueling discontent? The relationship between populist voting and political discontent. *Electoral Studies, 43*, 32–40.

Rottweiler, B., & Gill, P. (2020). Conspiracy beliefs and violent extremist intentions: The contingent effects of self-efficacy, self-control and law-related morality. *Terrorism and Political Violence.* DOI: 10.1080/09546553.2020.1803288

Schumacher, G., & Rooduijn, M. (2013). Sympathy for the 'devil'? Voting for populists in the 2006 and 2010 Dutch general elections. *Electoral Studies, 32*, 124–133.

Schumpe, B., Bélanger, J. J., Moyano, M., & Nisa, C. F. (2020). The role of sensation seeking in political violence: An extension of the significance quest theory. *Journal of Personality and Social Psychology, 118*, 743–761.

Sedikides, C., Wildschut, T., Arndt, J., & Routledge, C. (2008). Nostalgia: Past, present, future. *Current Directions in Psychological Science, 17*, 304–307.

Silva, B. C., Vegetti, F., & Littvay, L. (2017). The elite is up to something: Exploring the relationship between populism and belief in conspiracy theories. *Swiss Political Science Review, 23*, 423–443.

Smeekes, A., Jetten, J., Verkuyten, M., Wohl, M. J., Jasinskaja-Lahti, I., Ariyanto, A.,... & Butera, F. (2018). Regaining In-Group Continuity in Times of Anxiety About the Group's Future. *Social Psychology, 49*, 311–329.

Sprong, S., Jetten, J., Wang, Z., Peters, K., Mols, F., Verkuyten, M.,...., & Wohl, M. (2019). "Our country needs a strong leader right now": Economic inequality enhances the wish for a strong leader. *Psychological Science, 30*, 1625–1637.

Staerklé, C., Cavallaro, M., Cortijos-Bernabeu, A., & Bonny, S. (2022). Common sense as a political weapon: Populism, science skepticism, and global crisis-solving motivations. *Political Psychology.* https://doi.org/10.1111/pops.12823

Steenvoorden, E., & Harteveld, E. (2018). The appeal of nostalgia: The influence of societal pessimism on support for populist radical right parties. *West European Politics, 41*, 28–52.

Sternisko, A., Cichocka, A., Cislak, A., & Van Bavel, J. J. (2021). National narcissism predicts the belief and dissemination of conspiracy theories during the COVID-19 pandemic: Evidence from 56 countries. *Personality and Social Psychology Bulletin.* https://doi.org/10.1177/01461672211054947

Stojanov, A., & Halberstadt, J. (2020). Does lack of control lead to conspiracy beliefs? A meta-analysis. *European Journal of Social Psychology, 50*, 955–968.
Toner, K., Leary, M., Asher, M. W., & Jongman-Sereno, K. P. (2013). Feeling superior is a bipartisan issue: Extremity (not direction) of political views predicts perceived belief superiority. *Psychological Science, 24*, 2454–2462.
Van Boven, L., Kane, J., McGraw, A. P., & Dale, J. (2010). Feeling close: Emotional intensity reduces perceived psychological distance. *Journal of Personality and Social Psychology, 98*, 872–885.
Van den Bos, K. (2009). Making sense of life: The existential self-trying to deal with personal uncertainty. *Psychological Inquiry, 20*, 197–217.
Van den Bos, K. (2020). Unfairness and radicalization. *Annual Review of Psychology, 71*, 563–588.
Van der Bles, A.-M., Postmes, T., LeKander-Kanis, B., & Otjes, S. (2018). The consequences of collective discontent: A new measure of zeitgeist predicts voting for extreme parties. *Political Psychology, 39*, 381–398.
Van Prooijen, J.-W. (2016). Sometimes inclusion breeds suspicion: Self-uncertainty and belongingness predict belief in conspiracy theories. *European Journal of Social Psychology, 46*, 267–279.
Van Prooijen, J.-W. (2017). Why education predicts decreased belief in conspiracy theories. *Applied Cognitive Psychology, 31*, 50–58.
Van Prooijen, J.-W. (2020). An existential threat model of conspiracy theories. *European Psychologist, 25*, 16–25.
Van Prooijen, J.-W. (2022). Injustice without evidence: The unique role of conspiracy theories in social justice research. *Social Justice Research, 35*, 88–106.
Van Prooijen, J.-W., & Acker, M. (2015). The influence of control on belief in conspiracy theories: Conceptual and applied extensions. *Applied Cognitive Psychology, 29*, 753–761.
Van Prooijen, J.-W., Cohen Rodrigues, T., Bunzel, C., Georgescu, O., Komáromy, D., & Krouwel, A. P. M. (2022). Populist gullibility: Conspiracy theories, news credibility, bullshit receptivity, and paranormal belief. *Political Psychology*. https://doi.org/10.1111/pops.12802
Van Prooijen, J.-W., & Douglas, K. M. (2017). Conspiracy theories as part of history: The role of societal crisis situations. *Memory Studies, 10*, 323–333.
Van Prooijen, J.-W., Etienne, T., Kutiyski, T., & Krouwel, A. P. M. (2021). Conspiracy beliefs prospectively predict health behavior and well-being during a pandemic. *Psychological Medicine*. Doi: 10.1017/S0033291721004438
Van Prooijen, J.-W., & Krouwel, A. P. M. (2019). Psychological features of extreme political ideologies. *Current Directions in Psychological Science, 28*, 159–163.
Van Prooijen, J.-W., & Krouwel, A. P. M. (2020). Overclaiming knowledge predicts anti-establishment voting. *Social Psychological and Personality Science, 11*, 356–363.
Van Prooijen, J.-W., Krouwel, A. P. M., Boiten, M., & Eendebak, L. (2015). Fear among the extremes: How political ideology predicts negative emotions and outgroup derogation. *Personality and Social Psychology Bulletin, 41*, 485–497.
Van Prooijen, J.-W., Krouwel, A. P. M., & Emmer, J. (2018). Ideological responses to the EU refugee crisis: The left, the right, and the extremes. *Social Psychological and Personality Science, 9*, 143–150.
Van Prooijen, J.-W., Krouwel, A. P. M., & Pollet, T. (2015). Political extremism predicts belief in conspiracy theories. *Social Psychological and Personality Science, 6*, 570–578.

Van Prooijen, J.-W., & Kuijper, S. M. H. C. (2020). A comparison of extreme religious and political ideologies: Similar worldviews but different grievances. *Personality and Individual Differences, 159,* 109888.

Van Prooijen, J.-W., Ligthart, J., Rosema, S., & Xu, Y. (2022). The entertainment value of conspiracy theories. *British Journal of Psychology, 113,* 25–48.

Van Prooijen, J.-W., Rosema, S., Chemke-Dreyfus, A., Trikaliti, K., & Hormigo, R. (2022). Make it great again: The relationship between populist attitudes and nostalgia. *Political Psychology.* https://doi.org/10.1111/pops.12825

Van Prooijen, J.-W., & Van Vugt, M. (2018). Conspiracy theories: Evolved functionsand psychological mechanisms. *Perspectives on Psychological Science, 13,* 770–788.

Webber, D., Babush, M., Schori-Eyal, N., Vazeou-Nieuwenhuis, A., Hettiarachchi, M., Bélanger, J. J., Moyano, M., Trujillo, H., Gunaratra, R., Kruglanski, A. W., & Gelfand, M. J. (2018). The road to extremism: Field and experimental evidence that significance loss-induced need for closure fosters radicalization. *Journal of Personality and Social Psychology, 114,* 270–285.

Whitson, J. A., & Galinsky, A. D. (2008). Lacking control increases illusory pattern perception. *Science, 322,* 115–117.

Zuckerman, M. (1994). *Behavioral expressions and biosocial bases of sensation seeking.* Cambridge, UK: Cambridge University Press.

INDEX

academic radicalization 329–341
accessibility effects 58–61
achievement insecurity 7–8
affective synapse model 62–64
affirmation 65, 66, 68, 102, 103, 105, 134, 135
agriculture 29–30
alienation 9
ambiguity and uncertainty 56–58, 63, 112
ambivalence 37, 108, 115–117
animal behavior 85–86
anti-establishment sentiments and insecurity 368–382
anxiety 9, 134, 139, 188, 191, 192, 197, 198, 206, 211, 212, 216, 225, 228, 229, 230, 231, 232, 233, 268, 350, 369, 372; and insecurity 27–28; and mortality salience 35–37
attachment anxiety 191, 197, 225, 228, 229, 230, 232–237
attachment avoidance 234–237
attachment theory 17, 26, 187–200, 224–237
attitude change 111–113; and insecurity 111–113; and persuasion models 113–115
authoritarianism and insecurity 256, 270, 308, 309, 313, 314, 315, 323, 331, 334, 336–338, 357, 360, 363

benefits of insecurity 74–89
biology 83

bonds 205, 217, 225
brain 5–6, 22, 27, 169, 187, 311
business 58, 64, 80, 214, 368

cannibalism 23
catastrophizing 14–15
cognitive dissonance 17, 95–105; definition of 96–97; drive theory of 100–101; history of 97–98; and meaning 99–100, 101; and social reality 104–105; uncertainty and 103–105, 109–110
collective narcissism 256, 313, 315, 317, 319, 320, 321, 322, 378
collective narratives 12–13
collective victimization 274–275
consciousness 3–5, 8
consensual delusions 12–13
conspiracy theories 13, 17, 68, 175, 206, 246, 256, 258, 312, 315, 316, 322, 323, 349–363, 368, 369, 370, 372, 374, 375, 377, 381; and anti-establishment sentiments 369–382; and self-handicapping 362–365
COVID-19, 55, 95, 224–237, 352–354; and conspiracy theories 352–358

death 6, 16, 25–27, 35–48, 54, 55, 119, 170, 171, 172, 173, 175, 176, 178, 180, 191, 192–194, 197–199
delay of gratification 81, 87–88
democracy and populism 196, 276, 307, 309, 310, 315–323, 357, 359

democracy to autocracy 316–317; role of insecurity in 317
depression 8, 25, 36, 55, 196, 226, 340
deprivation 75, 86, 87, 89, 321, 355, 356, 373
drug abuse 115–119, 142
dyadic interdependence 206–219
dyadic relationships 181, 187, 192, 206–219

ecological basis of insecurity benefits 79–81
equality vs. inequality 13, 31, 36, 45, 173, 309, 310, 335, 339, 340, 341, 372
evolutionary functions of insecurity 74–89
evolutionary theory 21–32, 188, 191, 210, 265, 266, 269, 271, 307; and populism 309–323
existential concerns and attachment 191–200
existential insecurity 6, 17, 307, 356
existential threats 46–47; and attachment 187–200
existentialism 9
extremism and insecurity 329–340

fairness 173, 179, 293, 294, 295, 298–299, 330
fluency vs. disfluency 81, 82–83, 87–88, 311
foresight: evolution of 21–24; psychological costs of 24–25
free will 150, 151, 159–162, 168
freedom 7, 9, 13, 17, 31, 32, 44, 55, 150–164, 351, 357

games 153, 155, 156, 235
group cohesion 11, 12, 13, 310
group identification 193, 244–258, 313, 322
groups 11; and attachment 187–200; existential function of 266–268; survival perspective of 268–269; and terror management 45–47

health 16, 28, 31, 37, 40, 55, 74, 86, 88, 113, 115, 190, 205, 207, 224, 225, 226–229, 237, 245, 331, 351, 353
history 4, 5, 6, 9, 13, 18, 29, 30, 36, 41, 44, 59, 60, 62–63, 67, 68, 155, 170, 177, 180, 181, 228, 267, 271, 307, 309, 311–314, 354
hope and fear 58–60

human nature 307, 311, 323
Hungary 325–333
hunger 97, 100, 103, 105
hunter-gatherers 23–24, 27–29

ideologies 12, 13, 15, 17, 18, 42, 44, 46, 48, 60, 196, 218, 219, 246, 256, 265, 266, 275–277, 307–310, 312–314, 322, 323, 329, 330, 334, 337, 363
individual vs. collective security 265–278
individualism 9, 13, 15, 309, 310, 323, 357, 360, 361, 363
informational vs. personal uncertainty 291–293; and trust in institutions 293–294
insecure attachment 67, 190–192
insecurity and learning 84–85
insecurity trade-offs 80–82; taxonomy of 81
insecurity: and achievement 7–8; and ambiguity 56–58; and anti-establishment sentiments 368–382; and attachment 187–200; and attitude change 108–120; benefits of 74–89; and cognitive dissonance 95–105; and collective narratives 12–13; and commercial implications 16; and conspiracy theories 350–367; definitions of 3–4; evolutionary approaches to 16–17, 21–32; and group identification 255–263; management of 10–11; in modern nation states 30–32; philosophical views of 3; and politics 15–16; and populism 309–323; and public life 13–14, 16; and radicalization 329–341; and rationality 6–7; and relationships 9–10; role of consciousness in 3–6; and self-handicapping 130–144; and trust 150–164, 157–159, 287–299
intergroup relations 244, 251, 265, 266–278
internet 296–298, 323, 382

language 64, 99, 277, 317
leadership 44–45, 254–255, 256
liberalism 15, 309, 323
locus of control 134
loneliness 9, 199, 227, 228, 229, 311
love 37, 55, 192, 207, 216, 225, 228, 229, 231, 317, 358

marijuana use 116–117
marriage 74, 75, 171, 193, 212, 218, 226, 380

minority *vs.* majority influence 83–84, 112–113
misfortunes, explanation of 170–180
moral account 172, 173–176, 177, 178
moral certitude and populism 314–315, 322
moral meaning 170–183
moral responsibility 179–181
mortality salience 6, 25–27, 35–48
motivations 134–139; and self-handicapping 130–144
myth of the unknown 54–56

narcissism 256, 313, 315, 316, 317, 318, 319, 320, 321, 322, 378
narratives 12–13, 246, 256, 307, 309, 310, 311, 313, 314, 317–318, 331, 334, 337, 338, 340, 350, 351, 352, 354
needs 9, 21, 24, 36, 44, 46, 47, 98, 99, 103, 104, 105, 189, 191, 214, 237, 277–278, 287, 309, 330, 337
normal 15, 67, 332
normative 44, 74, 76, 77, 78, 80, 81, 84, 87, 88, 113, 114, 116, 118, 140, 163, 198, 198, 208, 355
nostalgia 368, 376, 379–380, 381, 382

obedience 11, 79, 197, 311, 360, 377
ostracism 112, 341, 373, 374
outsider 47, 136, 255, 156, 313, 317

pandemics 17, 225–237; and attachment experiences 228–230; and insecurity 229–231
perceived threats 42–44
personal responsibility 7, 162
personal uncertainty and trust 293–294
persuasion 108–120; models of 113–115
political bias 337–340
political consequences of insecurity 307–320, 330–331
political extremism 373–374; and insecurity 375–378
political insecurity and conspiracy theories 355–358
politics of uncertainty 307–323; and populism 309
politics 10, 12, 13, 14, 15–16, 17, 35–36, 42, 46, 192, 194, 196, 212, 213, 215, 218, 252, 272, 289, 307–320
populism 256, 307–323, 370–373; evolutionary perspective on 310–311; and insecurity 317–323; and tribalism 309

propaganda 311, 315, 316, 318, 319–321, 322

randomness 150, 155, 159, 161–163
rationality 6–7, 46, 47, 78, 156, 163, 323
relationships 9–10, 187–200

safety threats 209–212
self-enhancement 249, 368, 376, 377–378
self-handicapping 130–144; and conspiracy theories 362–365; motives for 137–138; paradox of 130–131; roots of 132–134; as self-protection 134–136; trade-off in 136–137
self-regulation 76–77, 251, 292
self-uncertainty 247–249; causes of 252–253; resources to deal with 251–253
social activism and insecurity 329–341
social identity 244–258; and self-uncertainty 247–249; and uncertainty 246–247
social influence 111–113, 254–255
social insecurity 8–9, 119, 357, 360
social institutions 287–299
social interaction 8–9, 151, 198, 248, 290
social media 11–12, 42, 43, 44, 61, 216, 254, 298, 323
social relationships and security 225–237
social safety system 207–209
social science radicalization 329–341
social support 11–12, 188–200
sociopolitical relationships 194–196
speed and accuracy trade–offs 77–79
strategic games and uncertainty 155–164
supernatural 175–176, 177, 182, 187, 199

Terror Management Theory 26, 35–48; empiric al evidence for 39–40
theory of mind 3, 5, 24, 151
theory of reasoned action 114–116
threat defense 209–212
time perspective and insecurity 8
tobacco use 116–117
tribalism 13, 14, 15, 18, 44, 190, 195, 270, 307, 309–323, 330, 349, 356
trust 150–164, 287–299

uncertainty and strategic games 155–164
uncertainty aversion 155
uncertainty challenge 54–68; and ambiguity 56–57; embracing uncertainty 66–68; escaping uncertainty 63–65; and strategy 151–164

unfelt needs 21, 24

values 13, 24, 36, 42, 43–47, 65, 66, 75, 78, 140, 190, 196, 248, 295, 309, 334, 343, 357, 360, 373

vigilance 61, 178, 265, 272, 273, 275, 276, 278, 330

violence 6, 18, 35, 36, 41, 42, 43, 45–46, 47, 48, 68, 227, 256, 270, 274, 277, 307, 308, 314, 377, 381

virtue signaling 314–315

WEIRD people 9, 178, 179, 298

woke activism 13, 17, 309, 310, 312, 314